Mammalian Semiochemistry

Mammalian Semiochemistry

The Investigation of Chemical Signals Between Mammals

ERIC S. ALBONE
University of Bristol
and
Clifton College, Bristol

With a contribution from
STEPHEN G. SHIRLEY
University of Warwick

A Wiley–Interscience Publication

JOHN WILEY & SONS LIMITED

Chichester · New York · Brisbane · Toronto · Singapore

Library of Congress Cataloging in Publication Data:

Albone, Eric S.
  Mammalian semiochemistry.
  'A Wiley–Interscience publication.'
  Bibliography: p.
  Includes index.
  1. Mammals—Physiology. 2. Mammals—Behavior.
3. Semiochemicals. 4. Animal communication.
I. Shirley, Stephen G. II. Title.
QL739.2.A38 1984   599'.01826   83-10231

ISBN 0 471 10253 9

British Library Cataloguing in Publication Data:

Albone, Eric S.
  Mammalian semiochemistry.
  1. Smell 2. Senses and sensation
I. Title II. Shirley, Stephen G.
  591.1'826   QP458
ISBN 0 471 10253 9

Typeset by Preface Ltd, Salisbury, Wilts.
Printed by Page Bros. (Norwich) Ltd.

知
不
知
上
不
知
知
病

Knowing ignorance is strength.
Ignoring knowledge is sickness.

Lao Tsu

Contents

Foreword ix

Preface xi

Chapter 1 **Mammalian semiochemistry** 1
    1.1 Semiochemistry and chemical ecology 1
    1.2 Terminology of chemical interaction 2
    1.3 Response guided and chemical image strategies . . 5
    1.4 Interdisciplinarity—some frustrations 8

Chapter 2 **Chemistry's contribution** 9
    2.1 Chemistry's contribution 9
    2.2 The impossibility of total chemical knowledge . . . 11
    2.3 The question of volatility 13
    2.4 Handling the volatiles 17
    2.5 The chemistry of an odour image 34

Chapter 3 **The skin** 40
    3.1 The uniqueness of skin 40
    3.2 The structures of skin 42
    3.3 The hormonal regulation of skin function 50
    3.4 Steroids in the skin 54
    3.5 Lipid chemistry of the skin surface 60
    3.6 Appendix 71

Chapter 4 **Scent glands** 74
    4.1 Scent glands; specialized skin glands 74
    4.2 Artiodactyla 74
    4.3 Lagomorpha 93
    4.4 Proboscidea 102
    4.5 Rodentia, including the Harderian gland 103
    4.6 Carnivora 115
    4.7 Primates and man 119

Chapter 5   Microorganisms in mammalian semiochemistry 135
      5.1   The role of microorganisms 135
      5.2   Microbial activity in skin 136
      5.3   Anal sacs; related structures and related scents . . 137
      5.4   The Hyaenidae 150
      5.5   The Mustelidae 151
      5.6   The Viverridae 159
      5.7   Rodentia 159
      5.8   Carnivore attractants as deer repellants 162
      5.9   Faecal odours; maternal pheromone in the rat . . 163

Chapter 6   Urine 165
      6.1   The importance of urine 165
      6.2   Urine display 166
      6.3   Some primer pheromone effects 168
      6.4   Gross composition of urine 174
      6.5   Urine volatiles 176
      6.6   Less volatile and involatile components 183

Chapter 7   Secretions of the reproductive tract 210
      7.1   Male secretions 210
      7.2   Female secretions 226

Chapter 8   Breath, saliva and the pig 235
      8.1   Oral odour and semiochemistry 235
      8.2   Chemical communication in the pig 237

Chapter 9   Mammalian chemoreception—*Stephen G. Shirley* . . . 243
      9.1   Sensory systems 243
      9.2   Properties of neurons 253
      9.3   Studies on the olfactory system 258
      9.4   Olfaction in practice; sensory strategy and com-
           munication 275

Appendix   Key chemical methods
      A.1   The analytical approach 278
      A.2   Chromatography 278
      A.3   Methods of identification 279
      A.4   Radioimmunoassay 286
      A.5   Interrelation of the major sex steroids 301

References 302

Subject Index 351

Foreword

L. G. Goodwin, FRS
Formerly Director of Science, Zoological Society of London

Chemical signals are essential to all living creatures. The pollen grain of the right species receives a signal to put forth a tube to grow down to the ovary of the flower; the free-swimming male gamete of the fern depends upon a signal to find the female cell on the underside of the prothallus. Parasitic schistosome miracidia locate and home in on a particular species of snail and the juvenile worms find their mates in the confusing chemical *milieu* of the tissues of their vertebrate host. Insects communicate by pheromones and, because these substances have an obvious practical application in the control of pests, they have been the subject of considerable research.

Dr. Albone, in his study of mammalian semiochemistry takes a wider view—the 'largely unexplored and unknown world of chemistry which underlies, in profound ways many of the biological interactions of mammals with members of their own and of other species in their shared natural environment'. As a chemist, he finds it difficult to understand why most of the research and literature on mammalian communication has, so far, been left largely to the biologist. The challenge and the fascination of the identification of the chemical structures of substances with such profound significance and potency, that act at a distance in unbelievable dilutions, might have been expected to generate much more enthusiasm for the work.

Against the odds, Dr. Albone has for many years carried out and championed research on chemical signals and his book will serve a useful purpose in bringing together what is known, and pointing out the great gaps in our knowledge of this essential area of biological communication. It will provide a valuable source for techniques and new ideas for chemical studies and it should certainly be read by all zoologists concerned with field studies and aspects of the behaviour and reproduction of mammals. The time is ripe for some serious cooperative research between the sciences.

The skin, its glands and its microorganisms, the urine, the secretions of the reproductive tract, and even the breath and the saliva all provide chemical signals that inform other individuals of the condition, attitudes and moods of their neighbours. Although the human sense of smell is poor compared with

x

that of most other mammals it probably plays a far greater part than many of us realize in influencing behaviour and the quality of our interactions with our own and other species. It is easy to guess that the smell of good cooking might attract a hungry man, but more difficult to understand why an inhalation at the entrance to the marquee at the Chelsea flower show should be, for most people, delectable whereas other odours should, to a varying degree, be repulsive. The multimillion-pound industry that seeks to reduce normal human smells by the application of deodorants and antiperspirants and to substitute them with alternative, manufactured scents and perfumes is surely built upon an important basic and intuitive human need.

We should perhaps pay more attention to the fine discrimination of Don Giovanni. 'Zitto!' he says, 'mi pare sentir odor di femmina. . .' And he was an expert.

Dr. Albone's book should awaken interest in this fascinating and poorly researched area of science.

Finchampstead, Berkshire

Preface

Mammalian semiochemistry is a science in bud, a science whose flowering is to come. Although biologists are increasingly aware of many interesting and important interactions between mammals which are mediated chemically, far too few chemists have yet taken up the challenge and have immersed themselves in this fascinating interdisciplinary quest. The reasons for the gulf are, no doubt, many. Certainly the rigidities of many of our institutions do not help. In addition, the immense complexity of mammalian systems is likely to deter some who are interested in rapid rewards for their efforts. But, perhaps principally, the problem is a manifestation of a gulf of perception which has ramifications on both sides of the interdisciplinary divide. Whatever its origin, it is a gulf which I hope this book will help to bridge.

Today, there are of course organizations and individuals who are themselves building very effective bridges in this area, and the names of many will be evident from the following pages, but the contributions of the *Journal of Chemical Ecology*, founded in 1975, of the European Chemoreception Research Organization and now of the International Society of Chemical Ecology merit special mention, even though their concerns are broader than the subject of this book. Conferences promoting collaboration between chemists and biologists in this area have also emerged, as for example, the conferences on chemical signals organized every few years by members of the State University of New York at Syracuse. In 1978, the NATO Special Programme Panel on the Eco-Sciences sponsored an Advanced Research Institute on Chemical Ecology, the formal conclusions and recommendations of which were published together with the proceedings the following year (Ritter, 1979). Among these were a number concerning mammalian studies, many of which echoed points which emerge in this work.

What I have written here arose out of my experience working as a chemist among biologists in this field. To many of my former colleagues I am indebted for the seeds of many of the ideas which I have developed here, and in this context I should particularly like to acknowledge the contributions of Professor Peter Flood, Dr. Graham Perry and Dr. Georges Ware.

In this book it is my ambition to provide those who are interested in moving into the area of mammalian semiochemistry, whether as biologists or as chemists, with facts and insights which will at once be interesting and practically helpful. At present no other chemically oriented text of this type exists in this area, and I hope that the reader will be tolerant if, in my effort to range widely, I have betrayed too blatantly a superficiality in some areas. I am all too aware of such weaknesses myself.

In the first two chapters, I look specifically at the contribution of chemistry, and at chemistry's limitations. After discussing the terminology of semiochemistry, and its appropriateness, I consider, at various levels, ranging from the conceptual to the practical, just what and how chemistry can contribute. Because it might be particularly useful to some, I have included an appendix outlining and referencing something of the key chemical techniques appropriate to the study of this area.

A fruitful approach to mammalian semiochemistry, I believe, necessarily involves a proper understanding of context, so that to unravel the semiochemical nature of any substrate would seem to require in most cases as full an understanding as possible of the biology and chemistry of that substrate. In the following six chapters, I have covered much of what is known of mammalian semiochemistry, but I have woven this into the matrix of a broader understanding of the major semiochemical sources, namely the skin, the specialized skin scent glands, microbial scent sources associated with mammals, urine, the secretions of the reproductive tract, and odours associated with breath and saliva.

In the final chapter, I am pleased Dr. Stephen Shirley of the Department of Chemistry and Molecular Sciences of the University of Warwick has provided matching insights into the field of mammalian semiochemistry from the standpoint of a molecular scientist concerned with mammalian chemoreception.

Finally, I should like to thank Professor D. L. Dineley and all those who have been so helpful to me in this task, but particularly to thank my wife who has contributed so much in encouragement and support in quite difficult times.

Bristol, England   ERIC ALBONE
January 1983

Chapter 1
Mammalian Semiochemistry

This book is concerned with the largely unexplored and unknown world of chemistry which underlies in profound ways many of the biological interactions of mammals with members of their own and of other species in their shared natural environment. It is my purpose to survey our present fragmented understanding of this subject and to suggest starting points for further scientific exploration.

1.1 SEMIOCHEMISTRY AND CHEMICAL ECOLOGY

The science of **semiochemistry**, derived from the Greek word for sign or signal—*semeion*—deals with the chemistry of those substances, known as semiochemicals, by means of which organism interacts with organism in the shared natural environment. The general term 'semiochemical' includes within itself the concept of pheromone. However, the definition of pheromone is rather precise and the discussion of whether or not this narrowly defined concept, or some modification of it, is or is not as appropriate to mammalian studies as it appears to be to the study of simpler organisms can lead to the unproductive if energetic discussions of semantics which not uncommonly surround such issues. For the most part it seems preferable to avoid these niceties except in those cases where their implementation genuinely advances understanding.

Closely linked with the science of semiochemistry is the field of **chemical ecology**. In essence, this relates to the same multidisciplinary study area, although the emphasis is a little different, and may be defined as the study of those *interactions* which occur between organisms in their shared environment which are mediated by chemicals they themselves produce. Clearly chemical ecology can make considerable biological progress in total ignorance of the underlying semiochemistry and, in fact, this has been the case in the mammalian field. The extensive literature which exists here has largely emerged from the work of students of animal behaviour and reproductive physiology while the contribution of chemists has been remarkably small.

The term **chemical communication** is sometimes used to denote much the same area of concern, although what chemical interactions constitute communication and what do not are matters for discussion. Marler (1977), for

1

example, feels that defensive chemical interactions do not fully come within the category of communication, communication in its fullest sense implying evolutionary specialization of a mutualistic cooperative nature.

Semiochemical effects occur between organisms of all types. Sensitivity to the chemical environment emerged at a primitive stage in evolution and all cell membranes respond to some degree to chemical stimuli (Koshland, 1980). Indeed chemical interaction between single-celled organisms probably played an important role in the evolution of multicellular organisms. Throughout the animal kingdom it is clear that the ability to sense and to respond to chemicals in the environment has major survival value. However, in spite of the ubiquity of this natural web of chemical interaction, our knowledge of the underlying chemistry remains very limited. Those semiochemical studies which have been undertaken have largely concerned insects. This arose from the economic imperative to seek more effective, less polluting means of controlling agricultural insect pests, for it seemed that an understanding of insect semiochemistry offered the possibility of simple chemical means of controlling, or at least of monitoring, populations of insect pests on a species-specific basis using relatively small quantities of natural substances. In the event, insect chemical communication has proved to be rather more complex than was once thought, responses such as attraction to a trap being commonly dependent on specific mixtures of compounds rather than on single compounds, as well as on other non-chemical factors, and pulsed chemical signals have even been reported in an arctiid moth (Conner *et al.*, 1980). Even so, implications for significant improvements in pest control methods are very real and the subject is under active study.

If insect semiochemistry is the subject of numerous articles and volumes, its mammalian counterpart, the subject of this book, is by contrast in its infancy. In the process of attempting to provide some starting points for the chemical exploration of this virtually uncharted subject, we shall see that mammalian semiochemistry, while benefiting from the achievements of insect studies, possesses many characteristics which are distinctively its own.

1.2 TERMINOLOGY OF CHEMICAL INTERACTION
(Nordlund and Lewis, 1976)

Semiochemical

A general term for single compounds or mixtures of compounds carrying information or otherwise mediating interactions between organisms in the shared natural environment (Regnier, 1971).

Pheromone

This term refers to semiochemical interactions occurring between organisms of the same species. As proposed by Karlson and Lüscher in 1959, the term described a substance secreted to the environment by an individual organism, which on being received by another individual of the same species, elicited a specific reaction, such as a definite behavioural response or developmental process. This concept has been widely used in studies of lower organisms.

In those cases where the pheromone concept is appropriate, two categories have been distinguished.

(1) **Releaser** pheromone. A pheromone to which the response is primarily behavioural and immediate.

(2) **Primer** pheromone. A pheromone to which the response is primarily physiological (e.g. endocrine) and longer term.

However Beauchamp *et al.* (1976) have noted that this precisely defined pheromone concept may not be so appropriate to mammals where chemical signals are less likely to lead to the reproducible and distinctive behavioural responses required by the definition. Mammalian responses commonly depend not only on the reception of a chemical signal but on intricate combinations of chemical and non-chemical (such as visual and tactile) cues, on the mammal's physiological state and on its past experience (Nyby *et al.*, 1978; Nyby and Whitney, 1980). In man, the situation is complicated even further by cultural factors.

An alternative approach is to broaden the definition of pheromone (Bronson, 1974; Müller-Schwarze, 1977; Thiessen, 1977a; Katz and Shorey, 1979) and this is encouraged by a realization that even with insects semiochemical complexity is much greater than was once thought. In response, Beauchamp *et al.* (1979) have pointed out that such a broad definition might not be very useful and could possibly be damaging and have argued in favour of its abandonment. If the term is to survive, it would seem better to use it in a narrow sense and to use other broader terms, such as semiochemical, for broader purposes. Martin (1980) has suggested such a narrow use to

'include those isolated chemicals shown to be relatively species-specific which elicit a clear and obvious behavioural or endocrinological function and which produce effects involving a large degree of genetic programming and influenced little by experience,'

and has proposed as the broader term **homeochemic** (contrast allelochemic, see below) to include substances responsible for semiochemical interactions of *all types* operating between members of the *same* species.

Rutowski (1981) has taken a different line and has anchored his definition of pheromone on a consideration of communication. He suggests that the term could be used profitably to describe compounds or mixtures of compounds which have a truly communicatory function between organisms (not necessarily of the same species), defining communication as occurring when a signal from one animal alters the behaviour or physiology of the animal which receives the signal in a way which promotes the fitness or reproductive success of both animals. The key here is that the interaction is of mutual benefit. For him, unlike Martin, it is of little consequence how far the interactions are learned and experience-dependent and how far they are innate.

Clearly there is considerable debate and disagreement between zoologists on how the term pheromone is to be used, if at all, and we seem to have the unfortunate circumstance of one term meaning different things to different people. However, it seems preferable to keep the definition of pheromone relatively narrow for use in those cases where the pheromone concept as originally conceived is appropriate and elsewhere to use the more general term, semiochemical.

Allelochemics

These are defined as semiochemicals mediating interactions between organisms of *different* species and have been classified in two categories.

(1) **Allomone.** A semiochemical adaptively favouring the emitting species, for example, a defensive secretion or a floral scent attracting pollinating insects. As far as mammals are concerned, some tropical flowers which are pollinated by bats emit odours which have been variously described as bat-like, musky, semi-foetid and 'like butanoic acid' and it is possible that these function to attract bats to the flowers (van der Pijl, 1961; Baker, 1963). Various species of bat are known to emit strong musky odours and to possess scent glands on the upper lip, on the forehead, on the throat or on the arm (Allen, 1967).

(2) **Kairomone.** A semiochemical adaptively favouring the receiving species, for example chemical cues by which a predator recognizes or locates its prey, or by which an insect is attracted to its food plant. Here, again, there has been discussion among zoologists concerning the value of this terminology (Weldon, 1980; Pasteels, 1982). These categories are biological rather than chemical and it is quite possible for a particular compound or mixture of compounds to fall within more than one category, to act both as a pheromone and as an allomone or kairomone depending on circumstances. For example, American bolas spiders attract their prey, male moths, by producing attractant allomones which also function as the sex attractant pheromone produced by the female moth (Eberhard, 1977). Indeed it has been suggested that kairomones may often be allomones or pheromones which have 'evolutionarily backfired' and come to be exploited by another species and act as kairomones secondarily.

There seems to be a general tendency for terminology of this type to proliferate and it could be asked how useful this is for it is not always clear under which category a particular chemical interaction should be included. In addition, our categories could have the disadvantage of limiting our perception. With Pasteels (1982), it would seem that a simple terminology would best provide the necessary flexibility to deal with a complex reality.

Perhaps also, Karl Popper, writing in his autobiography *Unended Quest*

(1976) has something important to say on this general matter of excessive terminology.

> Never let yourself be goaded into taking seriously problems about words and their meanings. What must be taken seriously are questions of fact, and assertions about facts: theories and hypotheses; the problems they solve and the problems they raise.'

1.3 RESPONSE GUIDED AND CHEMICAL IMAGE STRATEGIES

Even at our present stage of ignorance, we recognize that mammalian semio-chemical systems are characterized by a daunting complexity. One approach to their elucidation is to assert that before any useful chemical studies can be undertaken, a firm experimental bioassay procedure must be established. When this has been achieved chemical analysis of the signal substrate can be guided by repeated bioassay of substrate fractions until the biologically active components are first isolated and then identified. I label this approach a **'response-guided strategy'**. Such studies have proved themselves in insect semiochemical research and in other aspects of biological chemistry where precise bioassay procedures are possible. In this way, it has been possible to identifiy insect semiochemicals even if these are present in trace quantities in a mass of other biological material. Using a clearly defined behavioural assay, for example, the following pyrrole was identified as the major behaviourally active volatile component of the trail marks produced by the ant, *Atta texana* (Tumlinson *et al.*, 1972).

It was possible as the result of a complex sequence of fractionations to isolate and then to identify 150 μg of pure active component from some 3.7 kg of macerated worker ants. A number of other trail-active components were also present. However, the potency of this one substance was such that only 0.33 mg would be required to make the trail the length of the circumference of the globe!

Essential to this approach is the availability of a relatively simple, clear, behavioural or physiological bioassay of a type implied by the pheromone concept and, even then, procedures are complicated if activity is manifest not by single compounds but by relatively complex mixtures, for with fractiona-tion of the natural mixtures there is the likelihood of losing activity altogether. The simplicity required for a response-guided approach is more likely to be appropriate to the study of the simpler organisms, although it is now apparent that, even with insects, considerable complexity can occur.

The problem with this approach is that it is likely to be of only limited applicability for mammals. Clearly in those cases where it can be used, it must be fully exploited, but it is likely that large and important areas of mammalian semiochemistry exist where a simplistic pheromone concept is out of place for, as I have indicated, the neural structures and behaviour patterns of mammals are far more complex than those of insects. Definitive behavioural experimental assays are difficult to design because in general the mammal does not respond to a chemical stimulus in isolation but integrates the chemical stimulus with other non-chemical stimuli and with factors depending on its present internal physiological state and its past social and environmental experience. In addition, those responses which are noted in test situations, such as sniffing, licking and general investigation of the chemical source, are commonly non-specific, while a meaningful assessment of the mammalian response can often only be linked with a detailed study of the life of the mammal in its natural surroundings. Taken with the immense chemical complexity commonly encountered in the natural secretions and excretions which are the sources of chemical signals, these factors present profound obstacles to progress in mammalian semiochemistry. But they must be faced. In questioning the appropriateness of a response-guided approach to mammalian semiochemistry, Beauchamp *et al.* (1976) have raised a further related point.

> 'In a species that has a complex repertoire, it would be unprofitable to assume that a single note out of that repertoire carries all the information. It would also be highly unlikely that if all the notes in the repertoire were taken and randomly emitted, the signal would remain whole any more than randomization of the notes of a familiar tune would leave the tune intact. In fact, looking at single notes or rearranged notes would only succeed in "identifying" the components of the auditory signal if the signal were initially simple. Unfortunately these are exactly the procedures employed in pheromone identification work. The search for *the* chemical may often be as fruitless as would be the search for *the* note of a song.'

If the response-guided strategy is of limited value, what other complementary approaches are available? One might be a **'chemical image strategy'**. Instead of supposing that embedded in a mass of semiochemically meaningless material the mammal produces a limited number of substances which elicit very specific behavioural responses in a conspecific animal (the pheromone concept), a valuable alternative approach might be the holistic one in which the nature of the entire chemical image presented by one animal to another is surveyed. For, just as a mammal or a group of mammals presents a series of optical images from which a variety of information may be extracted, so they also exhibit to the chemical senses images differentiated with regard to

(1) chemical composition,
(2) spatial distribution, over their bodies and throughout their environment,
(3) time, both as the result of changes in the nature of the odour at source and as the result of environmental changes (ageing of scents).

These images may be perceived by olfaction, gustation (taste), and through the medium of the vomeronasal organ and, as an extension of the emitting animal, are clearly different in character from optical images. Thus they have a more diffuse, longer lasting, more slowly varying nature and persist in an area even when the mammal itself has moved away. Thus, they operate even when the communicating mammals do not encounter each other directly.

These considerations link with some of the ideas in Griffin's (1976) challenging little book on the subject of animal awareness in which he confronts some of the concepts which he believes modern ethologists are loath to countenance. He writes:

'The possession of mental images could well confer an important adaptive advantage on an animal by providing a reference pattern against which stimulus patterns can be compared. It is characteristic of much animal, as well as human, behaviour that patterns are recognized not as templates so rigid that slight deviations cause the pattern to be rejected, but as multidimensional entities that can be matched by new and slightly different stimulus patterns. This ability to abstract the essential qualities of an important object and recognize it, despite various kinds of distortion, is obviously adaptive. Mental images with a time dimension would be far more useful than static searching images, because they would allow the animal to adapt its behaviour appropriately to the probable flow of events, rather than being limited to separate reactions as successive perceptual pictures of the animal's surroundings present themselves one at a time. The concept of mental images that include both spatial and temporal dimensions tends to approach a working definition of conscious awareness.'

The implications of such a chemical image strategy for mammalian studies have yet to be explored. So far chemists working in the mammalian field have generally attempted to retain the response-guided strategy which has been so successful in work with insects. Certainly a chemical image strategy has its own problems as we shall see and is a recipe for a far more complicated life, but then mammals are complicated.

It is, then, the ambition of the chemical image holistic strategy to describe the chemical images which the mammal perceives. As with other modes of mammalian communication, particular importance will surround variations in images which correspond to changes or differences in important biological parameters in the animals concerned for if, for whatever reason, information is present, evolutionary pressures will tend to its exploitation. It will also be of considerable value to explore specializations both in scent- and other signal-producing organs and in chemoreception, for the fact that these specializations have evolved points to aspects of the semiochemical system which have acquired biological importance.

Today much is known concerning the chemical information which a mammal presents to its environment. This is less remarkable when it is recalled that most has arisen from concerns remote from semiochemistry. This book shall attempt to draw these understandings together and to view them as a springboard for future mammalian semiochemical studies.

1.4 INTERDISCIPLINARITY—SOME FRUSTRATIONS

Our subject is an interdisciplinary one and it is right to raise at this stage one of the real stumbling blocks to progress, a stumbling block which renders it all too obvious that scientists are people too, for it is a common frustration of interdisciplinary research that the collaborators within increasingly specialized component disciplines fail to understand the challenges, difficulties and opportunities, scope and limitations of their colleagues' specialties. It is not unknown, for example, for a biologist to regard a chemist's contribution as a routine technical service to the real scientific challenge of the biology, and of course some chemists can find for themselves justifications for analogous views. These are the rocks on which many an interdisciplinary enterprise has foundered for they imply unreasonable expectations of one's colleagues' contributions. As Mykytowycz (1979) has written:

'I suspect that many efforts to cooperate broke down in the early stages because of the difficulty in communication . . . Often the two specialists expect too much from one another. The chemist on the one hand hopes that, on the laboratory bench, mammals will behave like automata . . . and the biologist on the other hand believes that modern techniques and instruments available today make the routine analysis of odours a comparatively easy task.'

So it becomes a major importance that partners in such an enterprise possess the expertise of the specialist coupled with the awareness of the generalist, and both tempered with considerable humility. In the following pages, it would be presumptuous to begin to attempt to review the amassed knowledge of each component discipline from a specialist standpoint. Instead, my ambition is to attempt to shoot what I hope are some pertinent ideas across the disciplinary divides in the hope that they will be helpful to potential collaborators in the area of mammalian semiochemistry.

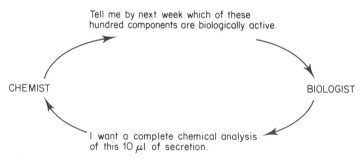

Figure 1.1   Naive optimism . . . a caricature of interdisciplinarity?

Chapter 2
Chemistry's Contribution

2.1 CHEMISTRY'S CONTRIBUTION

The substantial literature of mammalian chemical communication and chemical interaction continues to be largely the construct of biologists. In mammals numerous external secretory glands and other semiochemical sources have been described, scent-marking behaviour has been documented and behavioural, and physiological effects investigated, but linked chemical studies remain few. In addition to the daunting complexity of the subject, a major factor bearing on this situation is the fact that the intellectual terrain with which organic chemists feel familiar only rarely encompasses those areas in which an interest in mammalian semiochemistry has developed. If a proper understanding of mammalian semiochemistry is to arise, presently remote disciplines must interact. Here chemistry is one component among many, albeit a crucial one, for the concern is to provide an integrated understanding of the origins, nature and significance of the chemical medium by which mammals communicate (Figure 2.1). An early interest by perfumers and natural product chemists in certain external mammalian secretions was less fruitful than it might have been in this regard because in addition to the limitations of the chemical techniques then available, the studies were conducted with limited objectives and in disciplinary isolation.

The challenge presented to the chemist in attempting to apply his skills to the study of mammalian semiochemistry is considerable. From the findings both of insect pheromone research and of flavour chemistry we know that compounds present in a natural material even in trace quantitites can have a decisive effect on biological activity. We also know that the semiochemical substrates we are considering are frequently mixtures of great complexity and that biological significance may attach, not to individual compounds, but to mixtures of compounds.

However, the chemist does not enter this new area unprepared, for he has at his disposal an armoury of techniques and insights, both powerful and pertinent, which have been developed in such diverse fields as insect pheromone research (Brand *et al.*, 1979; Leonhardt and Beroza, 1982), clinical chemistry (Caprioli *et al.*, 1980), flavour and perfume chemistry (Teranishi *et al.*, 1971; Kolor, 1979), environmental chemistry and natural product chemistry (Young and Silverstein, 1975). Underlying these, and indeed

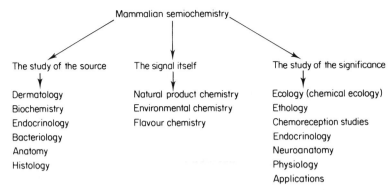

Figure 2.1   Mammalian semiochemistry in an interdisciplinary context

making progress in many of these areas possible at all, are relatively recent advances in modern microanalytical chemical methods. Even so, it should be noted that the chemical analytical sequence is biology dependent. The chemical analysis of a signalling substrate commonly involves some or all of the following steps:

(1) sample collection;
(2) preliminary fractionation; at this stage much of the sample may be discarded;
(3) separation of fractions of interest into components and, where necessary, the isolation of quantities of these components;
(4) classification and identification of components of interest;
(5) syntheses of these components and mixtures of components to confirm identification and/or for bioassay or application purposes;
(6) formulation and application.

Chemical considerations alone, however, do not provide a sufficient basis for following such a sequence. As a total chemical analysis can never be achieved, any particular analytical sequence involves a series of implicit or explicit decisions concerning priorities, concerning which compounds or groups of compounds present will be studied, and in what detail. For these, the chemist needs the guidance provided by the results of a linked biological programme.

Even at the stage of sample collection, we are involved in fractionation, in assigning priorities, for the sample we choose to collect cannot contain all the chemical information the animal presents to its environment. The choice of sampling procedure implies a selection of a part of this total chemistry. This selection must be consciously guided by biological insights.

One interesting approach used by Goodrich *et al.* (1978, 1981a,b) and Hesterman *et al.* (1976, 1981) in their studies on rabbit semiochemicals has been to present rabbits sequentially with the separated components of complex signalling secretions and to focus attention specifically on those components which produce an instantaneous change in heart rate. The limitations of the method are at once obvious for it is by no means self-evident that other components might not, either individually or in combination, be semiochemically active. On the other hand it does provide a rapid and very convenient biological pointer to components of potential semiochemical importance. Their technique has been to examine signalling secretions by gas chromatography and to split the effluent from the gas chromatograph, which delivers a train of separated components, and direct part to the gas chromatograph detector and part to a restrained rabbit. Components which elicit a momentary elevation or depression of heart rate of greater than 15% as they emerge from the gas chromatograph are deemed worthy of more detailed semiochemical study. Maximal effects were not necessarily associated with the maxima of gas chromatogram peaks, as might be expected, since the olfactory properties of even a single compound are known to be capable of dramatic variation with ambient air concentration. Also the possibility that highly active components present in trace quantities emerge from the column simultaneously with inactive components present in large quantities adds a further complication, for the inactive component is likely thereby to be assigned, at least initially, spurious semiochemical activity.

Although it might be convenient to select for study a particular secretion in isolation, in reality not only does the animal present its environment with compounds deriving from a multiplicity of body sources, but it also frequently mixes these secretions or excretions while marking its own or another animal's body surface or objects in the environment. And, while we know that the biological significance of a mixture of substances may be different from that of the sum of its parts, even if we work with an entire animal we are bound to select certain types of compound (e.g. volatile substances) for study. Further, it has to be remembered that the entire animal is itself not a constant source, so that even the investigation of the *total* information obtainable from one animal at a particular time represents a selection. As a result, some selection is inevitable.

2.2 THE IMPOSSIBILITY OF TOTAL CHEMICAL KNOWLEDGE

Although most attention has so far been directed to olfactory signals and so to the consideration of the volatile components associated with a signalling material, involatile components also possess signalling potential so that ideally one would wish to take into account the total chemistry a mammal presents to its environment. This is impossible, for as Zlatkis and his colleagues have written (1979a):

'A complete profile of all the constituents of biological fluid is at present impossible with available analytical techniques. The goal at which most workers aim is the development of a complete profile of a selected group of substances (such as steroids, amino acids, etc.) or of compounds with similar physical properties (for example, volatiles).'

12

Nor are new analytical advances likely to alleviate the situation. In an impressive article, Blumer (1975) contemplates the inherent complexity of the chemistry of nature. Pondering just one class of organic compound (the porphyrins) present in sediments, he has contrasted the state of knowledge in the 1930s when investigators using classic organic analytical techniques such as crystallization, column chromatography, and visible spectrophotometry inferred 'the presence of a few easily anticipated and explainable components that were thought to have been recovered in pure form' while, just four decades later:

'The combination of new separating methods . . . in joint application . . . demonstrated the existence of individual components probably numbering in the 10^5 range even in a single sediment.'

Blumer goes on to comment that:

'For someone who worked in the field at that time (1930s), it is humbling to think of the tens of thousands of pigments which he handled daily, but to whose existence he was totally oblivious!'

It seems that as analytical methods improve, analyses do not become simpler, but instead nature reveals a matching complexity.

And yet, even in the midst of this complexity, trace components cannot be

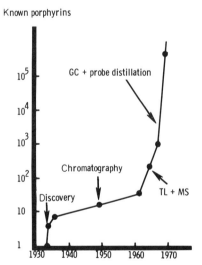

Figure 2.2 Porphyrins in sediments; an example of the complexity of nature matching advances in chemical analytical methodology. (Reproduced by permission of Verlag Chemie from Blumer, 1975)

neglected. Reviewing the chemical methodology of pheromone studies Claesson and Silverstein (1977) have drawn attention to the fact that the presence of even a fraction of a per cent of a highly active trace contaminant in an isolated component presumed to be a single pure compound can lead to the mistaken conclusion that the isolated compound is active when in fact it is not.

It follows that whatever the limitations of response-guided strategy, the **total** description of the chemical image a mammal presents is itself impossible, for the description is unending. Faced with this situation, the best that can be achieved in the advancement of knowledge is to use whatever imperfect tools are at hand, to pursue a chemical image strategy knowing that our descriptions must be limited and to pursue a response-guided strategy knowing that large areas of mammalian semiochemistry will be beyond its scope.

2.3 THE QUESTION OF VOLATILITY

In the context of mammalian chemical communication, the volatiles emitted by a signal source have attracted, and are likely to continue to attract, considerable attention. In this chapter, a range of techniques appropriate for collecting and examining these volatiles is discussed in some detail. The contributions of flavour chemists and of insect pheromone research are particularly influential here. However, it is appropriate to note that involatile components are likely to have substantial semiochemical importance and that it is expected that research will be increasingly directed to these *as well as to* volatile components in the future.

Because in certain instances olfactory thresholds can be extremely low, compounds possessing vapour pressures too low to register in many of the analyses of profiles employed in examining scent sources can still be important odorants. This is not unexpected. But what of the semiochemical potential of components which are totally involatile?

Involatile compounds can mediate semiochemical effects in various ways.

(1) Where physical contact occurs between a mammal and a signal source, there is no reason at all why the signal should be volatile. Mutual licking and grooming sequences occur very commonly in mammals. The male hunting dog sniffs and licks the female vulva increasingly as oestrus is approached and will even vigorously nibble at the female's skin from the ear to the groin (van Heerden, 1981). The asiatic elephant transfers chemical information from its mate to its mouth using the tip of its trunk (Eisenberg *et al.*, 1971), black-tailed deer sniff and lick each other's tarsal scent organs (Müller-Schwarze, 1977) and the male golden hamster is motivated to lick its mate's vaginal secretion (Johnston, 1974). A complete list of such behaviour would be very long, but in very few cases has the nature of involatile semiochemical components of the signal source been defined with any precision, and in only a few cases has precise evidence been presented concerning the semiochemical contribution of such components. It seems, for example, that involatile components present in guinea pig urine are involved in sexual recognition in that

species (Berüter *et al.*, 1973). The puberty accelerating pheromone present in male mouse urine also seems to lack volatility (Vandenbergh *et al.*, 1976) and bioassay procedures have demonstrated that components in female mouse urine which reduce aggression in male mice are so poorly volatile that they are only partially removed by evaporating the urine to a viscous oil either at atmospheric pressure or under vacuum (Evans *et al.*, 1978). It seems also that the compounds in male mouse urine which evoke the well known Whitten and Bruce effects (Section 6.3) are associated with the protein fraction (Marchlewska-Koj, 1977), while in the hamster vaginal secretion an apparently involatile 'mounting pheromone' component appears to be associated in some way with high molecular weight protein fractions (Singer *et al.*, 1980). In such studies it is always difficult to be sure the behavioural or physiological responses are due to involatile components rather than to trace quantities of volatiles which have not been completely removed. Thus Novotny *et al.* (1980a) have observed that even the most rigorous measures to remove volatiles from mouse urine peptide fractions yield materials which still retain a distinctive mouse odour to the human nose.

On licking the source, the mammal is able to detect chemicals present not only by taste and by olfaction, but also via the separate neural pathway provided by the chemoreceptors of the vomeronasal organ (see Section 9.1). This organ, present in most terrestrial species, with the exception of the higher primates and certain bats, comprises a pair of elongated fluid-filled sacs, which are lined with receptor cells and which open anteriorly to connect with the palate via the nasopalatine duct. Commonly the animal responds after making mouth contact with the source by displaying a stereotyped grimace or **flehmen response** (a lifting of the upper lip, closing the nostrils, deep breathing and possibly head movement) which is generally thought to be associated with the use of the vomeronasal organ (Estes, 1972; O'Brien, 1982). Flehmen is known in a wide range of mammals. In the asiatic elephant, referred to above, a male responds to an oestrous female by sampling female urine or the female anogenital area with its trunk, transferring substrate directly to the openings to its vomeronasal organ in the roof of its mouth, and exhibiting flehmen (Rasmussen *et al.*, 1982).

In most cases, dissolved material taken into the mouth is directed to the organ by the action of the tongue and by the pumping action of the organ itself (Meredith *et al.*, 1980). Using rhodamine, an involatile fluorescent dye, it has been shown quite clearly that involatile components present in hamster urine very efficiently reach the vomeronasal organ of a hamster inspecting this material (Beauchamp *et al.*, 1980; Wysocki *et al.*, 1980). Here it seems that the involatile substances reach the vomeronasal organ via the external nares either by direct contact of the nose with the urine or by transfer via the tongue. The vomeronasal organ appears to be particularly important in relation to reproduction (Bailey, 1978; Keverne, 1979; Müller-Schwarze, 1979; Johns, 1980; Meredith, 1980; Ladewig *et al.*, 1980; Mossing and Damber, 1981).

Figure 2.3   Flehmen in a black tailed deer, *Odocoileus hemionus columbianus*. Note the curled upper lip exposing the palate in which is located the entrance to the vomeronasal organ. (Photograph courtesy D. Müller-Schwarze)

(2) The question of volatility can play no part in the transmission of water-borne signals between aquatic mammals, unless of course it acts at the water–air interface. However such chemical signalling remains little studied (Lowell and Flanigan, 1980).

(3) Involatile compounds can also play a part in chemical signalling by modifying the release of volatile substances with which they are associated in a scent source. Studies on model scent marks (Regnier and Goodwin, 1977) using the polar lipid odorant, phenylacetic acid, in radiolabelled form, has shown that the volatility of this substance decreases with the increasing polarity of the involatile lipid in which it is dissolved (in this case, volatility is retarded in the sequence DEGS > squalene > mineral oil) but is increased by ambient humidity. Such effects differ significantly from odorant to odorant, the volatility of a less polar lipid odorant being less affected by substrate polarity or by ambient humidity. As a result, the volatile profile will vary significantly with the nature of non-volatile components of the scent mark, as well as with environmental factors.

Many mammalian scent materials, such as the anal sac secretions of certain carnivores, consist of two phase liquid (aqueous–oil) mixtures, each phase modifying the odour of the other depending on the distribution of odorants between them. The presence in an aqueous scent source of such dissolved involatile material as salts and proteins or of solid particles also modifies the

Table 2.1. Differences in odour threshold measured in water and in vegetable oil (ppb). Reproduced with permission of Pudoc, Wageningen

Compound	Threshold concentration		Ratio oil to water
	in water	in oil	
Methylmercaptopropanal	0.2	0.2	1 : 1
2,5-Dimethylpyrazine	1800	2600	1 : 1.4
2-Phenylethanal	4.0	22	1 : 5
Hexanal	4.5	120	1 : 27
Hept-2-enal	13.0	1500	1 : 115
Nonanal	1.0	1000	1 : 1000
Non-2-enal	0.08	150	1 : 1875
Dec-2-enal	0.3	2100	1 : 7000

Guadagni *et al.*, 1972; Rothe, 1975.

headspace volatile composition as the result of physical and/or chemical interactions (Maier, 1970, 1975) (Table 2.2).

(4) Finally, volatiles could exist in a scent source as involatile scent precursors, to be released subsequently as the result of enzymatic or microbial action. Once again, mammalian studies are lacking but we know that many vegetable aroma compounds exist as involatile precursors in the intact vegetable cell, to be released enzymatically only when the cell is disrupted (Grosch *et al.*, 1975; Tressl *et al.*, 1975). Thus many organosulphur volatiles associated with onion and garlic aroma occur in the intact cell as S-alkyl and S-alkenyl-L-cysteinsulphoxides while thioglucosides fulfil a similar function in radish and other plants and the unsaturated C_{18} acids, linoleic and linolenic acids, act as precursors for the range of saturated and unsaturated aldehydes which dominate the aromas of such vegetables as tomato and cucumber.

Table 2.2. The binding of volatiles by protein. A buffered (pH 6.9) aqueous solution of soy protein (5%) differentially binds organic volatiles compared with protein-free solution (Gremli, 1974)

Volatile[1]	% retention (reversible plus irreversible) by protein[2]	% irreversibly bound by protein[3]
Hexanol	0	0
Hexan-2-one	5–16	0
Hexanal	37–44	<5
Hex-2-enal	68–75	28–34
Decan-2-one	59–68	0
Decanal	94–97	13–22
Dec-2-enal	100	38–48

[1] 10 mg per 100 ml solution.
[2] Determined by headspace analysis at 22 °C.
[3] Bound to protein after removing water and volatiles under vacuum.

Saliva enzymes could liberate volatiles from involatile precursors taken into the mouth by licking a signal source. In one case, saliva enzymes are known to function as part of a chemical signal source, although details of the system are at present unresolved (Harderian gland secretion, Section 4.5).

In general, however, the communicatory significance of involatile components remains unexplored even in the case for aquatic mammals where water solubility replaces volatility as a physical parameter of major semiochemical importance. Experimental evidence seems to suggest that cattle are able to detect at a distance by olfaction the presence of inorganic salts in solution, although by what mechanism this occurs remains unclear (Bell and Sly, 1980).

2.4 HANDLING THE VOLATILES

In mammalian semiochemistry, the volatiles emitted by the odoriferous substrate are of considerable biological interest so that while not underestimating the signalling potential of other components, a detailed discussion of techniques appropriate to their study is in order. It is important to note that the chemical compositions determined are technique dependent and are as dependent on the details of the experimental techniques employed as they are on the composition of the substrate itself.

Techniques have been developed by flavour chemists for the study of volatiles associated with food materials and involve steam or vacuum distillation and solvent extraction, with accompanying procedures to minimize thermal decomposition or air oxidation of labile components. Some work of this kind on plant aroma volatiles has direct relevance to mammal studies, as in cases where a mammal is attracted to a major plant food source by olfactory cues (Bullard and Holguin, 1977). However, the quantities of starting material available to the mammalian semiochemist are usually very much smaller than those available to the food flavour chemist.

The chemical analysis of these volatile substances relies heavily on gas chromatography (GC) based techniques. But while the application of GC methods in general have been explored in a very wide variety of analytical contexts so that there now exists a vast and rapidly growing literature on their application, one aspect of the analytical procedure does merit special attention. This concerns the methods by which the volatiles associated with a scent source are collected prior to analysis. This crucial procedure is less simple than it seems. Jennings and Filsoof (1977) state that:

'Although gas chromatography has permitted major advances in the field of volatile analysis, it has not been an unmixed blessing. All too frequently the resultant chromatogram is regarded as a *true representation* of the composition of the starting material. Such an assumption overlooks the facts that not all compounds in the injected sample are stable to the gas chromatographic process and that sample preparation procedures can exercise profound effects on the quantitative and qualitative composition of the injected sample.'

These authors go on to show how a model system consisting of a synthetic mixture of ten volatile organic compounds yields quite different gas chromatographic profiles depending on which sample preparation procedure is used.

However, little analytical work has yet been conducted specifically on

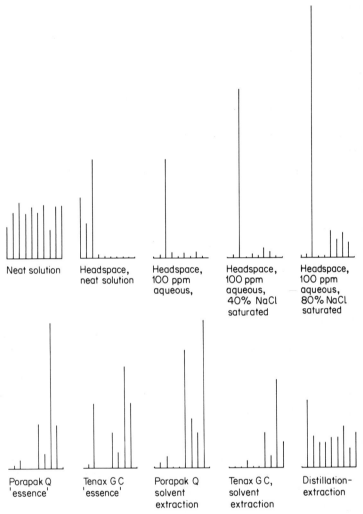

Figure 2.4   Different gas chromatographic profiles obtained from the same mixture of volatile compounds as the result of using different sample preparation techniques. In order of increasing GC retention time (to the right), the mixture consists of ethanol, pentan-2-one, heptane, pentan-1-ol, hexan-1-ol, hexyl formate, octan-2-one, limonene, heptyl acetate and γ-heptalactone. (Reprinted with permission from Jennings and Filsoof, 1977. Copyright 1977 American Chemical Society)

mammalian scent components. The following approaches derived largely from the concerns of perfumery, insect pheromone studies and clinical chemistry, provide important bridgeheads which are now beginning to be exploited in mammalian studies. It will be convenient to refer to work in this area under the following headings.

(1) Headspace analysis, in which the vapour above an odour source is analysed directly.

(2) Enfleurage techniques, in which volatiles emitted by a scent source are taken up in an involatile, inert lipid placed in proximity to the source.

(3) Microdistillation methods, in which the scent source is heated and the volatiles are collected as they distil off.

(4) Trapping methods in which volatiles are trapped from a large volume of air or carrier gas.

(5) Microtrapping methods, adapted for very small samples.

(6) Whole organism studies.

These are in no way exclusive categories and there is no absolute division between any of them. However, they present a useful grouping for our ideas.

Headspace analysis
(Hachenberg and Schmidt, 1977)

Since, with scents we are ultimately concerned with the relative proportions of components in the vapour phase, it is well to remember that these proportions are not the same as those present in the scent-emitting substance. Quantitatively quite different profiles of volatiles are likely to be obtained if we examine the headspace above a secretion sample and compare this result with the proportions of these compounds observed on analysing the secretion directly.

Techniques of headspace analysis provide chemical data concerning the composition of a vapour which is in equilibrium with a scent-emitting substance maintained in a closed container. The material under study is allowed to stand in a sealed container at a constant temperature for sufficient time for equilibrium between the gas and condensed phases to be achieved. A vapour sample then taken from the sealed container, for example by a gas-tight syringe via a septum, is analysed directly by injecting the vapour into a gas chromatograph. To facilitate the attainment of the necessary equilibrium conditions, the volume of the sealed container must not be excessive (<10 ml) while the volume of the scent-emitting substance it contains must not be too small relative to this volume (~3 ml) for otherwise erroneous values for the headspace equilibrium concentrations would be obtained (particularly of the minor volatile components). Of course the container should be strong enough to withstand the pressure generated where an elevated temperature is used.

Provided the system is in equilibrium, the relative proportions of each component in the vapour phase so determined is related to its partial vapour pressure in the system. For component 'i' this partial pressure, $p(i)$, is proportional to $p_0(i)$, the vapour pressure which would be exerted by a pure sample of 'i' at the same temperature, to $c(i)$, the concentration (mole fraction) of 'i' in the liquid phase under study, and to its activity coefficient, $\gamma(i)$, a quantity which expresses mathematically the fact that the contribution to the vapour phase of each component in the liquid phase is modified by interaction with the other liquid phase components (except in what are called ideal mixtures where such interactions are negligible and $\gamma = 1$):

$$p(i) = p_0(i) \cdot c(i) \cdot \gamma(i)$$

The value of $\gamma(i)$ depends on the nature and relative proportions of all components present in the liquid phase, on temperature, and to some extent on pressure. For the dilute solutions of odorants commonly encountered in a natural scent, $\gamma(i)$ can be assumed, as an approximation, not to vary with $c(i)$.

In such a complex situation, it is often difficult to work back from an analysis of headspace composition to determine the composition of the scent emitter. For this purpose, calibrations would be required using standard mixtures having a composition which in all regards is as close as possible to that of the scent-emitting material under study. This is because $\gamma(i)$ depends on the total mixture composition. A peak enhancement method can be used in which headspace analysis is performed on a sample and then on a second sample identical save that it contains an added known quantity of the particular volatile component being determined with precision.

Although such matters should be borne in mind when considering the differences between vapour and liquid phase compositions of a natural scent, the above situation represents a considerable simplification of natural conditions in a number of important regards. For example, a scent-emitting substance is not generally in equilibrium with its vapour. Even so, it is generally to be expected that vapour and liquid compositions will differ in the kinds of ways indicated.

A major problem with headspace analysis centres on the need to increase the overall sensitivity to determine components which may be present in the vapour phase in trace amounts. One possibility is to inject larger sample volumes into the gas chromatograph, but this is limited by the drastic loss in chromatographic resolution which occurs when the vapour sample size approaches the volume of the column itself, which may be as low as a millilitre for high efficiency capillary columns. A second method is to raise the temperature of the sample, thus increasing the saturated vapour pressure terms, $p_0(i)$, but this is also subject to practical limitations. A third convenient method is to increase the activity coefficient, $\gamma(i)$. This has been achieved by adding salts such as ammonium sulphate or potassium carbonate to about 30% for the headspace analysis of such aqueous scent sources as urine. Quite substantial improvements are thus obtained, the magnitude of the salting-out effect depending on the particular salt employed. Non-electrolytes may also be used in certain cases. Thus, the partial vapour pressures of volatile organic substances dissolved in a water-miscible organic solvent, such as dimethylformamide or acetone, can be increased by the addition of water, the greater increase being noted in the more hydrophobic components.

Important adjuncts to headspace analysis are the various concentration methods by which substances present in the vapour phase; even at very low levels are trapped out prior to chromatographic analysis. Continuous non-equilibrium methods will be described in which a stream of inert gas is passed through the system and the volatiles taken up in it are trapped out. These methods by which substances present in the vapour phase; even at very low levels are trapped out prior to chromatographic analysis. Continuous non-equilibrium methods will be described in which a stream of inert gas is passed

Enfleurage techniques

A range of methods are available for the study of natural odour sources which derive ultimately from the method of 'enfleurage' employed in perfumery. In enfleurage, flowers are left sealed at room temperature in small containers the interiors of which are coated with involatile oil or fat. Scent components emitted by the flowers dissolve in the oil or fat and this transferred scent is subsequently examined further. Bergström (1973) has described a modified enfleurage method in which parts of orchid flowers were maintained for some days in contact with a preconditioned preparation of Chromosorb G loaded with 10% silicone high vacuum grease. The flower scent substances, which continued to be evolved during storage, were slowly transferred to the high vacuum grease and subsequently examined by gas chromatography.

The perfume chemist, in order to be able to formulate a satisfactory perfume, is less concerned to know the vapour composition of a particular scent than he is to know the composition of the liquid mixture which will yield that vapour, for as we have noted the composition of volatiles in the vapour and in the liquid phases will generally be different. In this context, enfleurage methods have been further elaborated as follows.

If a scent-emitting material is allowed to stand in a sealed container in close proximity to a small quantity of an inert, involatile oil, a situation will eventually be achieved in which scent volatiles will be present both in the oil and in the scent source such that their levels in both these condensed phases result in equilibrium with the same scent vapour. Although the proportions of the odorants which are then present in the oil may differ from those in the original substance, the oil drop will yield the same vapour and have the same odour so that gas chromatography of the oil drop will yield the information the perfume chemist requires.

Experiments of this type, reported by Sully (1971), employed dibutyl phthalate as the odour solvent. Phthalate esters are themselves odourless substances having excellent lipid solvent properties, and are readily purified from low-boiling impurities. Raoult's law indicates that uptake of vapour will increase as the molecular weight of the absorbing liquid decreases. Dibutyl phthalate has the advantage of a relatively low molecular weight (278) coupled with a longer GC retention time than most odorants (b.p. 340 °C). Diethyl phthalate is also used (mol. wt. 222; b.p. 296 °C). In his experiments, Sully studied the uptake of odorants by a microdrop ($\sim$3 μl) of the phthalate ester suspended in a platinum wire loop (6 mm $\times$ 1 mm) over the odour source in a sealed container. Different components required different times to reach equilibrium and equilibration times of between 18 hours and 4 days were employed. The rate-limiting step appeared to be the diffusion of the odorant through the air and the rate of equilibration was found to be greatly increased under a partial vacuum. As water is commonly a major component of many natural materials, the vacuum employed did not normally exceed the saturated vapour pressure of water at the temperature of the experiment. Following an experiment on thyme odour using diethyl phthalate, peaks of gas chromatographic retention time in excess of that of diethyl phthalate were noted, suggesting that all likely odoriferous components had probably been transferred.

The method has been applied more recently by Bullard et al. (1978a,b) to the chemical analysis of a fermented egg product which has mammal pest control potential, being both a repellant to species of deer which cause browse damage to fir seedlings of importance to forestry, and also attractive to coyotes which constitute an important predator of livestock in the USA. Seventy-six odour components were identified by this method and a further 22 by other techniques. A synthetic product was then formulated for field testing.

Microdistillation methods

A further approach to the study of volatiles is to employ microdistillation techniques in conjunction with gas chromatography. Thus, following the technique described by Bergström (1973) for use with packed gas chromatography columns, and by Ställberg-Stenhagen (1972) for capillary columns, volatile substances were similarly transferred directly from insect or plant tissue samples to the gas chromatography column. The tissue sample was loaded into a specially designed 'precolumn tube' which was inserted into a modified gas chromatograph injector port. The volatiles were then evaporated from the samples by raising the temperature of the modified injector port while passing a stream of inert gas through the loaded precolumn tube. The volatiles were condensed from the gas stream by passing through a section of empty gas chromatography column cooled in dry ice/ethanol immediately before the analytical column in the gas chromatograph. When distillation was complete, the precolumn tube was replaced by a clean tube (to reduce the risk of pyrolysis product formation from the organic residue), the coolant was removed from the trapping tube, and gas chromatography was allowed to proceed in a temperature-programmed mode. The risk of thermal artefact formation again arises and the baking temperature must be as mild as possible consistent with adequate sample collection. Bergström recommended heating the modified injector port up to 170 °C over 10 minutes. Bergström's method has been modified for use in many subsequent studies.

A closely similar precolumn technique has been used to study the volatile substances associated with the tarsal gland hair tufts of the reindeer (Andersson *et al.*, 1975).

A related micromethod has been reported by Morgan and Wadhams (1972a,b). Insect tissue was injected directly into a gas chromatograph and the volatiles baked off into the carrier gas and analysed directly. Using this method it was possible to analyse the mixture of some 20 volatile compounds (total mass, 1 μg) associated with a single insect gland, so that it becomes feasible to consider analysing the volatiles from, for example, a single mammalian sebaceous follicle. A typical procedure is as follows.

Individual insect glands, or batches of glands (tissue volume range 1–100 μl) are sealed in glass capillary tubes and introduced into the heated (210 °C) injection port of the gas chromatograph using a solid sample injector device produced by Hewlett-Packard. The sealed capillary is retained in position until temperature and gas flow conditions have stabilized (about 5 minutes) and then crushed by depressing the plunger of the solid sample injector, thus releasing the volatiles onto the analytical column. This method clearly raises concern that some of the products observed may be thermal artefacts of the system, but in the example quoted this did not seem to be so. More recently the method has been used with capillary GC columns (Morgan *et al.*, 1979).

The method has been used to study the volatiles associated with the supracaudal scent gland of the red fox (Albone, unpublished).

A simple apparatus for vaporizing volatiles from a biological matrix and injecting them onto a gas chromatograph has also been described by Karlsen (1972).

Trapping methods
(Zlatkis and Shanfield, 1979)

The volatiles associated with biological fluids have been collected by sweeping a stream of inert gas over the surface of the fluid under examination and subsequently trapping the organics taken up in the gas stream. Cold traps have been used for this purpose in the analysis of urine and breath volatiles (Teranishi *et al.*, 1972). In addition to questions concerning the efficiency of this process major difficulties are the tendency of the traps to plug with ice and the large quantities of water which are retained in the final product. Ten exhalations of human breath alone will contribute 0.8 g of water to such a collecting system. A more satisfactory system employs adsorbent trapping materials (Zlatkis *et al.*, 1973c; Liebich and Al-Babbili, 1975).

Many common sorbents, such as silica gel, alumina and charcoal, suffer the disadvantage of binding volatiles too strongly and/or possessing strong affinities for water. Adsorbents employed have usually been selected from among the materials in use as gas chromatography column packings, for which compound retention characteristics are already known (Butler and Burke, 1976). Some of these have been evaluated in relation to the environmental sampling of airborne pollutants (Bertsch et al., 1974; Parsons and Mitzner, 1975; Russell, 1975). Trapped organics are subsequently recovered by heating the adsorbent in an inert gas stream (purging) although solvent elution has also been used.

An adsorbent now in common use in this context is Tenax GC, a porous polymer derived from 2,6-diphenyl-p-phenylene oxide, which has the merit of allowing the passage of water vapour while efficiently adsorbing organic vapours at room temperature (Novotny et al., 1974, McConnell and Novotny, 1975). Tenax has high thermal stability (to 350–400 °C) and trapped organics are satisfactorily recovered by purging at 300 °C. There remains concern that any thermally labile compounds present may be destroyed in the process and lower purging temperature (to 200 °C) have been used, although this leads to a less complete recovery. Thus C_6 and C_7 ketones have been found to desorb well at 300 °C but poorly at 220 °C. The Tenax technique appears not to be satisfactory for the often quite odorous compounds of molecular weight higher than about 200 although reports vary and volatiles of considerably greater molecular weight have been reported desorbed from certain other adsorbents (Chromosorb 105) at high gas flow rates. Quantitative studies on insect semiochemical alkenyl acetates have shown that Tenax can be used effectively to trap and subsequently to release, either by thermal desorption or by solvent (hexane/diethyl ether, 8/2) elution, low levels of C_{12} to C_{18} acetates (Cross, 1980; Cross et al., 1980). A typical procedure is as follows.

(1) Trapping. Urine (100 ml) containing ammonium sulphate (30 g), added to increase the volatility of low molecular weight organics, is contained in a sample bottle immersed in a 90 °C water bath and fitted with a water condenser (coolant temperature 12 °C), and stirred magnetically while helium gas is passed over the urine surface at 20 ml/min for one hour. The gas stream emerges from the apparatus via the condenser and passes subsequently through a trapping tube containing 2 ml Tenax GC (35/60 mesh) which had been carefully preconditioned in an inert gas flow for 30 min at 375 °C. At the end of the procedure, the trap may be purged with dry nitrogen for a short time to remove any residual water.

(2) Recovery of volatiles. The Tenax trapping tube is designed to fit as a precolumn tube into a modified gas chromatograph injector port. The tube is positioned in the port so that inert carrier gas flows through the Tenax in the reverse direction to that employed during trapping. Volatiles are desorbed by heating to 300 °C and are subsequently condensed out in a cold (liquid air or dry ice/acetone cooled) section of capillary tubing placed before the main analytical column of a gas chromatograph, possibly also coated

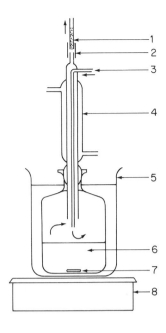

Figure 2.5   Apparatus for trapping urine volatiles on Tenax. 1 Trapping tube containing Tenax. 2 Teflon sleeve. 3 Helium flow inlet. 4 Condenser. 5 Hot water bath. 6 Urine sample. 7 Magnetic stirring bar. 8 Hot plate-magnetic stirrer. (Redrawn with permission from Zlatkis *et al.*, 1973c)

with the GC stationary phase. The coolant is then removed and gas chromatography proceeds in a temperature-programmed mode. Gas chromatographic analysis of the complex mixtures obtained from urine samples requires the use of high efficiency capillary columns. If a suitable gas chromatograph is not immediately available, the trapping tubes containing Tenax plus adsorbed volatiles may be sealed and stored and analysed at a later date without sample deterioration.

The profiles of volatiles thus obtained are dependent on the precise experimental procedure used, and this must be carefully standardized if comparative studies of different urine samples are to be undertaken. In addition to such factors as the gas chromatographic conditions used and the nature of the adsorbent employed, variables in the procedure which influence the profile characteristics obtained are the temperatures of the heated water bath and of the condenser coolant water, the time period during which helium is passed in the trapping procedure and the desorption conditions subsequently used. The profile obtained is not merely a function of the sample, but very much of the procedure employed also.

Other adsorbents in addition to Tenax have been used. These include Chromosorb 101 (styrene-divinylbenzene copolymer) in a similar procedure for the analysis of urine volatiles (Matsumoto *et al.*, 1973) and Chromosorb 105 (desorption at 150–170 °C) (Murray, 1977) for rabbit anal gland secretion and faecal pellet volatiles (Goodrich *et al.*, 1978, 1981a). Porapak (an alkylstyrene copolymer) is also an effective adsorbent, but like Chromosorb

Table 2.3. Adsorbent traps for headspace volatiles: some examples pertinent to mammalian semiochemistry. Tenax used except where stated otherwise

Substrate	Remarks, including GC column used	References
Human urine	100 ml samples; Emulphor ON 870 capillary	Liebich and Al-Babbili, 1975; Zlatkis *et al*., 1973a,b,c
Human urine	Large scale (use of Chromosorb 101 trap) SF-96-50 plus 5% Igepal CO-880 capillary	Matsumoto *et al*., 1973
Mouse urine	15 ml samples; Emulphor ON 870 capillary	Liebich *et al*., 1977
Red fox urine	0.5 ml samples; SF-96-2000 plus 5% Igepal CO-880 capillary	Bailey *et al*., 1980
Red fox urine	Sampled from wild fox marks in fresh snow; Ucon 50-HB-2000 capillary	Jorgenson *et al*., 1978
Hamster oestrous (vaginal secretions)	Collected on filter papers/50 samples bulked; 10% Carbowax 20M on 80/100 mesh Chromosorb packed column	Singer *et al*., 1976
Human axillary (volatiles)	Directly and via cotton wool pads; 2% Carbowax 20M packed column	Labows *et al*., 1979c
Human breath	50 exhalations; Emulphor capillary	Zlatkis *et al*., 1973c
Tissue homogenate volatiles	Rat liver, lung and brain tissue separately homogenized; volatile profiles from each tissue type characteristically different, probably reflecting distinctive functions and metabolic activities; GE-SF-96 plus 10% Igepal CO-880 capillary	Politzer *et al*., 1975
Human blood component	5 ml serum/plasma; Emulphor ON 870	Zlatkis *et al*., 1974
Rabbit faecal pellets	30 g (use of Chromosorb 105 trap); Carbowax 20M or FFAP glass SCOT column	Goodrich *et al*., 1981a,b Hesterman *et al*., 1981
Ambient environment	Air pollution studies; several hundred compounds noted; Emulphor ON 870 capillary	Bertsch *et al*., 1974
Rabbit anal gland (homogenate volatiles)	20 glands (use of Chromosorb 105 trap); Carbowax 20M glass SCOT column	Goodrich *et al*., 1978
Mouse urine	Ucon 50-HB-2000 glass capillary	Novotny *et al*., 1980a

101, lacks the thermal stability of Tenax. Having a maximum operating temperature of only 230 °C, 'bleeding' (volatile artefact formation from the thermal breakdown of the adsorbent itself) occurs significantly above 150 °C before many compounds of importance are desorbed. Porapak Q has, however, been recommended for trapping the more volatile organics and less volatile compounds can be eluted with solvent. Carbosieve (carbon molecular sieve) is an effective adsorbent of high thermal stability, but here adsorption is so strong that a desorption temperature is required in excess of the pyrolysis temperature of most organic substances.

A method (Parliment, 1981) which has been developed in food science for the extraction, concentration and fractionation of low levels of volatile aroma compounds from quite large volumes of aqueous medium (and here the semiochemist might think of urine) is to pass the dilute aqueous solution through a column containing a reverse-phase packing material. Such a material consists of, for example, silica gel to which octadecyl groups have been bonded chemically. As the aqueous material passes through the column the apolar groups selectively extract the aroma volatiles from the solution and finally the extracted organics are selectively desorbed in small volumes of solvent by eluting the column with solvent gradient.

Applications to small samples

A disadvantage of the method described centres on the large samples generally required (50–450 ml), although profiles have been recorded using this method from 5 ml samples of plasma and serum (Zlatkis *et al.*, 1974; Liebich and Wöll, 1977) and from small volumes of red fox urine (0.5 ml) (Bailey *et al.*, 1980) and of hamster vaginal secretion (Singer *et al.*, 1976). A micromethod (Zlatkis and Andrawes, 1975) has now been developed and applied to the analysis of the volatiles associated with samples of human urine (1 ml), plasma (100 μl) and breast milk (20 μl) (Stafford *et al.*, 1976).

Table 2.4. Volatile components of human breast milk. Tentative identification based on GCMS analysis. Reproduced by permission of Elsevier Scientific Publishing Company from Stafford *et al.* (1976)

Compound	Compound
Hexanal	Methylhexanal
Heptanal	2-Methyl-4-hexene
6-Methyl-2-heptanone	2-Methylamylnitrile
2-Pentylfuran	Hexanol
Ethylcyclopentanone	4-Methyl-1-pentene
Acetonylfuran	Pentylcyclohexene
Ethylcyclohexanone	Aminopentanylfuran
Ethylcyclohexenone	

Using this method, body fluid samples are saturated with ammonium carbonate and organics are extracted in diethyl ether by vortex mixing followed by centrifugation. A precolumn designed to fit into a modified gas chromatograph injector port is previously packed with 0.5–1.0 g pyrex glass wool and preconditioned in a helium gas stream at 350 °C for one hour, and the diethyl ether extract is now injected into this tube so that liquid spreads evenly over the surface of the glass wool. The ether solvent is stripped off by passing helium for a short time at room temperature through the tube and the residual volatile organics analysed by thermal desorption from the glass wool (240 °C) into the gas chromatograph as in the precolumn method described in the previous section.

The method depends on the fact that glass wool has good adsorptive properties for many volatile organic compounds, but low affinity for such polar compounds as diethyl ether and water. The ether extract contains higher molecular weight lipids and although the desorption temperature is relatively mild and no evidence has been obtained of pyrolysis products in the analysis results, the possibility of the pyrolysis of these higher molecular weight substances yielding artefacts cannot be totally excluded.

The efficiency of this method, and of the extraction step in particular, has recently been improved very considerably by the use of a 'transevaporator' (Zlatkis and Kim, 1976; Lee *et al.*, 1978; Zlatkis *et al.*, 1979a,b). 2-Chloropropane (b.p. 35.7 °C) is now recommended as the extraction solvent of choice and the use of glass beads has replaced the use of glass wool.

The procedure is in two steps.

Step 1 The profile of the most volatile components

(1) The precolumn tube (d, Figure 2.6) is loaded with Tenax and the water condenser is inserted at g.

(2) Between 25 and 200 μl of biological fluid is adsorbed on to the lower end of the Porasil tube, a. Porasil E is a strongly hydrophilic porous silica adsorbent.

(3) The apparatus is assembled and helium is passed for 5–10 minutes to transfer the most volatile organics from the Porasil to the Tenax on which they become strongly adsorbed.

(4) Tube d is disconnected, flushed with helium at room temperature to remove any traces of water vapour (which could block the precolumn through ice formation during desorption), and the volatiles are then desorbed at 280 °C into the gas chromatograph for analysis as described previously.

Step 2 The profile of less volatile components

(1) The condenser at g is removed and a second precolumn tube containing glass beads is connected (d).

(2) 2-Chloropropane (0.8 ml) is placed in c and swept through the Porasil

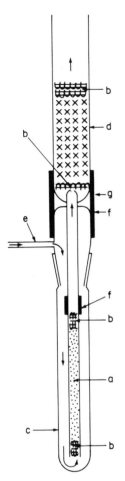

Figure 2.6 The transevaporator. a Tube containing Porasil E (about 0.3 ml). b Glass wool plug. c Outer tube. d Precolumn tube loaded with (for step 1) Tenax, or (for step 2) 80–100 mesh glass beads. e Helium flow inlet. f Teflon sleeve. g A short water cooled condenser inserted here in step 1 to prevent excess water reaching the Tenax. Porasil, Tenax and glass beads are all preconditioned. (Redrawn from Lee *et al.*, 1978 and reproduced by permission of Elsevier Scientific Publishing Company)

column, a, by applying helium pressure at e. This efficiently transfers soluble organics to the glass beads, d, while leaving water, hydrophilic and high molecular weight substances on the Porasil.

(3) Column a is removed and the apparatus reassembled with c in a warm (50 °C) air bath. Helium is passed, evaporating the 2-chloropropane and any traces of water from d.

(4) Tube d is disconnected and its volatiles thermally desorbed into the gas chromatograph for analysis as described previously.

Capillary gas chromatograms containing some 200 peaks were obtained in this way from small samples of human serum. Human breast milk was found to be even richer in volatiles and profiles from human saliva were also recorded. The method was not suitable for similar small samples of human urine as the levels of volatiles were not sufficiently high.

Although the profiles obtained were highly reproducible, percentage recoveries were found to vary (reproducibly) from compound to compound, being particularly low for hydrophilic substances such as propan-1-ol.

The direct injection of crude sample into the gas chromatograph has been used in the qualitative study of hydrophilic and therefore less easily extracted low molecular weight volatile components of blood serum, such as acetone and ethanol. Dangers of artefact formation in the heated gas chromatograph injection port exist. Here the acetone observed is probably in part formed from the decomposition of acetoacetate in the gas chromatograph (Liebich and Wöll, 1977).

Similarly, direct injection methods have been used in the analysis of the volatile fatty acids associated with mammalian communicatory secretions such as the anal sac secretions of dogs and coyotes (Preti et al., 1976), of red fox (Albone and Fox, 1971) and of mink (Sokolov et al., 1980). However, the problem of artefact formation remains (which is not to suggest that it is non-existent for other methods of sample preparation). For example, it is very possible that the 2-piperidone observed in such gas chromatograms is an artefact. It is interesting to note also that the headspace volatile profile may differ quite substantially from the profile obtained by the direct injection method. These points are discussed also in relation to anal sac secretion in Section 5.3.

Studies of whole organism volatiles

Attention has been given to the analysis of the volatiles pervading the air in the vicinity of living organisms. A simple approach is to trap out volatiles from an air stream passed over the organisms by means of a liquid nitrogen trap. This has been employed with large groups of insects. To collect a sufficient quantity of material for analysis, collection has to be continued over an extended time period, and this leads to such problems as ice clogging the traps. More serious, however, is the fact that such collections tend to be highly inefficient for the very low concentrations of compounds encountered. An ingenious condensation trap has, however, been described which overcomes this problem and enables the total organic content of the air to be trapped by condensing the entire air (Browne et al., 1974). After collection, the liquid air is slowly evaporated in a low temperature environment and the residue extracted in an organic solvent.

A modified trap, based on this method, makes possible continuous collection of volatiles over 48–96 hours by permitting condensed liquid air to partially evaporate from the system when the trap becomes full.

An alternative approach is to use a solid adsorbent trapping material. For example, in studies on the pheromones produced by *Trogoderma* beetles, Porapak Q was used as a solid adsorbent to trap organics present in the ambient air surrounding a large number of these insects. The trapped organics were subsequently stripped from the Porapak, not by temperature desorption, which suffers the major limitations described previously, but by elution with organic solvent. In this way C_{17} unsaturated aldehydes were collected and identified as major pheromonal substances. In one study (Cross *et al.*, 1977), (*Z*)-14-methyl-8-hexadecenal was identified as the major airborne pheromone of the beetle *Trogoderma variabile* (emission rate, 37 ng per beetle per day), even though this substance was undetected in the extracts of macerated beetles. The authors speculate that this might indicate the enzymatic destruction of the compound during maceration, or, more probably, the production and release of the chemical by the beetle only 'on demand'. In either case, this finding indicates the value of studying, where possible, the substances emitted into the environment by a living organism to supplement studies on the chemicals which can be extracted from its tissues. The principle of this method, then, is to aerate living organisms for an extended period of time and to trap out the organic volatiles they produce using a solid adsorbent. Advantages of the method are that collection of volatiles can be continued for long periods of time with the minimum of attention (contrast the use of cold trap) and that water vapour passes through the trap and so causes no problems.

This method clearly has potential for the study of the total volatiles produced by mammals also. For example, a chamber has been developed at the US Department of the Army (Ellin *et al.*, 1974) in which human subjects sit

Table 2.5. Method for trapping organic volatiles from the air surrounding living insects (Cross *et al.*, 1976)

Stage 1: Trapping insect volatiles

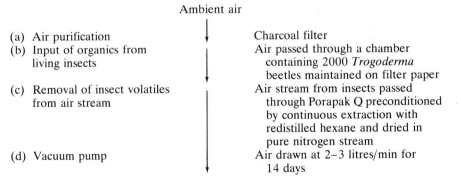

Ambient air

(a) Air purification — Charcoal filter
(b) Input of organics from living insects — Air passed through a chamber containing 2000 *Trogoderma* beetles maintained on filter paper
(c) Removal of insect volatiles from air stream — Air stream from insects passed through Porapak Q preconditioned by continuous extraction with redistilled hexane and dried in pure nitrogen stream
(d) Vacuum pump — Air drawn at 2–3 litres/min for 14 days

Stage 2: Stripping organics from adsorbent
Insect volatiles collected by continuous extraction of charged Porapak adsorbent with hexane

Notes
(1) In spite of precautions, artefacts derived from the solid adsorbent occur in the final extract. In this case 1-(ethylphenyl)-1-phenylethane, derived from Porapak, was noted.
(2) No solid adsorbent will be equally efficient at trapping every class of volatile compound.
(3) Pre-extracted Amberlite has also been used for air purification (Cross *et al.*, 1980).

Table 2.6. Volatile compounds indentified in human effluvia. Adapted with permission of Elsevier Scientific Publishing Company from Ellin *et al*. (1974)

Hydrocarbons
Methane
Ethane
Propane
Pentane
Hexane
Heptane
Decane
Dodecane
2-Methylpropane
2,2,4-Trimethylpentane
2,3-Dimethylhexane
2,2,4-Trimethylhexane
Cyclopropane
Cyclopentane

Unsaturated hydrocarbons
Isoprene
Butene
Hex-1-ene
Hex-2-ene
Oct-1-ene
Oct-2-ene
Nonene
2,5-Dimethylhexa-1,5-diene
Decene
Buta-1,3-diene
Propadiene
2-Methylbut-2-ene
2-Methylpent-1-ene
4-Methylpent-1-ene
2-Ethylbut-1-ene
Propyne
Hex-1-yne
Hex-2-yne
Hept-2-yne
Nonyne
But-2-en-1-ol
Butene-1,4-diol
Pinene
Camphene
Cyclohexene

Alcohols
Methanol
Ethanol
Propanol
Butan-1-ol
Isobutanol
Butan-2-ol
2-Methylpropan-2-ol

Pentan-1-ol
Pentan-2-ol
Hexanol
2-Methylpentan-1-ol
4-Methylpentan-2-ol
2-Ethylbutanol
Heptanol
2-Ethylhexanol
Ethan-1,2-diol
2-Ethoxyethanol
3-Methylcyclopentane-1,2-diol

Acids
Lactic acid
Pyruvic acid

Amines
Ammonia
Methylamine
Ethylamine
Dimethylamine
Diethylamine
Propylamine
Butylamine
Amylamine

Ketones
Acetone
Butan-2-one
Pentan-3-one
4-Methylpentan-2-one
Heptan-4-one
Allylacetone
Mesityl oxide
Cyclohexanone
Octan-2-one

Aldehydes
Methanal
Ethanal
Propanal
Pentanal
Heptanal
2-Ethylhexanal
Crotonaldehyde

Esters
Methyl furoate
Pentyl acetate
Allyl acetate
Methyl butyrate

Table 2.6.—*continued*

Methyl acetate	Pyrrolidine
Butyl butyrate	Pyrrole
	Butylpyrrole
Nitriles	Benzothiazole
Acetonitrile	Benzothiophene
Butylnitrile	2,3-Dimethylpiperidine
Hexylnitrile	Benzofuran
Allylnitrile	Tetrahydrofurfuryl alcohol
	Thiophene
Aromatics	Vinylpyridine ← Vinylpyridine
Toluene	Methylpyridine
p-Cumene	
p-Cymene	*Sulphur compounds*
Benzene	Methanthiol
Aniline	Butanthiol
Mesitylene	Pentanthiol
Ethylbenzene	Hexanthiol
Styrene	Heptanthiol
o- and *p*-xylene	
Cinnamyl alcohol	*Ethers*
Phenylhexane	Dimethyl ether
	Diallyl ether
Heterocyclics	
Furfuraldehyde	*Halogenated hydrocarbons*
Furfuryl alcohol	Dichlorobutane

for periods of an hour while their volatile effluents are collected using cryogenic traps or solid adsorbents (Porapak Q; Chromosorb 102). In this way three to four hundred different compounds were noted of which 135 were identified (see Table 2.6). The authors remark that although they had expected to observe a large number of alcohols and ketones in the effluvia of man, they had not expected the large number of unsaturated and branched-chain hydrocarbons they observed nor had they expected isoprene to be one

Table 2.7. Rates of emission of five volatile human effluents. Reproduced by permission of Elsevier Scientific Publishing Company from Ellin *et al*. (1974)

	Rate of emission (μg/h)		
Compounds	Subject 1	Subject 2	Subject 3
Ethanol	25	58	100
Isoprene	425	251	270
Acetone	360	240	470
Butanol	16	26	41
Toluene	0.6	14	13

of the most abundant human volatiles. Five compounds commonly found in the effluvia of all subjects were quantitated (see Table 2.7).

This approach is now beginning to be applied to problems of mammalian chemical communication and Tenax traps have been used for the study of volatile effluvia generated by certain domestic species.

2.5 THE CHEMISTRY OF AN ODOUR IMAGE

The difficulty of defining an odour chemically is well known to flavour chemists. Although a number of cases are known where an aroma impression is dominated by a single or a small number of 'character impact compounds' (Ohloff, 1978), the characteristic chemical signatures by which we recognize most natural scents and flavours are commonly the property of highly com-

Table 2.8. Some character impact food aroma compounds. Reproduced by permission of Springer Verlag from Ohloff (1978)

Natural occurrence	Character impact compound dominating food flavour
Pear	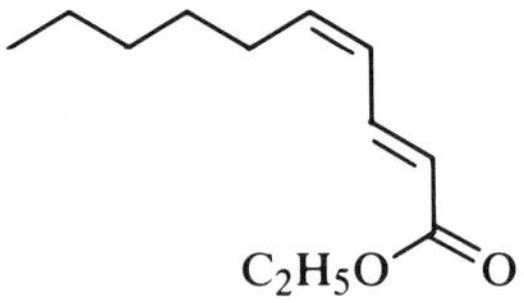
Grapefruit	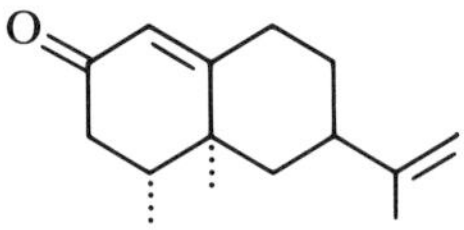
Cucumber	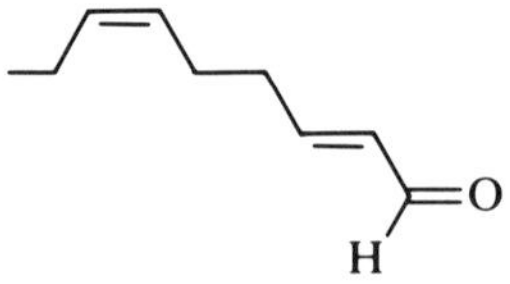
Raspberry	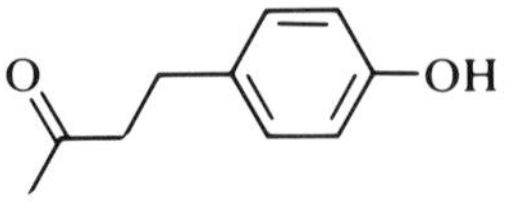

Even a single chemical will generally possess a complex odour quality exhibiting a number of distinct odour notes, probably corresponding to different types of odorant molecule–chemoreceptor interaction.

Table 2.9. Known flavour compounds (1975) classified by compound class

	Number
Aliphatic/aromatic hydrocarbons	170
Isoprenoid hydrocarbons	130
Functionalized isoprenoids	170
Alcohols/phenols	190
Acetals/ethers	140
Carbonyls	310
Acids	230
Esters	450
Lactones	90
Furans/pyrans	110
Amino acids	40
Other nitrogen compounds	290
Thiazoles/oxazoles	60
Other sulphur compounds	220
Total	2600

It has been estimated that the total number of flavour compounds existing is between 5000 and 10,000. Reproduced by permission of Pudoc, Wageningen, from Rijkens and Boelens (1975).

plex mixtures of chemicals. Thus the characteristic aroma of strawberry is associated with a mixture of volatiles which occurs in whole strawberry at the level of 10 ppm and which includes hundreds of different compounds of which more than 300 have so far been identified. Advances in flavour chemistry reveal increasing chemical complexity so that now flavour chemists estimate (Rijkens and Boelens, 1975) that each of the 150 commonly recognized aroma types are likely to be composed of between 300 and 800 components. Nor can gas chromatography be employed in a naive fashion for it is most unlikely that compounds exhibiting the largest peaks on the aroma gas chromatogram will be the substances of greatest importance in determining aroma quality. This arises because odour potency can vary from compound to compound by many orders of magnitude. An indication of this can be obtained by comparing odour threshold values, the levels at which the compounds in question, presented usually in very dilute aqueous solution, can just be detected by the human nose (Stahl, 1973).

Thus, the major determinant of the odour of California green bell peppers is a compound of extremely potent odour which occurs as a minor component of the complex mixture of pepper volatiles. This is 2-methoxy-3-isobutylpyrazine ($T = 0.002$ ppb) (Buttery et al., 1969). The very

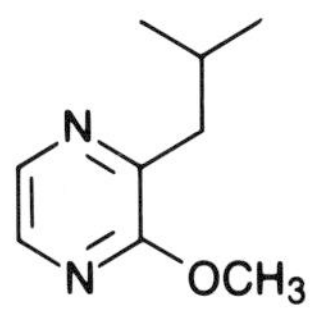

Table 2.10. Some odour threshold values (*T*) in ppb of water. (Ohloff, 1978; adapted by permission of Springer Verlag; Buttery *et al.*, 1981)

Compound		*T*
Acetone		450,000
Butan-1-ol		10,000
Butanoic acid	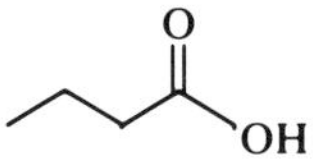	3000
α-Terpineol	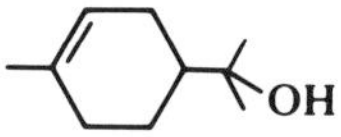	350
γ-Decalactone	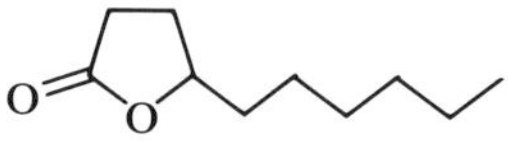	88
Butanal	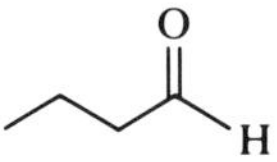	60
Limonene	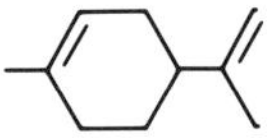	10
Hexanal	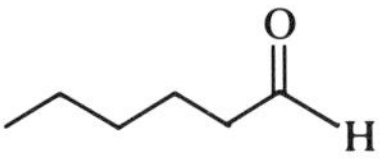	4.5
α-Ionone	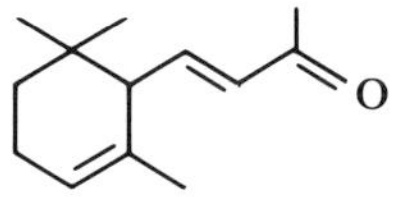	0.4
Ethyl butanoate	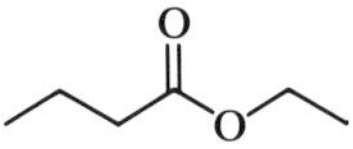	0.1
Geosmin	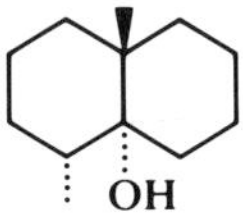	0.021
β-Ionone	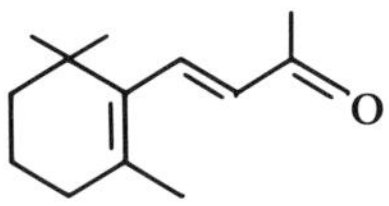	0.007
Thiamin odour compound	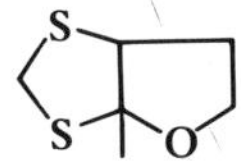	0.0004
2,3,6-Trichloroanisole	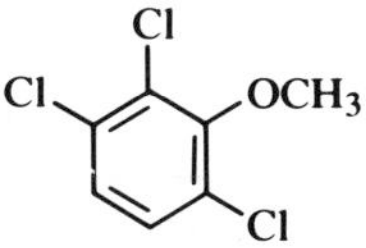	0.0000003

increasing odour potency

low olfactory threshold of this compound suggests its tight binding to its olfactory receptor and, interestingly, this substance has now been found to bind selectively to cow olfactory mucosa homogenate and only slightly to other tissue in comparison (Pelosi *et al.*, 1981, 1982).

Such considerations alert us to the limitations of making simple comparisons between gas chromatograms when examining odour changes accompanying variation in some source parameter. Comparisons would be more meaningful if gas chromatographic data were modified by incorporating some factor to take into account these differing odour potencies. One approach is to annotate the gas chromatogram with odour descriptor terms as the result of splitting the GC effluent so that part can be continuously assessed by a trained observer (Dravnieks, 1975). A different approach has been attempted by flavour chemists using the concept of aroma value (or odour unit value) such that (Rothe, 1975):

$$\left(\begin{array}{c}\text{the aroma value of a} \\ \text{component of a mixture}\end{array}\right) = \frac{\text{(the concentration of the component in the mixture)}}{\left(\begin{array}{c}\text{the odour threshold concentration of the same} \\ \text{component determined in a closely similar medium}\end{array}\right)}$$

Clearly a profile incorporating aroma values will appear quite different from the corresponding conventional gas chromatogram. To illustrate this point

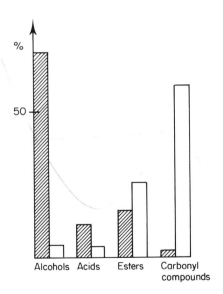

Figure 2.7 Percentage by mass (hatched) of various compound classes represented in whisky aroma, compared with their odour unit percentage contributions. (Reproduced with permission from Salo, 1975)

38

Table 2.11. Relative percentage and odour units of aroma components associated with potato-crisp oil (Guadagni *et al.*, 1972). Reproduced by permission of Pudoc, Wageningen

Compound	Relative % in potato-crisp oil	Threshold value (parts/10^9 parts of oil)	$10^{-3} \times$ number of odour units
Methylmercaptopropanal	2.0	0.2	100,000
2-Phenylacetaldehyde	18.0	22	8180
3-Methylbutanal	5.0	13	3850
2-Ethyl-3,6-dimethylpyrazine	6.5	24	2720
Deca-2,4-dienal	7.5	135	560
2-Methylbutanal	7.4	140	530
2-Ethyl-5-methylpyrazine	6.0	320	190
Pent-1-en-3-one	0.1	5.5	180
Hexanal	2.1	120	175
2-Methylpropanal	0.5	43	120
Non-2-enal	1.5	150	100
2,5-Diethylpyrazine	1.0	270	37
2,5-Dimethylpyrazine	6.5	2600	25
Heptanal	0.6	250	24
Oct-2-enal	0.7	500	14
Hept-2-enal	1.8	1500	12
Dec-2-enal	1.2	2100	6
Nonanal	0.1	1000	1

From Rothe, 1975.

note for example how although the alcohols dominate the composition of the imitation whisky organic volatiles, the carbonyl compounds dominate the aroma (Figure 2.7). Note also how the odour unit profile of potato crisp oil aroma (Table 2.11) differs from the composition profile.

But although raw gas chromatography data may be inadequate, aroma value profiles are themselves open to the criticism and are not available for non-human species anyway. Thus their use assumes that the sensation is the simple sum of the sensations produced by the components taken separately. The situation is, however, not simple and a vector model had been used to sum perceived intensities of odorants in mixtures (Berglund *et al.*, 1973). Synergistic and antagonistic effects can occur between odorants (Köster, 1969). For example, a mixture of isobutyric acid, butyric acid and hexanoic acid, each present at levels which would be individually subthreshold (odour values 0.3, 0.2 and 0.3), together produce a detectable odour considerably in excess of the sum of their individual odour values (odour value, 3.3). This means that the threshold of the mixture is substantially less than would be expected by adding the individual contributions of the different acids (Salo, 1975).

Further, their use implies that for each odorant, the same constant increase in odour intensity occurs for each multiple of its threshold concentration and

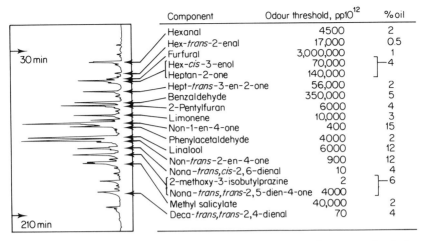

Component	Odour threshold, pp10^{12}	% oil
Hexanal	4500	2
Hex-*trans*-2-enal	17,000	0.5
Furfural	3,000,000	1
Hex-*cis*-3-enol	70,000	}4
Heptan-2-one	140,000	
Hept-*trans*-3-en-2-one	56,000	2
Benzaldehyde	350,000	5
2-Pentylfuran	6000	4
Limonene	10,000	3
Non-1-en-4-one	400	15
Phenylacetaldehyde	4000	2
Linalool	6000	12
Non-*trans*-2-en-4-one	900	12
Nona-*trans*,*cis*-2,6-dienal	10	4
2-methoxy-3-isobutylprazine	2	}6
Nona-*trans*,*trans*-2,5-dien-4-one	4000	
Methyl salicylate	40,000	2
Deca-*trans*,*trans*-2,4-dienal	70	4

Figure 2.8 Composition of steam volatile components (10 ppm of whole pepper) from California green bell peppers. Gas chromatogram annotated with odour threshold data. (Adapted with permission from Buttery *et al.*, 1969. Copyright 1969 American Chemical Society)

it also requires an adequate knowledge of odour thresholds. Yet an odour threshold varies considerably as a function of the person tested, as a function of the test procedure employed (Rothe, 1975), and as a function of the composition of the medium in which the odorant is dissolved.

Although we can be sure that considerations of this kind are pertinent to odour perception in other mammals also, knowledge of threshold data in non-human species is minimal.

A detailed discussion of links between odour perception (and other types of chemoreception) and the chemical nature of the odorant is given in Chapter 9 (see also Beets, 1978; Moskowitz and Warren, 1981; Theimer, 1982).

Chapter 3
The Skin

3.1 THE UNIQUENESS OF SKIN

Much of the chemical ambience of a mammal arises from its skin. The skin is unique among tissues in occupying the interface between the internal physiological and biochemical world of the mammal and the external world of social and ecological relationships in which it lives. The chemistry of the skin surface mirrors this intermediate role. Its products guard the mammal against infection and dehydration and, by the evaporation of sweat, provide a mechanism for temperature regulation. And since the skin represents to the outside world aspects of the internal physiological state of the mammal, it is not surprising that it plays an important part in signalling, both visually and chemically (Mykytowycz and Goodrich, 1974). Many of the characteristic body odours of mammals are, at least in part, associated with their skins, as has been reported, for example, in the case of the African hunting dog, *Lycaon pictus* (van Heerden, 1981). Sometimes the most noticeable skin odours derive from skin microorganisms, and this aspect is discussed in Chapter 5, but more commonly nothing is known of the nature of the odorants or of their precise origins. In this chapter we examine something of the surface chemistry of skin, of the origins of the materials which make up its surface film and of the factors which regulate their production.

The substances which the skin presents to the environment include the secretions of the three classes of skin glands—sebaceous, apocrine and eccrine—as well as materials contributed by the epidermis itself. It is often difficult to distinguish precisely between the contributions of these different sources as their products are generally mixed together and with other substances on the skin surface in the form of a partly emulsified aqueous–lipid film deriving from a multiplicity of sources. In addition, skin surface chemistry exhibits further modifications as the result of the action of commensal skin microorganisms and of various environmental factors. Patterns of chemical production vary over the body surface and are regulated closely by hormonal and other physiological factors so that, in spite of the complicating factors, information concerning the mammal's internal state becomes encoded in its skin chemistry.

The distribution of skin glands in man

Sebaceous glands

Man is exceptionally well endowed with sebaceous glands, these being best represented on the forehead and in the scalp and being absent from the palms of the hands and the soles of the feet, according to the following order of abundance:

scalp, forehead > chest > back > axillae > anogenital region >
abdomen > arm, leg > palm, sole (absent)

The reason for such a profusion of sebaceous glands is not understood. Specialized or enlarged sebaceous glands, in many cases devoid of hairs, occur around major points of entry into the body, including the lip surface, around the nostrils, at the edge of the eyelids (meibomian glands), in the external ear canal, around the nipples, in the labia minora and the prepuce and on the mucocutaneous surface of the penis and clitoris.

Apocrine glands

In man these are limited to the axillae, the nipples and areolae, the anogenital area, the edge of the eyelids and the external ear canal.

Eccrine glands

These are widely distributed over the entire body surface except for the lips, labia minora, prepuce and the mucocutaneous surface of the penis and clitoris. They are most numerous on the sole of the foot and the palm of the hand. Especially large eccrine glands occur in the axillae and groin.

Mixed gland type areas

The secretions observed on the body surface are mixtures of materials from various sources. Apocrine/sebaceous mixed secretions seem to be semiochemically important in the axillae, and possibly in the groin. In the nipple (Montagna, 1970), compound sebaceous glands and apocrine glands open directly to the skin surface and may contribute, together with milk components, to the nipple odour by which infants seem to be able to recognise their mothers. Clusters of large sebaceous glands and some eccrine and apocrine glands are present in the areolae. Cerumen (ear wax) is a mixture of secretions of four gland types: cerumenous specialized sebaceous, typical sebaceous, apocrine and eccrine.

Although the majority of studies in the area covered by this chapter have related to the human species, those few studies which have been undertaken on the general skin chemistry of other mammals lead us to suppose that as a general rule considerable interspecies variation will be observed. Unfortunately, very few of the chemical studies reported here arose out of a semiochemical interest so that trace constituents of possible major communicatory significance have not been sought. Even so, present understandings of the chemistry of the skin surface provide an important starting point for future research in mammalian semiochemistry. And even where skin surface components lack volatility sufficient for olfactory communication, they may yet play an important part in chemical communication (see Section 2.3).

In this chapter we consider in turn those structures of the skin which, because they influence the chemistry of the skin surface film, must play some part in any production of skin-derived semiochemicals, the ways in which these structures interact with circulating hormones, and finally something of what is known of the chemical composition of the skin surface film. In the following chapter these ideas are extended to include those skin structures which are the specialized scent gland complexes which many species possess.

3.2 THE STRUCTURES OF SKIN
(Rothman, 1954; Montagna and Parakkal 1974)

Skin consists of three major layers. The outer layer, or epidermis, contains no blood vessels and receives nutrients only by diffusion from the underlying dermis. Beneath the dermis lies an inner layer of subcutaneous adipose (fatty) tissue.

The epidermis, the outer layer of skin

In man the epidermis is 0.1–0.3 mm thick, except on the soles of the feet and the palms of the hands where it is considerably thicker. The epidermis itself consists of a number of layers. The inner layers are living and these give rise to the outer horny layer, the **stratum corneum**, which is dead. This tough, impervious, protective outer structure is composed principally of inert, fibrous proteins called keratins.

Within the epidermis, new cells arise in the innermost, basal layer, and as basal cell proliferation continues, the epidermis cells once formed undergo a slow process of degeneration and keratinization while progressing through the various layers of the epidermis, ultimately to be compacted into the inert stratum corneum. Later, as more keratinized cellular remnants are added to this layer, they move into the upper stratum corneum where adhesion is progressively lost and finally their material is sloughed off as tiny flakes or **squames** from the skin surface. A person may lose up to a gram of material per day from his body surface by this desquamation process. These squames have a considerable absorptive capacity for water and other fluids and provide one avenue for the distribution of material from the skin surface. For each epidermal cell, the whole sequence takes from two to four weeks.

The process of keratinization is itself a source of lipids for the skin surface. During keratinization, the cytoplasmic globular proteins of the epidermal cells are transformed into the substance of the inert, fibrous and amorphous keratins while the nucleus and all intracellular membranes disintegrate, the

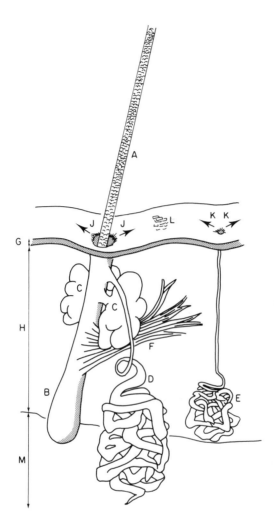

Figure 3.1   Structures of human skin. Considerable variations occur in other species.
A   Hair, covered with scales the pattern of which varies with species. On occasions,
these may be adapted to retain secretion (osmetrichia or scent hairs). B   The hair
follicle, from the base of which the hair grows. In primates, the follicles are typically
arranged in clusters of from 20–30 in the prosimians to 2–3 in the apes. In such species
as the dog and the sheep, compound follicles each supporting a number of hairs occur.
C   Sebaceous glands discharging into the pilosebaceous canal surrounding the hair in
the upper hair follicle. D   Apocrine gland discharging into the pilosebaceous canal
above the sebaceous duct. E   Eccrine gland discharging directly to the skin surface.
F   The piloerection muscle (arrector pili) which contracts involuntary in response to
cold stimuli and emotional stress, causing the hair to be elevated and tending to expel
secretion from the pilosebaceous duct, although how effective this expulsion is is
uncertain. Absent in some species (e.g. lemurs, seals). G   The epidermis. H   The
dermis. J   Discharge of sebaceous and apocrine secretions. K   Discharge of eccrine
sweat. L   Desquamation (flaking) of the surface layer of the epidermis (stratum
corneum). M   Subcutaneous fatty tissue

final residue being a flattened, dehydrated, keratinized remnant. In the process, considerable quantities of non-polar lipids are released. However, it is not always easy to distinguish this lipid from that of sebaceous origin with which it is mixed and conversely it is almost impossible to collect a sample of human sebum from the skin surface which we can be sure is free of epidermal lipid. This is because sebum diffuses into and through the absorbent layers of the stratum corneum following secretion. In man, the contribution of epidermal lipid to the skin lipid film is estimated at 5–10 $\mu g/cm^2$ which, unlike sebaceous lipid, is formed evenly over the entire body surface. In most species it is often very uncertain how much skin or fur lipid is of epidermal and how much is of sebaceous origin.

The dermis

Unlike the epidermis, the dermis contains blood vessels and nerves. It is also rich in materials which provide physical protection for the underlying tissues. These substances include microfibres of collagen and elastin as well as a protective aqueous gel of proteoglycans (mucopolysaccharide combined with protein) and also of the free mucopolysaccharide, hyaluronic acid. Interestingly, the hyaluronic acid is the basis of a visual chemical signal for oestrus in certain primates. In response to blood oestrogen, the hyaluronic acid content of the sex skin of female baboons and some macaques increases. This leads to a considerable increase in osmotic pressure, to the swelling of the sex skin and to the visual signal denoting oestrus.

Hair

Hair is characteristic of the mammal skin. Its prime function is to provide thermal insulation by trapping a layer of air within the pelt. In some mammals air spaces even occur within the medullae of the hairs of the winter coats. This insulation is increased by the erection of the hairs (piloerection) and is protected from the effects of water by an oil film produced by sebaceous glands associated with the hair follicles. As will be seen, both hair and the associated sebaceous glands can fulfil a semiochemical role also, for the hairs can function both to accumulate and disperse chemical signals derived from such glands. In some cases, skin hairs have evolved structures to this end. Such hairs which are specifically adapted for chemical communication have been called **osmetrichia**.

Sebaceous glands

Sebaceous glands are present in the skins of all mammals except whales and porpoises and are generally associated with hair follicles. Of the mammals studied only lemurs have sebaceous glands which are as numerous and as large as those in man. Their distribution over the body surface is far from

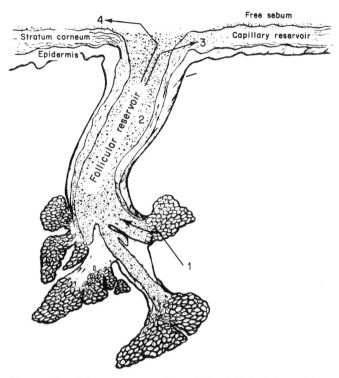

Figure 3.2   Sebum transport through the follicle (adapted from Kligman and Shelley, 1958 © 1958 The Williams and Wilkins Co., Baltimore). 1   Sebum enters follicle following the death and disintegration of lipid-filled sebaceous cells. 2   Sebum in the follicular reservoir where it is commonly subjected to the action of resident microorganisms. 3   The horny outer layer of the epidermis (the stratum corneum) which holds sebum in its interstices, rather like a wick, as well as contributing some lipids of its own. 4   Free sebum, mixed with apocrine and eccrine secretions, is readily distributed over the body surface and on objects in the environment by movement and physical contact

uniform. In man they range from a maximal density of 900 cm^{-2} in the scalp to total absence from volar surfaces (the palms of the hands and the soles of the feet). In most mammals the largest sebaceous glands are commonly found in the skin of the muzzle, in the external auditory meatus and in the anogenital area. In man the forehead is a major sebaceous secretory area, the large follicles there having an enhanced capacity compared with those of the scalp as the result of the absence of more than a rudimentary hair.

Sebaceous glands produce an oil liquid called **sebum** which arises as the result of the holocrine secretory process which characterizes these glands. In this, secretory cells which are formed at the periphery of a glandular lobule undergo a specific form of degeneration. As they are displaced from the periphery by the formation of more sebaceous cells and travel towards the centre of the lobule, they slowly fill with lipid and after about a week (in man) eventually become so enlarged with lipid that they die and burst yielding their totality to the central secretory duct, which becomes filled with a mixture of lipid, other cellular debris and microorganisms.

Sebum is almost totally lipid. In man whose skin is exceptionally rich in sebaceous glands, sebum provides the bulk of the lipid of the skin surface, which despite the uneven sebaceous gland bodily distribution, readily covers the entire body surface as the result of normal muscular activity. Even so, the distribution of lipid remains uneven, ranging up to 300 μg/cm^2 depending on the site examined (Greene *et al.*, 1970). It has been estimated that in the scalp at a given instant, each pilosebaceous follicle may contain about 10 μg of lipid (Nicolaides and Kellum, 1965). In a young man a major secretory area such as the scalp may yield 200 mg lipid in 24 hours (Nicolaides, 1965) while production rates in forehead skin in man have been found to range up to 150 μg/cm^2/h (Strauss and Pochi, 1961). We are, then, considering quite substantial quantities of material. The 1.4 m^2 of the average human body surface is believed to yield between 1 and 2 g sebum per day. On the other hand, in man sebum does not appear to fulfil any essential function.

As always, our knowledge of other species is more limited. Acetone washings produce about 250 mg lipid from the fur of the guinea pig and about 100 mg from a rat (Wheatley and James, 1957) while sheep's wool may contain about 20% by weight of wool wax. Clearly considerable quantities of sebum coat the fur in many species so that fur functions as a sebum reservoir. Precise measurements of sebum secretion rates are, however, often difficult to obtain as individual animals commonly groom for considerable periods of time, transferring fur lipid about their bodies and to objects in their environments (Ebling, 1974).

Apocrine glands

This term encompasses a variety of different structures which were once thought to share a common secretory mechanism in which the apical portions of secretory cells are shed to form secretion leaving the remainder of the cell intact. Although these sweat glands are now commonly believed to secrete by the merocrine process typified by eccrine sweat glands, it remains convenient to group these structures together and to contrast them both with sebaceous and with eccrine tissues. However, the complexity of secretory mechanisms should be emphasized. An electron microscopic study of human axillary apocrine sweat glands has, for example, brought forward evidence for the formation of secretion in this one tissue by all three mechanisms, apocrine, eccrine (merocrine) and holocrine (Schaumburg-Lever and Lever, 1975).

Apocrine glands occur throughout the hairy skins of most carnivores and many other mammals. In higher primates, however, their distribution is more limited and in man their occurrence is highly localized. Local apocrine sweating can be induced, for example in the dog, by the local application of heat (Aoki and Wada, 1951), and in many species, although not in man, there is evidence that apocrine sweating aids thermoregulation. This is the case in such large-bodied tropical species as the rhinoceros. What is said to be apocrine sweating occurs widely among bovine and equine species in response to heat stress and adrenaline infusion. However, in view of the copious quantities of sweat produced in many species it is probable that merocrine processes in the epitrichial (hair follicle associated) sweat glands are at least partly responsible (Carter and Dowling, 1954; Findlay and Jenkinson, 1960; Waites and Voglmayr, 1963; Allen and Bligh, 1969; Jenkinson and Robertshaw, 1971). In the black bedouin goat of the Negev, such sweat production provides a major means of evaporative heat loss to the extent that cutaneous moisture loss from the skin of the side of the head may range up to half a litre per square metre per hour (Dm'iel et al., 1979).

Such sweat glands are, however, uncommon in very small mammals as these species have a need to conserve water, and instead rely for evaporative heat loss on panting and on the cooling effect of distributing saliva over the body surface. Urine washing can also have a thermoregulatory effect. As we shall see, such activities can have a semiochemical dimension also (see Section 4.5).

In contrast with eccrine glands, human apocrine glands are usually associated with hair follicles, discharging into the pilosebaceous canal at a point above the sebaceous duct. However, this is not generally the case with apocrine glands in other species. For example, apocrine glands secrete directly to the skin surface in most other primates. In the rhinoceros, referred to previously, abundant apocrine sweat glands occur in the absence of hair follicles (Care, 1969).

Typically, an apocrine gland consists of a tube, the inner coiled part of which is lined with secretory tissues and embedded deep in the dermis, even extending into the subcutaneous fatty tissue. The remainder is a relatively straight duct which conveys the secretion to the pilosebaceous canal.

Human apocrine glands become active only with the approach of puberty, although thereafter sex hormones seem to have little effect on their activity. In certain other species, apocrine gland activity is more obviously under sex hormone control. They are, however, under neurohumoral control. In contrast with eccrine glands, apocrine glands respond to catecholamine stimulation, indicating adrenergic innervation. They also respond to cholinergic stimulation, although in man any response is masked by the profuse eccrine sweating which also results.

After puberty, human apocrine glands secrete but sparsely and irregularly, each gland yielding only about 10 μl of secretion every one to two days. Perhaps for this reason above others, our knowledge of the chemical composition of apocrine sweat has remained primitive. However, it is also unfortunate, since apocrine glands seem to be of considerable semiochemical importance in a number of species including man. More recently some attention has been given to the composition of human axillary apocrine sweat (see Section 4.7), which contains among its major lipid components, free cholesterol (76% lipid) and triglyceride (19% lipid), the remaining lipid being wax ester, cholesteryl ester, squalene and free fatty acids (Leyden *et al.*, 1981). Histochemical observations suggest that major secretion components include lipoproteins.

Human apocrine glands secrete in response to emotional stress stimuli, such as fear and pain, the excretion of a small quantity of fluid being followed by a lengthy refractory period during which the gland is unresponsive to further stimulation.

Eccrine glands
(Kuno, 1956; Sato, 1977)

Although important differences in structure occur between eccrine glands in different species, all consist of a simple coiled secretory tube embedded in the dermis and possibly extending into the subcutaneous fatty tissue, and a duct leading directly to the skin surface and *not* associated with a hair follicle.

The dominance of eccrine sweat glands is a characteristically human feature. In other primates their occurrence is sporadic while in non-primate species their distribution is very limited. Eccrine sweat glands occur on the friction surface (e.g. volar surfaces—palms, soles) of all primates and marsupials, on the knuckle pads of gorillas and chimpanzees, on the underside of the prehensile tails of certain monkeys, and on the digital pads of very many other species. However, their histological and histochemical similarity to human eccrine glands depends on the taxonomic position of the species concerned. They also occur in such specialized regions as the snout of the pig and the muzzle of the rabbit.

The entire human body is covered with up to four million eccrine sweat glands, each of an average mass of 30–40 μg. Together, under extreme circumstances, these glands can yield up to 10 litres of eccrine sweat in a day. More commonly, the human body loses about one litre of water each day through the skin, although some of this evaporates through the epidermis itself and does not involve the sweat glands.

Eccrine sweat is the most dilute of body fluids, consisting of more than 99% water. In man, its principal, although not its only, function is to prevent the rise of body temperature following heat exposure or vigorous exercise. Although human eccrine glands are distributed over almost the entire body surface, this distribution is uneven. In addition to this, the pattern of sweat production varies considerably from individual to individual and is influenced substantially and somewhat unexpectedly by posture. Visualization of sweating patterns (eccrine or apocrine) achieved by painting the body with a preparation of iodine, allowing it to dry, and treated with a preparation of fine starch powder (Wada, 1950) prior to heat exposure has shown that pressure on the soles of the feet (or hips) inhibits sweating on the lower limbs, so that the lower limbs are relatively free of eccrine sweat on standing (Takagi and

Sakurai, 1970). In a similar way pressure in the axilla or on the side of the chest inhibits sweat production on the same side of the upper body while increasing sweat on the opposite side.

Eccrine sweating also occurs in response to emotional stress. However, such sweating exhibits characteristics which distinguish it from sweating due to exercise and heat stress. It affects only certain parts of the body—for example eccrine sweating on the palms and soles occurs only in response to emotional stress—and exhibits no latency of onset as does heat stress eccrine sweating.

Eccrine sweat glands are under involuntary nervous control and are innervated principally by cholinergic nerves although there is a minor adrenergic component also. It has been suggested that adrenaline released from the adrenal glands as the result of exercise may increase eccrine sweating.

Human eccrine sweat is a clear, very dilute, aqueous solution, the principal organic components of which are urea (0.3–0.6 g/l) and lactate (0.9–3.6 g/l). The major inorganic components are sodium ions and chloride ions. The eccrine apparatus is richly supplied with blood vessels and appears to be surrounded by a capillary net which closely follows the form of the secretory coil and duct. *In vitro* studies using single excised glands have shown that the product which forms in the secretory coil closely resembles an ultrafiltrate of plasma in composition, the fluid appearing to be formed by a process of active sodium ion transport accompanied by the passive diffusion of chloride and other plasma constituents. On subsequent passage through the duct, reabsorption processes occur, so that in some ways eccrine gland function resembles that of the kidney. In view of these considerations, it is not surprising that the precise composition of the eccrine sweat delivered to the skin surface depends substantially on the rate of sweat production.

A wide range of concentrations of inorganic ions is observed in eccrine sweat. Sodium ion concentrations are below (20–80 mM) and potassium ion concentrations are above (5–20 mM) those of plasma (Na^+, 145 mM; K^+, 5 mM), while the principal anion is chloride. Although it appears that the urea present in eccrine sweat is largely of plasma origin, it seems that the lactate is not, and is probably the product of glycolysis in the eccrine gland itself. The trace organic composition of eccrine sweat is clearly quite complex and has not been studied in great detail. Components which have been studied include creatinine, uric acid, histamine, prostaglandins, amino acids (which occur at secretion: plasma concentration ratios ranging from 10 for serine and aspartic acid to <0.2 for glutamine and cystine), reducing sugars (~30 mg/l) and protein (200–800 mg/l). This latter includes glycoproteins and acid mucopolysaccharides, albumin, α- and γ-globulins, immunoglobulins and enzymes (esterases). The characteristic odour associated with some schizophrenic patients was for some time thought to be associated with low levels of *trans*-3-methyl-2-hexenoic acid which was detected in their sweat, until detailed observations using mass fragmentography showed that low levels of this compound (median 9 ng/ml sweat) were present in all sweat and that its

Table 3.1. Skin surface pH

	Mean pH	pH range	n
Human, white, male	4.85	4.04–6.26	51
Human, white, female	5.50	4.70–6.66	52
Guinea pig	5.50	4.26–7.17	93
Cat	6.43	5.57–7.44	27
Rat	6.48	5.74–7.51	60
Rabbit	6.71	5.97–7.50	18
Dog	7.52	5.18–9.18	40

Draize, 1942.

occurrence was not related to schizophrenia. The origin of the substance is unclear (Gordon et al., 1973).

Ammonia occurs in eccrine sweat (0.5–8 mM), although some of this may arise by the bacterial degradation of secretion urea. The pH of human eccrine sweat is determined in a complex way by the carbon dioxide/bicarbonate and lactic acid/lactate systems, as well as by amino acids, ammonia and other sweat components, and varies from pH 5 at low sweat rates to pH 7 at high sweat rates. On standing in air, eccrine sweat may become slightly alkaline (pH 7–8). The pH of the human skin surface is, however, always acid, pH 4–6, that of men tending to be slightly more acid than that of women. The skin surface film exhibits a substantial buffering capacity resisting change in pH. The normal skin pH of a number of laboratory animals has been reported to be less acid than that of man, and even to be slightly alkaline. Of course, in these non-human species, sweat is largely 'apocrine', not eccrine. The skin pH may have an important influence on skin surface-chemistry and odour production.

3.3 THE HORMONAL REGULATION OF SKIN FUNCTION
(Strauss and Ebling, 1970; Ebling, 1974, 1977a,b;
Pochi and Strauss, 1974; Shuster and Thody, 1974)

Hormones exert a profound effect on the biology of mammalian skin. Seasonal changes in hair growth, which are so marked in many species—indeed seasonal variations in human body hair growth rates have been detected—are, like the moulting cycles and breeding cycles themselves, under endocrine control (Johnson, 1977). And of course, the growth of human facial, axillary, pubic and other body hair which commences at puberty is a reflection of internal endocrine changes.

But changes in hair growth are not the only manifestation in the skin of hormone action. In many species, sebaceous and apocrine gland activities exhibit similar seasonality. Sebaceous glands are characterized by a sensitivity to hormonal influence, and while both eccrine and apocrine glands are under neurohumoral control, apocrine glands unlike eccrine glands are also influenced by hormonal action.

Sebaceous function

Observations made on a variety of mammalian species, including man, of changes in sebaceous activity at puberty, following castration and following hormone administration, and also of differences between the sexes, have shown that **androgens** markedly increase sebaceous gland size and the rate of sebum production, except in such cases as mature males where sebum production is already maximal. There are also some indications that the gross chemical composition of sebum could be influenced by androgen administration although effects of this type seem not to have been studied in detail or in relation to trace components. It is reported for example that the fatty acid residue ratios nC_{16}/nC_{18} and $nC_{18:1\Delta9}/nC_{18}$ increase in the sebum of actively secreting, androgen treated sebaceous tissue.

The effects of **oestrogen** are more variable. Generally sebum production and gland size are reduced, although the effect is not the exact opposite of androgen administration in that sebaceous cell proliferation (mitosis) is not reduced. Rather, oestrogens are presumed primarily to reduce intracellular lipid synthesis. However, in some cases, for example in relation to the sebaceous side glands of the shrew, *Suncus murinus*, and the sebaceous abdominal gland of the gerbil, *Meriones unguiculatus*, activity seems to be stimulated by oestrogen. **Progestogens** seem to have little effect on sebaceous function, although the literature on this point is rather confused. In some cases, such as the sebaceous side glands of the shrew and the abdominal gland of the gerbil just mentioned, progesterone exerts a stimulatory effect.

The responses of the sebaceous side glands of the shrew and the abdominal gland of the gerbil illustrate the general observation that it is quite common for specialized sebaceous glands to exhibit different hormone sensitivities from the sebaceous glands of the general skin surface. An example of such a specialized sebaceous gland, which although not a skin gland, is of semiochemical importance, is the rat preputial gland (see Section 7.1).

Pituitary homones (Ebling *et al.*, 1975; Thody and Shuster, 1975) have profound biological effects which extend to a major influence on skin gland function. Studies on the *rat* illustrate how the pituitary exerts important but complex influences on sebaceous functions. They also show that different sebaceous tissues may be influenced in different ways.

It is shown, for example, that following removal of the pituitary gland (hypophysectomy), sebaceous activity is greatly reduced. It appears not only that some pituitary hormones act *indirectly* on sebaceous tissue as the expected secondary consequence of their effect on other endocrine organs—for example pituitary ACTH (adrenocorticotropin) acts as a sebaceous stimulant as the result of its action on the adrenal cortex—but also that other pituitary hormones act *directly* on sebaceous tissue *inter alia* influencing their responses to steriod hormones.

Experiments comparing intact, castrated and hypophysectomized-castrated rats have shown the following.

(1) The low levels of skin sebum production in hypophysectomized-castrated rats are unresponsive to treatment with testosterone except in the presence of such pituitary hormones as prolactin, α-MSH (melanocyte stimulating hormone) and growth hormone. α-MSH and growth hormone stimulate skin sebum production even in the absence of testosterone and it appears that these two pituitary hormones each exert two different effects on sebaceous tissue, one which is *synergistic with testosterone* and one which is *independent* of testosterone. Prolactin appears to act only in the former way.

In the lower animals, α-MSH is concerned exclusively with pigmentation through the formation of melanin, a dark coloured protein complex of a macromolecular substance derived from the tyrosine metabolite, indole-5,6-quinone.

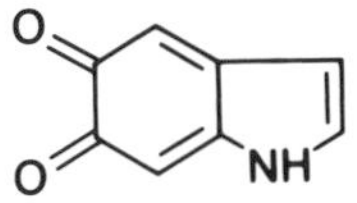

It is of some interest to note that with the evolution of the mammalian hairy skin its function has broadened to include a regulation of sebaceous function, and so to have some bearing on semiochemistry (Nowell *et al*., 1980a,b).

Melanocytes occur in the skin at the interface between the dermis and the epidermis where they contribute melanin granules to the epidermal cells. They also occur in sebaceous tissue and melanin may become so abundant in certain seba (e.g. of the black lemur) that the secretion is rendered dark brown. In man, melanocytes are distributed unevenly over the body surface, higher than average densities occurring for example on the face and forehead, and around the genitals and nipples and on the areolae. They occur in association with human sebaceous glands, for example in the eyebrows, in the glands of Zeis (eyelash) and on the adult female nipple.

(2) Although the administration of testosterone requires the presence of one of these pituitary hormones in order to achieve sebaceous stimulation, this is not the case for certain testosterone metabolites (Ebling *et al*., 1973) notably 5α-dihydrotestosterone, and to a lesser extent androstenedione and 5α-androstane-3β,17β-diol, and it seems probable that one effect of these pituitary hormones is to facilitate the formation of such testosterone metabolites in sebaceous tissue.

Clearly a full account of the endocrine control of skin sebaceous function is complex and many questions await resolution. However, the fact of endocrine control is clear. In addition, the endocrine system controls scent marking behaviour and may itself be modified by the reception of chemical signals (see Section 6.3).

Apocrine function

Much less attention has been directed to an understanding of apocrine glands. As indicated, the development of human apocrine function at puberty sug-

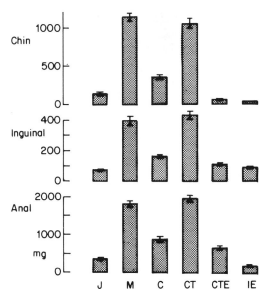

Figure 3.3 Masses of submandibular (chin), inguinal and anal glands from 13 week juvenile (J) and 21 week mature (M) rabbits. C, castrated; CT, castrated and treated with testosterone; CTE, castrated and treated with testosterone and oestradiol; IE, intact and treated with oestradiol. Means/s.e. for groups of 6–8 animals. (Reproduced by permission of Plenum Publishing Corp. from Ebling, 1977a)

gests a sensitivity to hormonal influence; however, no effects, functional or histological, have been observed in these glands following hormone administration in adults.

In other species the situation may be rather different. In the wild rabbit, *Oryctolagus cuniculus,* the submandibular or chin gland, the inguinal or groin gland and the anal gland are all apocrine, although the inguinal gland also includes a sebaceous component. All show responses to hormone administration similar to those noted in sebaceous glands, the submandibular gland showing the greatest sensitivity being considerably larger in the male, particularly in the dominant male, and having its growth both inhibited by early castration and restored by testosterone administration. Oestrogen administration leads to a decrease in gland size and antagonizes the effect of testosterone.

Such effects are not unusual in mammalian apocrine glands. Where apocrine and sebaceous glands occur together in a specialized glandular complex, both gland types are commonly affected in a similar way by androgen. For example, the forehead gland of the male roe deer, *Capreolus capreolus,* contains apocrine and sebaceous components, both of which develop and regress with seasonal changes in size of the testes, reaching maximal size and producing a copious red, rancid-smelling secretion in the summer breeding season. Both glandular components regress following castration and are stimulated by subsequent testosterone administration. In a similar way, the musk-scent producing post-auricular apocrine glands of the shrew, *Suncus murinus,* regresses after castration and stimulated by testosterone—and also by progesterone and by oestradiol (Dryden and Conaway, 1967). These glands are located on the neck and throat and behind the ear, and discharge directly to the skin surface. Small sebaceous glands are also associated with the hair follicles in this area.

3.4 STEROIDS IN THE SKIN
(Sansone-Bazzano and Reisner, 1974)

The skin is one of the major sites of steroid metabolism in the body, and as just indicated, steroids exert a major regulatory effect on skin function. They and their metabolites also occur in skin secretions as the result of steroid metabolism in the skin and elsewhere in the body. Extracts of sweat from an absorbent shirt worn by a young man revealed a range of C_{18} and C_{19} steroids in microgram quantities (Oertel and Treiber, 1969). Some 90% of these steroids were present in free form, while solvent extraction of the skin surface itself revealed a similar profile of steroids, here largely present as sulphoconjugates (sulphates and sulphatides).

To investigate this further, radiolabelled dehydroepiandrosterone sulphate was injected intravenously and its excretion pattern and that of its metabolites were noted by tracing the fate of the radioactivity. This weak androgen is secreted internally from the adrenal cortex of both men and women at the

Table 3.2. Steroids in human sweat. Reproduced by permission of FEBS from Oertel and Treiber (1969)

	Free steroid (μg)	Sulphoconjugate (μg)
Dehydroepiandrosterone	29.5	6.4
Androstenediol	12.8	2.6
Androstenetriol	8.9	1.5
Androstenedione	1.1	0.2
Testosterone	0.1	0.1
Androsterone	71.4	9.0
Etiocholanone	44.7	6.2
Etiocholandione ⎫ Androstandione ⎭	37.5	3.3
Oestrone	3.1	0.3
Oestradiol	2.5	0.2
Oestriol	0.4	0.1

Extracted from a shirt worn experimentally by a young man.

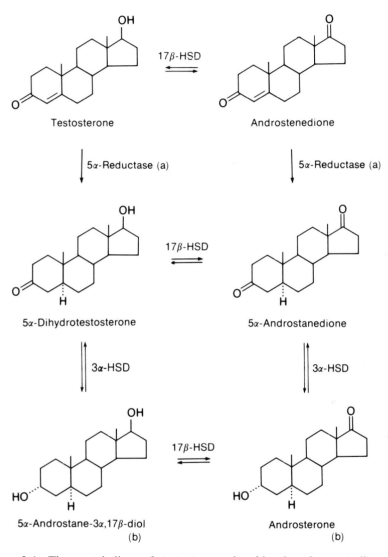

Figure 3.4 The metabolism of testosterone in skin, based on studies using homogenates of human skin, rat skin, hamster flank organs, and other tissues. Dehydroepiandrosterone, the principal androgen secreted (as sulphate) by the adrenal cortex, also stimulates sebum production in man, being converted to androstenedione and testosterone in skin. (a) The 5β-epimers, observed as urinary metabolites of testosterone, are not formed in skin or sebaceous tissue incubations. (b) The 3β-epimers, 5α-androstane-3β,17β-diol and 3-epiandrosterone, are also formed by 3β-HSD action

rate of about 20 mg each day. This experiment showed that the skin represented a major excretion route for this substance and its metabolites and that only a relatively small proportion of the radioactivity was excreted in urine. The profile of radioactive metabolites noted in the skin was similar to that recorded above. The results suggested that the free steroids present in the sweat were formed by hydrolysis of the sulphoconjugates immediately prior to excretion.

Thus it is possible that the steroid composition of the skin surface may reflect aspects of the internal endocrine state, and so have some semiochemical significance. It is possible also that the steroid metabolites of major significance in that regard are present only at very low levels measurable only by such highly sensitive methods as radioimmunoassay.

The skin's capacity to metabolize steroid hormones has been studied most closely in relation to the androgens. Investigations of the fate of radiolabelled testosterone following intravenous injection in the rat show that while this androgen is metabolized more rapidly per unit mass of tissue in the accessory sex organs than in other tissues, the skin by virtue of its great mass provides the principal site for testosterone metabolism for the organism as a whole. It appears that testosterone sensitivity is related to the tissue's ability to transform this androgen to the potent metabolite, 5α-dihydrotestosterone.

These and other aspects of skin steroid chemistry have been investigated by incubating radiolabelled steroids with tissue preparations, commonly whole skin homogenates, or with preparations of specialized sebaceous tissues, such as those derived from rodent preputial glands, and by following the formation of metabolites, for example, by radio-TLC and radio-GC. These techniques complement histochemical methods.

In man, the rate of 5α-dihydrotestosterone formation from testosterone in

Table 3.3. The androgen content of tissue 45 min after intravenous administration of ^{3}H-testosterone to castrated rats. Reproduced with permission from Sansone-Bazzano and Reisner, © 1974 The Williams and Wilkins Co., Baltimore

	Androgen content measured by radioactivity (cpm × 10^{-5})		
	Total skin	Total preputial gland	Total prostate
Testosterone	113.9	0.9	3.6
Dihydrotestosterone	18.8	0.7	4.0
Androstenedione	8.9	0.1	0.2
Androstanedione	8.9	0.2	0.1
Androstanediol	6.4	0.1	0.2
Relative radioactivity per unit mass of tissue	Skin tissue	Preputial gland tissue	Prostate tissue
	1.0	3.0	0.6

skin varies with the region of the body examined (Wilson and Gloyna, 1970). Skin samples from the prepuce, scrotum and labia majora when compared with samples from the thigh, chest and back, showed a scatter of elevated activities up to some 20 times those of the comparison samples. Preputial skin also exhibited a similar range of values, although here a clear age dependence was noted, highest values being observed in infants and lowest values in adults.

Histochemical insights

The tissue sites of certain steroid-metabolizing enzymes can be conveniently located histochemically (Baillie *et al.*, 1966). Hydroxysteroid dehydrogenases (HSDs) are involved in the interconversion of hydroxy- and ketosteroids, in the presence of an appropriate hydrogen-transfer cofactor (NAD or NADP). A range of different HSDs exist, exhibiting specificities for hydroxy-groups at different positions in and with different configurations in regard to the steroid nucleus. Other factors which may affect specificity are the position of any unsaturation in the steroid nucleus and whether the steroid is C_{18}, C_{19} or C_{21}.

Hydroxysteroid dehydrogenase activity within a tissue may be located simply and rapidly by incubating a tissue section in the presence of an appropriate hydroxysteroid substrate, an appropriate hydrogen transfer cofactor, and a tetrazolium salt. At the site of HSD activity, hydroxysteroid oxidation is coupled with the reduction of the tetrazolium salt to deposit an insoluble coloured dye.

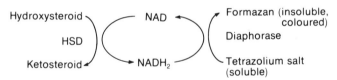

The level of HSD activity is indicated qualitatively by the intensity of tissue staining. Particular HSDs are localized by the selection of the appropriate hydroxysteroid substrate.

Although the method has its limitations—for example, end products are not isolated, and quantitative data concerning metabolites are not obtained—it does provide a most valuable adjunct to biochemical incubation methods, by indicating at a subcellular level the location, and to some extent, the nature of, and variations in, steroid metabolizing activity. Unfortunately a number of important steroid-metabolizing enzymes, such as the 5α-reductases, cannot be localized by an analogous histochemical method.

Even so the method does provide some interesting insights, and will no doubt be used more extensively in a semiochemical context in the future. It does reveal substantial HSD activity in skin—confined to the sebaceous tissue—and points to differences between different regions of the body. In human skin, it is apparent (Baillie *et al.*, 1966) that HSD activity varies

58

Figure 3.5 Anatomical distribution, irrespective of sex, of human sebaceous glands exhibiting histochemically demonstrable HSD activity. (Reproduced with permission from Baillie, Ferguson and Hart (1966), *Developments in Steroid Histochemistry*. © Academic Press Inc. (London) Ltd.)

markedly over the body surface, ranging from high HSD levels (mainly Δ^3-3βHSD, 16β-HSD and 17β-HSD) on the face and neck, to total absence of activity in certain other areas.

Observations of enzymes do not give direct information concerning the levels of particular steroids in particular sebaceous tissues and their secretions, even if they do direct us to tissues of possible interest. Levels of particular steroids depend on a variety of factors which include the outcome of competition between different steroids for particular enzymes (for example, between oestradiol and testosterone, both of which are 17β-hydroxy, for 17β-HSD), and also between different enzymes for particular cofactors (NAD or NADP). Biochemical studies show that cofactor dependencies may vary around the body, the predominant 17β-HSD of the forehead being NAD dependent while that of the axilla is NADP dependent.

To take another example–the supracaudal (tail) gland of the red fox (Albone and Flood, 1976) consists largely of highly developed sebaceous glands together with a few apocrine glands. These sebaceous glands show intense HSD activity of a type which distinguishes this scent gland from other

59

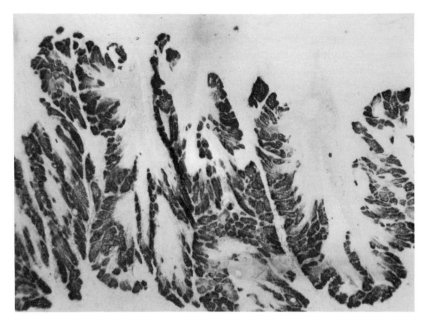

Figure 3.6 Histochemical evidence for HSD activity in red fox (*Vulpes vulpes*) supracaudal scent gland: substrate 5β-androstan-3β-ol-17-one. The skin surface is uppermost. Staining is restricted to the cytoplasm of the sebaceous cells and is absent from the degenerate cells at the centres of the glandular lobules. (Reproduced by permission of Plenum Publishing Corp. from Albone and Flood, 1976)

sebaceous tissue on the fox's body. The steroid chemistry of this specialized skin gland complex has, however, yet to be investigated.

Other scent glands which have been studied in this way include the chin gland of the male cuis, *Galea musteloides* (Holt and Tam, 1973), the submaxillary gland of the boar (Flood, 1973) and the antebrachial organs of certain lemurs (Sisson and Fahrenbach, 1967).

Table 3.4. Histochemically demonstrable HSD activity in the supracaudal gland as compared with other sebaceous tissue from the red fox (Albone and Flood, 1976)

| HSD sought | Intensity of reaction | |
	Supracaudal tissue	Other skin
α-3α	++	++
β-3β	++++	++(1)
Δ^5-3β	++	++
17α	+	−
17β	+	+++

(1) β-3β reaction for back skin was higher than other tissue (+++), save supracaudal tissue.

3.5 LIPID CHEMISTRY OF THE SKIN SURFACE

The chemistry of the human skin surface lipid has been studied in considerable detail. Comparative chemical studies have also been conducted on vernix caseosa, the greasy sebaceous material which covers newborn infants. This substance, as well as exhibiting certain characteristic chemical features of its own, is further differentiated from adult human skin surface lipid by being free of environmental contaminants and of the secondary products of chemical and microbial degradation present in adult human surface lipids. The squalene component of human sebum, for example, forms oxidation products on exposure to light.

Sebum is unique among animal lipids in the complexity of its composition and in the presence of components which are normally encountered only in trace quantities if at all in internal tissues. Although studies on non-human seba are not extensive, we do have evidence that major differences in sebum chemistry occur even between closely related species (Nicolaides et al., 1968, 1970). And even within a species, the composition of the lipid mixture produced by specialized sebaceous glands may be different from that of the general skin surface. Sebum lipids are synthesized in the sebaceous glands themselves, so that sebum lipid chemistry does not reflect directly the composition of circulating lipid or changes in the lipid composition of the diet, except where a major dietary deficiency is encountered (Nikkari, 1965). However, skin surface lipid composition, for example with regard to squalene content, does vary with fasting as overall sebaceous activity is affected (Downing and Strauss, 1974).

Lipid classes

Assays using TLC (Nicolaides et al., 1968; Downing et al., 1969) and more recently HPLC (Aitzetmüller and Koch, 1978), show that the major compound classes represented in skin surface lipid include

> hydrocarbons
> sterol esters
> aliphatic (wax) esters
> glycerides (glycerol esters)
> free fatty acids
> free sterols

The gross composition of the largely sebaceous human skin surface lipid is shown in Table 3.5. When these observations are compared with the results of studies with other species, it appears that man is unusual in producing a sebum rich in **glycerides**. The presence of high levels of **free fatty acids** in skin surface lipid is also highly characteristic of the human species and is linked to the production of sebaceous triglyceride since these free fatty acids are derived from triglycerides by the action of skin microorganisms. Although published data of the type given in this table are somewhat variable, levels of

Table 3.5. Human skin surface lipid: gross % composition by weight

	Adult			Foetal
	Surface lipid (1)	Sebum (1)	Epidermal lipid (1)	Vernix caseosa (2)
		Contribution to total surface lipid		
		Major	Minor	
Hydrocarbon (squalene)	10 (7)	12	<0.5	9
Sterol esters (cholesterol esters)	2.5	<1	10	33 (5)
Aliphatic (wax) monoesters	22	23	0	19 (4)
Triglycerides	25 (3)	60	10	26
Mono- and diglycerides	10 (3)	0	10	0
Free fatty acids	25 (3)	0	10	0
Free sterols (mainly cholesterol)	1.5	0	20	9 (5)
Other	4	5	40 (6)	

(1) Based on various sources summarized in Nicolaides, 1974.
(2) From Kärkkäinen et al., 1965.
(3) These data are highly variable between individual samples and represent different degrees of microbial hydrolysis of sebum triglyceride: surface lipid triglycerides in the range 20–50% and free fatty acids in the range 8–40% are reported (Downing et al., 1969; Marples et al., 1972).
(4) Including 7% wax esters (mainly type 2a), not present in more than trace quantities in adult sebum (Ansari et al., 1969; Fu and Nicolaides, 1969).
(5) 76% cholesterol plus 24% lathosterol.
(6) Including 30% glyco- and phospholipids.
(7) Man excretes between 100 and 500 mg/day of this hydrocarbon (Nikkari et al., 1974).

glycerides and free fatty acids are especially so as the result of this action, triglycerides yielding diglycerides, monoglycerides, free fatty acids and free glycerol.

Although the seba of all species of mammals examined contain esters of some kind, these are generally not triglycerides, but other classes of ester which are less readily hydrolysed by microbial activity. Thus, the seba of rodents, the rabbit and sheep are low in triglyceride and also in free fatty acid, while neither of these compounds classes are observed by TLC in the hair lipids of the chimpanzee, baboon, hamster, guinea pig, cat, dog or cow (Nicolaides et al., 1968).

A further distinctive component of human sebum is the unsaturated C_{30} hydrocarbon, **squalene**. Other hydrocarbons which are commonly encountered in skin surface lipid are generally considered to be environmental contaminants, such as petroleum hydrocarbons.

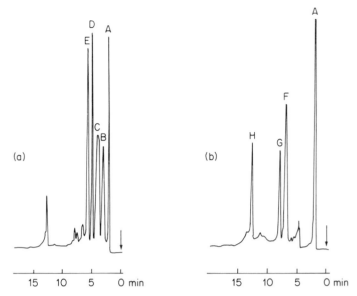

Figure 3.7   Gradient elution HPLC of lipids extracted from human hair. (a) Chromatogram of hair lipid following methylation with diazomethane. (b) Chromatogram of unsaponifiable fraction of hair lipid following alkaline hydrolysis and extraction into an organic solvent. A   hydrocarbons (squalene); B   cholesterol esters/wax esters; C   free fatty acids (as esters); D   triglycerides; E   internal standard; F   wax alcohols; G   cholesterol; H   air oxidation products. (Adapted from Aitzetmüller and Koch, 1978, by permission of Elsevier Scientific Publishing Company)

In addition to the triglycerides already mentioned, human sebum also contains **sterol esters** and aliphatic esters (**wax esters**). These latter are present in the seba of all species studied, although in some cases such as the mouse, rabbit, goat and cattle, their levels are relatively low. Apart from hydrocarbons, the **wax monoesters** comprise the least polar components of sebum. Human sebum (but not vernix caseosa) and the sebum of the chimpanzee appear to be unusual in containing a higher proportion of wax monoester than of sterol ester.

However, it is in the **wax diesters** that a greater species diversity is manifest, even at the compound class level. For example, TLC data show that the skin surface wax diesters of cattle, the rabbit and the cat are exclusively of type 1, being 35, 66 and 66% of the total surface lipids in these species, while those of the dog, mouse, guinea pig and baboon are almost exclusively of type 2a, and account for 50, 67, 41 and 21% of the total surface lipids respectively. However, rat and sheep seba contain wax esters of both types (Table 3.7) while adult human sebum contains neither (Nicolaides *et al.*, 1968). Although relatively few species have been studied by skin chemists, from those cases which

Table 3.6 Wax esters of the skin surface. Among the tissues of land mammals, wax esters commonly occur in more than trace quantities only in the skin. The multitude of wax esters present are grouped in the following major structural classes separable by TLC

Class	General formula (where R, R', R″ are long-chain alkyl groups)
Wax monoester	$R-CO-O-R'$
Wax diester, type 1 (diesters of 2-hydroxyacids)	$R-CH-CO-O-R'$ $\mid$ $O-CO-R''$
Wax diester, type 2a (diesters of alkane-1,2-diols)	$R-CH-CH_2-O-CO-R'$ $\mid$ $O-CO-R''$
Wax diester, type 2b (diesters of alkane-2,3-diols)	$\quad\quad\quad CH_3$ $\quad\quad\quad\mid$ $R-CH-CH-O-CO-R'$ $\mid$ $O-CO-R''$

The average molecular mass of wax esters in the seba of a variety of mammalian species has, like that of the triglycerides of human sebum, been found to lie in the range 800–900 (Nicolaides *et al*., 1970; Nikkari, 1974).

Table 3.7. Lipid classes represented in some non-human sebaceous lipids (adapted from Nikkari, 1974). SL, surface lipid; P, preputial gland lipid

Lipid compound class	Rat SL (1)	Rat P (2)	Mouse SL (3)	Mouse P (4)	Sheep SL (5)
Hydrocarbon	1	2			trace
Sterol ester	27	14 (7)	10	5	25
Wax monoester	17		5	48	
Triglyceride	8 (6)	60	6	17	6
Free fatty acid	1	2		trace	
Free alcohol	7	2	13 (9)	3	12
Glycerol ether diesters		0		14	10
Diester waxes type 1	10	0			9
Diester waxes type 2	11	0	67 (8)		7
Other	19	20		13	31

(1) Nikkari, 1965.
(2) Nicolaides, 1965.
(3) Wilkinson and Karasek, 1966.
(4) Snyder and Blank, 1969.
(5) Nikkari, 1974.
(6) Possibly includes some glyceryl ether diester.
(7) Including some wax monoester (Sansone and Hamilton, 1969).
(8) Nicolaides *et al*., 1970.
(9) 5% C_{30} sterol + 8% C_{27} sterol.
Many of the identifications are based on chromatographic mobility only.

Table 3.8. Compositions of the hydrolysis products obtained on hydrolysing seba of various species (adapted from Nikkari, 1974)

	Adult human	Vernix caseosa	Rat	Mouse	Guinea pig	Rabbit	Sheep
Hydrocarbon (squalene)	12	3	~0	0	0	0	~0
Sterols	3	20	25	13	21	4	39
Monohydric alcohols	12	5	17	6	5	31	5
Alkane-1,2-diol	0	~0	6	28	11	2	3
Fatty acid	68	44	50	44	55	53	48
2-Hydroxyacid	0	4	6	0	0	~0	14

have been documented it is clear that the skin chemistry is complex and that species variations also manifest themselves in other ways, for example in the nature of the acids which are combined in the wax esters. Even at a gross level we can note such differences, not only between species when we compare lipids of a particular class, but also between the acid moieties combined in different classes of lipid in the same species.

Type 2b wax esters have yet to be reported in mammalian skin lipids but occur as major components of the sebaceous preen gland secretions of certain birds (chicken, turkey). While preen glands themselves are outside the scope of this text, a consideration of their lipid chemistry has some bearing on mammalian semiochemistry, not least providing an example of the application of pertinent analytical techniques (Haahti and Fales, 1967; Hansen et al., 1969; Jacob and Poltz, 1974).

What is unusual about the chemistry of sebum?

We have already noted the presence of wax esters, compounds not commonly encountered in quantity elsewhere in mammalian systems, as important

Table 3.9. Acid types represented in various skin lipid fractions

		Acid type		
Species	Lipid type	Saturated straight chain	Saturated branched chain	Unsaturated
Cow	Wax diester type 1	100	0	0
Cat	Wax diester type 1	73	19	8
Dog	Wax diester type 2	33	64	3
Rat	Wax diester type 1	41	30	29
Rat	Wax monoester	6	30	64
Rat	Sterol ester	12	53	35

From Nikkari, 1974 and references cited there. For other related work, see Nicolaides et al., 1970 and Nikkari and Haahti, 1968.

sebum constituents. The distinctive features of sebum chemistry do not end here.

(1) A major feature of sebum chemistry resides in the diversity of the fatty acids and alcohols represented in the esters referred to above. Thus, although triglycerides occur widely in organisms, the range of the fatty acid moieties present in human sebum triglyceride make this material unique. A detailed chemical analysis of the fatty acids obtained on hydrolysing human skin surface lipid shows that although the biochemically common fatty acids, palmitic acid (nC_{16}), myristic acid (nC_{14}), stearic acid (nC_{18}), oleic acid ($nC_{18:1\Delta9}$) and linoleic acid ($nC_{18:2\Delta9,12}$) are present, this accounts for only 37% of the acids by weight. The remaining 63% is composed of more than 200 different fatty acids not normally encountered in more than trace quantities in animal tissue. This has prompted a leading dermatological chemist to ask (Nicolaides and Apon, 1977):

'Why does the adult human skin synthesize so many acids in such minute amounts? Perhaps these acids serve a pheromonal function and contribute to our distinctive "chemical signature" just as the skin gives each of us a distinctive fingerprint. It seems highly improbable that any two individuals would produce exactly the same amounts of each of these numerous acids.'

The same author has also suggested that the unusual chemistry of skin surface lipid could assist the mammal by inhibiting the growth of pathogenic microorganisms (Nicolaides, 1974).

(2) A second feature of skin lipid chemistry is the accumulation by many species of intermediates formed on the biosynthetic route from acetate to cholesterol. In internal tissues, in contrast, these intermediates are generally observed only in trace quantities, biosynthesis progressing efficiently to the biologically valuable end product.

Fatty acid and fatty alcohol composition

By far the most study has been directed to the study of human skin surface lipid acids.

The major acids (free plus esterified) of human skin surface lipid are (Nicolaides, 1974):

nC_{16}	25.3% total acid
$nC_{16:1\,\Delta6}$	21.7
$nC_{18:1\,\Delta8}$	8.8
nC_{14}	6.9
$isoC_{16:1\,\Delta6}$	4.0
nC_{15}	4.0
nC_{18}	2.9
$nC_{18:1\,\Delta6}$	1.9
$nC_{18:1\,\Delta9}$	1.9

plus more than 200 acids at lower levels.

Table 3.10. The contribution of fatty acids (free plus esterified) of different types of human skin surface lipid (Nicolaides, 1974). Copyright 1974 by the American Association for the Advancement of Science

Structural type	Saturated acids		Monoene acids		Diene acids	
	range	%	range	%	range	%
Straight chain, even carbon number	$C_{10}-C_{30}$	35.8	$C_{14}-C_{24}$	36.4	$C_{16}-C_{22}$	2.8
Straight chain, odd carbon number	$C_{11}-C_{27}$	5.4	$C_{15}-C_{25}$	3.9	$C_{17}-C_{19}$	0.1
iso-Acids	$C_{12}-C_{28}$	4.0	$C_{14}-C_{26}$	5.3	—	
anteiso-Acids	$C_{11}-C_{27}$	1.4	$C_{15}-C_{25}$	1.3	—	
Monomethyl, internally branched (1)	$C_{11}-C_{26}$	2.6	$C_{17}-C_{25}$	0.2	—	
Dimethyl, branched (2)	$C_{13}-C_{24}$	0.8		—		—
Total		50.0		47.1		2.9

(1) Branching mainly at C_4.
(2) Including trace quantities of trimethyl branched acids.
(3) For information relating specifically to the sterol ester and wax ester fractions, see Nicolaides et al., 1972.

Here total acid (free plus esterified) composition differs from that noted in internal sources in the following regards.

(1) An unusually wide range of alkyl chain lengths is represented. The findings given in Table 3.10 probably represent an underestimate as the techniques used tend to be insensitive to acids of very long and very short chain lengths. In vernix caseosa, mouse skin and rat skin longer chain lengths have been noted, in the latter case up to C_{38}. In the mouse, 30% of the fatty acids present in the skin surface lipid sterol ester fraction are monoene acids in the range $C_{30}-C_{35}$ (Wilkinson and Karasek, 1966).

(2) An unusually high proportion of odd carbon number fatty acids is present.

(3) An unusually high proportion of iso- and anteiso-acids is present.

(4) Even though they constitute a small weight percentage of the total fatty acid fraction, an unusually large number of different internally branched mono- and polymethyl fatty acids is present. Close similarities between the composition of this fraction in adult skin surface lipid and in vernix caseosa confirm their endogenous nature (Nicolaides and Apon, 1977).

(5) An unusual pattern of unsaturation is shown. Although the 1Δ6 carbon–carbon double bond is uncommon in internal sources, it is a major contributor to human skin surface lipid, mainly as the $nC_{16:1\Delta6}$ acid, and to certain other mammalian sebaceous secretions. In contrast, the 1Δ9 carbon–carbon double bond is common in other sources as $nC_{18:1\Delta9}$ (oleic acid). However, in skin lipid acids this double bond is represented chiefly by $nC_{16:1\Delta9}$, a minor constituent of internal tissues.

These $C_{16:1\Delta6}$ and $C_{16:1\Delta9}$ acids, and acids related biosynthetically to them by chain extension and degradation, namely acids of the general formulae

$$C_{(16+2a):1\ \Delta(6+2a)}$$

and

$$C_{(16+2a):1\ \Delta(9+2a)} \qquad \text{(where a is an integer)}$$

together with other families of acids derived from $\Delta6$ and $\Delta9$ acids of other chain lengths and branching patterns, make up the bulk of the monoene acids of adult human skin surface lipid and vernix caseosa.

In contrast to sebaceous tissue where such atypical fatty acids are synthesized, the major fatty acids of the epidermis are the ubiquitous, biologically valuable fatty acids common to other tissues (e.g. oleic acid, $C_{18:1\Delta9}$). However, it seems unlikely that the epidermis contributes these to the surface lipid layer to the extent one might expect since they provide a source of energy for use in the latter stages of keratinization when external nutrients are no longer available. In contrast, the more uncommon fatty acids present in epidermal tissue are more likely to survive to reach the surface layer. In human skin surface lipid wax ester and sterol ester fractions, the acid moieties of 25 and 36 different *families* of biosynthetically related monosaturated acids respectively have been identified. Among these, families based on $\Delta6$ and $\Delta9$ parent acids respectively accounted for 98 and 1 mole % of wax ester monounsaturated acids and 89 and 11 mole % of sterol ester monounsaturated acids (Nicolaides et al., 1972).

The patterns of unsaturation in the fatty acid moieties of vernix caseosa lipids also consist predominantly of the $C_{16:1\Delta9}$ families, although their relative contributions differ in the different lipid classes (Ansari et al., 1969; Nicolaides et al., 1972). The relative contributions to the unsaturated acid content of the sebaceous and epidermal tissues are a matter of debate. It has been suggested that the higher proportion of $\Delta9$ acids in the sterol esters and type 2 diester waxes of vernix caseosa indicate a major epidermal contribution here, while the large proportion of $\Delta6$ derived monounsaturated acid moieties in the wax esters are evidence of a largely sebaceous origin (Downing and Strauss, 1974).

The primarily sebaceous skin surface lipids of the rat appear to derive principally from the $C_{16:1\Delta9}$ acid while in the mouse the precursor unsaturated acid appears to be largely $C_{18:1\Delta9}$ (Nicolaides and Ansari, 1968; Wilkinson, 1970).

A detailed study of the small proportion of diene acids of human skin surface lipid showed the major components to be

$C_{18:2\ \Delta5,8}$ 39 mole % of total diene acids
$C_{18:2\ \Delta9,12}$ 18 mole % of total diene acids
$C_{20:2\ \Delta7,10}$ 18 mole % of total diene acids

together with a large number of minor diene acids.

A similarly complex pattern of occurrence, including unsaturation of a predominantly Δ6 type, is found among the fatty alcohols. Since these alcohols occur as wax ester components, they are of sebaceous origin. Although their structures are closely similar to those of the fatty acids already discussed and from which they are thought to derive biosynthetically, there is a difference between the alcohols and the acids as far as chain length distribution is concerned. In general the chain lengths of the predominant alcohols are greater than those of the predominant acids, presumably reflecting selectivity of the enzyme(s) responsible for reducing acid to alcohol. Thus, although the C_{16} monoene is the most abundant unsaturated acid, the C_{20} monoene is the most abundant unsaturated alcohol in human sebaceous wax ester (Downing and Strauss, 1974).

Cholesterol intermediates as skin lipid components

The biosynthesis of the ubiquitous sterol, cholesterol, from acetate involves a sequence of reactions passing through a series of intermediate compounds which include (in order of formation) geranyl pyrophosphate, farnesyl pyrophosphate and the aliphatic hydrocarbon, squalene, which goes on to cyclize to the sterol lanosterol and thence through a number of steps to form cholesterol itself. Now, lanosterol contains a carbon–carbon double bond in its side chain, whereas the side chain of cholesterol is saturated. In internal tissues, such as the liver, it seems that the predominant pathway from lanosterol to cholesterol involves a series of intermediates possessing unsaturated side chains, and that this unsaturation is reduced only at the final biosynthetic step (Bloch pathway). In the skin and in certain other sebaceous tissues (rodent preputial glands) an alternative pathway is important (Kandutsch–Russell pathway). Here reduction of the side chain occurs at an earlier stage and a series of intermediates (as esters) between lanosterol and cholesterol containing saturated side chains are observed. In some species, certain of these intermediates accumulate in skin lipid.

These sterols occur variously in free and in ester form. It is known that intermediates of the Kandutsch–Russell pathway are mainly in the sterol ester form while those of the Bloch pathway are mainly free. This might suggest that esterified sterols, such as are encountered in rat surface lipid, are likely to be of sebaceous origin while free sterols are epidermal. These considerations are unreliable, however, as the epidermis itself can promote sterol esterification. By incubating radiolabelled cholesterol with epidermal homogenates from a number of species, including man, esterification was observed to proceed apparently enzymatically in preparations of the stratum corneum, and even in the surface film itself (Freinkel and Aso, 1969).

Human sebum contains high levels of squalene, sebaceous sterol biosynthesis being largely blocked at this stage. The relatively low levels of sterol noted in human skin lipid are principally of epidermal origin. The major sterol present in adult human surface lipid is cholesterol (93% total sterols),

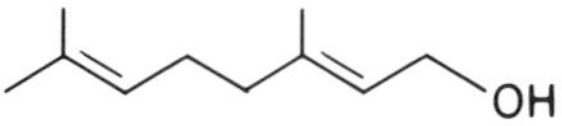

Geraniol (C$_{10}$) man (postulated)
(Nicolaides, 1974)

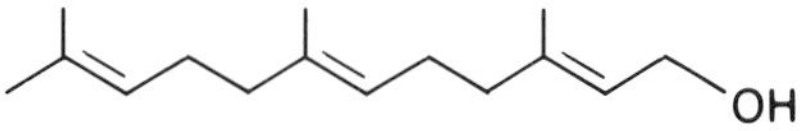

Farnesol (C$_{15}$) man (Nicolaides, 1965);
also detected by odour in
incubations of skin slices
with acetate

Squalene (C$_{30}$) man

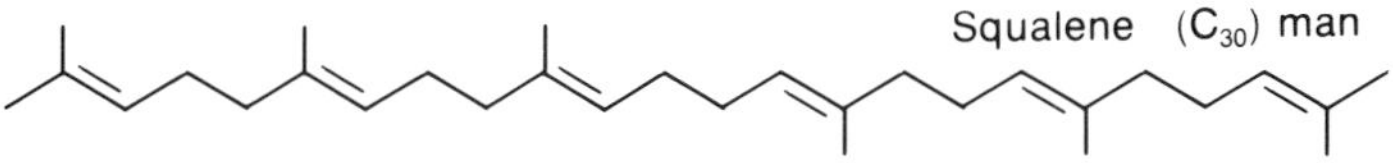

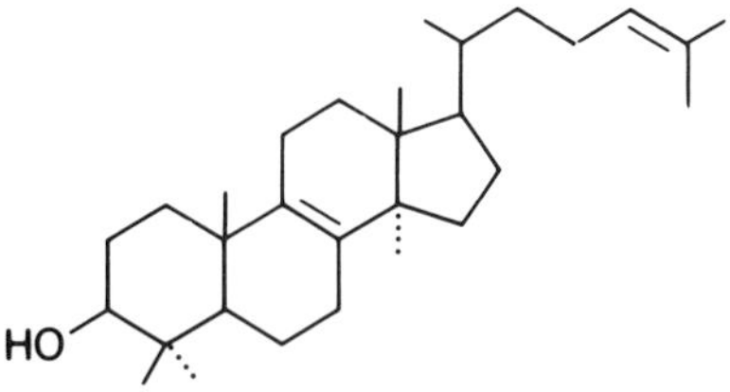

Lanosterol (C$_{30}$) sheep

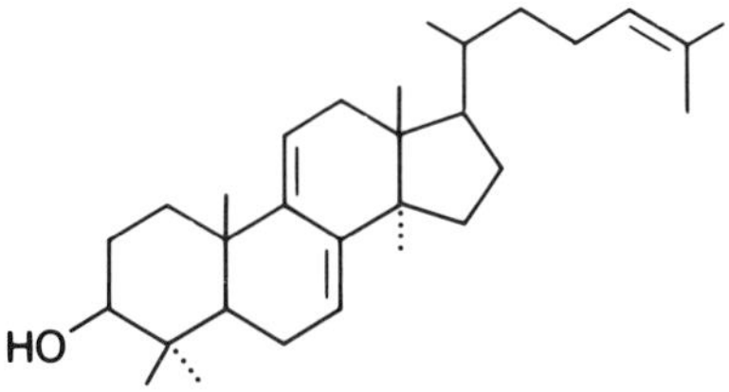

Agnosterol (C$_{30}$) sheep

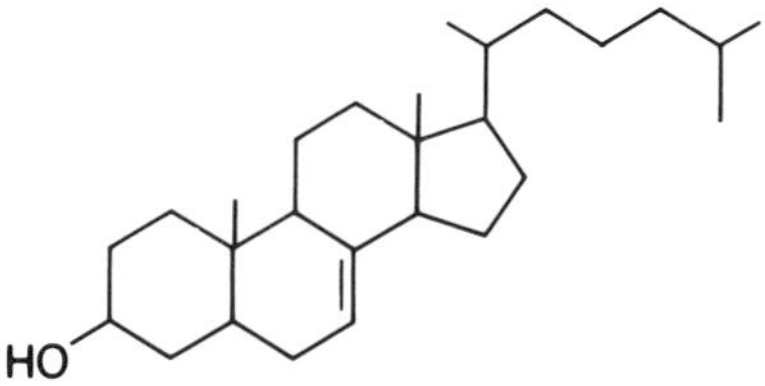

Lathosterol (C$_{27}$) rodents

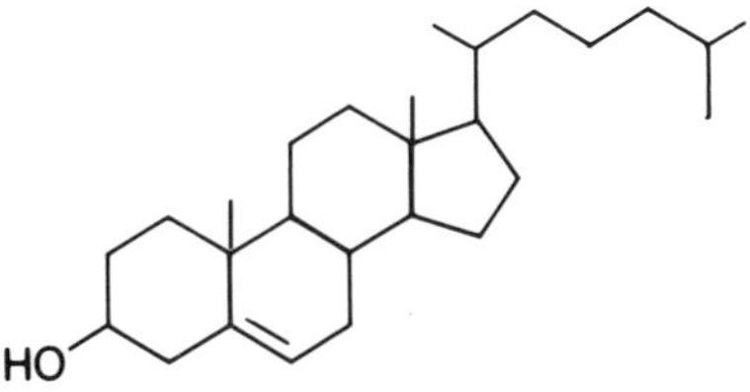

Cholesterol (C$_{27}$) ubiquitous

Figure 3.8   Intermediates in cholesterol biosynthesis noted in skin surface lipids

with lathosterol accounting for 3%, C_{30} sterols (e.g. lanosterol) for 2% and C_{28} and C_{29} sterols for 1.5% (Nikkari, 1974). 7-Dehydrocholesterol ($\Delta^{5,7}$-cholestadienol) is a minor (0.1% total sterols) but physiologically important sterol formed in skin, for in sunlight this substance forms vitamin D_3. This latter substance has been detected in human skin by mass spectrometry and gas chromatography at levels of the order of 3 μg/g (Rauschkolb *et al.*, 1969). It occurs principally in the lower layers of the epidermis, only trace quantities reaching the stratum corneum.

In contrast, the sterol content of the surface lipid of other species is generally higher than in man since sebaceous sterol biosynthesis is not blocked at the squalene stage. Sterol compositions have been tabulated in a number of reviews based on earlier reports, but many of these studies are not easily comparable since the techniques employed and the rigour of the identifications vary considerably from report to report. It does emerge, however, that major species variations do exist. For example, rat surface lipids contain unusually high levels of lathosterol (40% of total sterol), the proportion relative to the other major sterol present, cholesterol, varying with such factors as age and sex (Wheatley and James, 1957). Other sterols present included desmosterol derivatives, methosterol, agnosterol and lanosterol (Nicolaides 1965). Mouse surface lipids also contain high levels of lathosterol and histochemical studies indicate that in both species its origin in sebaceous (Brooks *et al.*, 1956).

A more detailed chemical study of rat skin lipids based, not on surface washings, but on an examination of whole skin revealed the presence of 15 related sterols of which the major components are listed in Table 3.11. Of the rat skin sterols intermediate between lanosterol and cholesterol, 87% by weight were of the Kandutsch–Russell pathway and only 13% of the Bloch pathway.

It is of interest that human vernix caseosa, which possesses a higher sterol content than adult human sebum, contains 17% lathosterol. It appears that the block in sterol biosynthesis which occurs at squalene in adult sebaceous glands is not complete in the foetus.

Sheep surface lipid (wool wax) in contrast accumulates (as ester) an earlier intermediate, lanosterol, which with related C_{30} sterols, accounts for 50–60% of wool wax sterols, the other major sterol being cholesterol (Downing *et al.*,

Table 3.11. Rat skin sterols (Reproduced with permission from Clayton *et al.*, 1973)

	% total sterol
Cholesterol	32
Lathosterol	21
4,4-Dimethylcholest-8-enol	10
4α-Methylcholest-7-enol	7
4α-Methylcholest-8-enol	7

1960). A small quantity (2%) of agnosterol and 24, 25-dihydroagnosterol is also present.

The relative contribution of sebaceous and epidermal tissue to the sterol composition of skin surface lipid remains confused and probably varies considerably with species. As we have already indicated, the sterols of human skin surface lipid probably arise principally from the epidermis, sebaceous sterol biosynthesis being blocked. This is reflected in the varying proportion of sterol (epidermal) to squalene (sebaceous), (or more reliably to wax ester which is also entirely sebaceous) present in human skin surface lipid taken from different body regions possessing different sebaceous gland densities (Greene *et al.*, 1970). For a similar reason, a relatively high proportion of sterol to wax ester is observed in the surface lipid of children of 3–6 years of age, reflecting the enhanced sebum production rate (hormone control) of adults compared with children. Interestingly, young babies show an enhanced sebum production rate compared with children perhaps as the result of the effects of residual maternal hormones (Ramasastry *et al.*, 1970).

3.6 APPENDIX

Abbreviated notation of lipid components

C_{18} a compound or group containing 18 carbon atoms.

nC_{18} a compound containing 18 carbon atoms in a single unbranched chain.

$isoC_{18}$ a compound containing 18 carbon atoms in a single chain, unbranched save that it terminates with the group

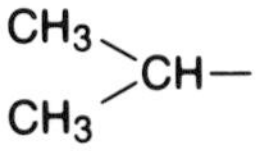

$$\begin{matrix} CH_3 \\ \diagdown \\ \diagup \\ CH_3 \end{matrix} CH-$$

anteisoC$_{18}$ as $isoC_{18}$, save that the chain terminates

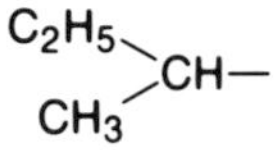

$$\begin{matrix} C_2H_5 \\ \diagdown \\ \diagup \\ CH_3 \end{matrix} CH-$$

Such compounds are commonly encountered in skin lipids. For other hydrocarbon chains, the compound is named from its longest carbon chain, the positions of branching, substitution or unsaturation being indicated by numbering the carbon atoms from one end (invariably from the end of the COOH group in a fatty acid).

Δ a compound containing a C=C double bond. Unless specifically stated, lipid double bonds are assumed to be in the (Z)- or *cis* configuration. Low levels (0.3% total acids) of (E)- or *trans*-momounsaturated acids have been detected in hydrolysed human skin lipids (Morello and Downing, 1976) but are thought to be of dietary origin or the result of microbial action.

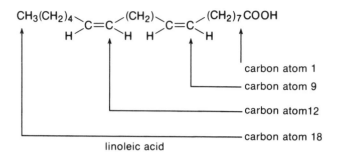

linoleic acid

The nC$_{18.2\Delta9,12}$ carboxylic acid (linoleic acid) is shown above

      └─positions of double bonds (lower numbered carbon atom)
    └─double bonds (*cis*) present
   └─number of double bonds
  └─number of carbon atoms
 └─straight chain compound

Biosynthesis of sebum components (Downing and Strauss, 1974)

Although the detailed biochemistry is complex, it is sufficient to our understanding of the patterns of lipid components encountered here to note the following.

Saturated fatty acids are biosynthesized by sequential addition of $-CH_2CH_2-$ units derived from malonic acid to a suitable starter acid, such as acetic acid. The acids do not react in their free forms, but combined with coenzyme(E). The first step (in fact a composite of many steps) can be represented

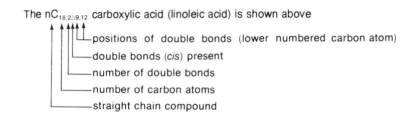

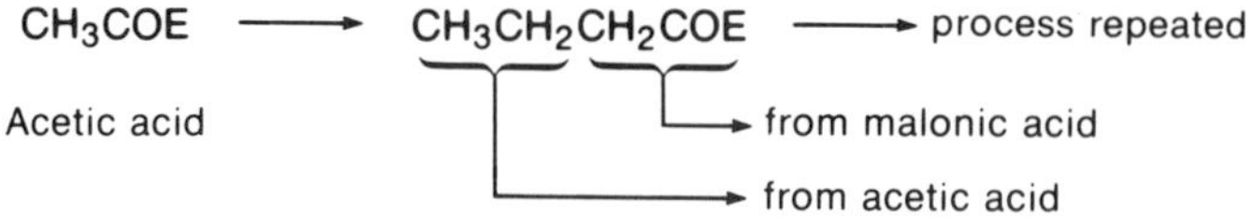

In this way, the carbon chain, which starts with the two carbon atoms which were initially present in the acetic acid, may be lengthened by sequential addition of C$_2$ units to make a whole family of nC$_{\text{even}}$ carboxylic acids. In fact, the enzyme systems involved do not release these intermediate acids, but repeat the process a number of times sequentially releasing finally the nC$_{14}$ acid. This may then undergo further chain elongation, by the addition of individual C$_2$ units, or chain shortening by the removal of C$_2$ units at the

COOH terminus (β-oxidation), so that families of acids are formed. The different families of acids are formed thus (Downing and Strauss, 1974).

starter acid			products
Acetic acid	CH_3COOH	$\longrightarrow$	nC_{even} acids
Propionic acid	CH_3CH_2COOH	$\longrightarrow$	nC_{odd} acids
Isobutyric acid	$CH_3-\overset{\displaystyle \underset{\displaystyle CH_3}{\mid}}{CH}-COOH$	$\longrightarrow$	$isoC_{even}$ acids
Isovaleric acid	$CH_3-\overset{\displaystyle \underset{\displaystyle CH_3}{\mid}}{CH}-CH_2-COOH$	$\longrightarrow$	$isoC_{odd}$ acids
2-Methylbutyric acid	$CH_3CH_2-\overset{\displaystyle \underset{\displaystyle CH_3}{\mid}}{CH}-COOH$	$\longrightarrow$	$anteisoC_{odd}$ acids

Internal methyl branching occurs where methylmalonic acid replaces malonic acid as an extender acid.

Unsaturated fatty acids are formed from the saturated acids just discussed. It is a characteristic of human sebaceous lipid which distinguishes it from lipid formed elsewhere in the body that this unsaturation is introduced preferentially into the $\Delta 6$ position and into the C_{16} acid. The introduction of a double bond at $\Delta 9$ also occurs, as in other tissues, but again the preference is for C_{16} and not C_{18} substrates.

After a double bond has been introduced at $\Delta 9$, it is not unusual for a further double bond to be introduced at $\Delta 6$. It is generally the case for a second double bond to be introduced one CH_2 unit from the first on the side toward the carboxyl terminus.

The unsaturated acids thus formed can undergo further chain elongation and further desaturation, and in this way the numerous families of unusual sebum fatty acids which characterize sebum may arise.

Fatty alcohols probably arise by reduction from the fatty acids first formed. 2-Hydroxy-fatty acids also arise from fatty acids by hydroxylation, and alkane diols by their subsequent reduction. Finally, esterification processes lead to the lipids observed in sebum.

Chapter 4
Scent Glands

4.1 SCENT GLANDS; SPECIALIZED SKIN GLANDS

Scent glands, specialized integumentary organs from which semiochemicals emanate, are nothing more than elaborations of the basic skin structures described in the previous chapter. As a result, the insights developed in that chapter are pertinent to a general understanding of these organs. The degree of specialization, of course, varies considerably from case to case, ranging from the occurrence of little more than an above average concentration of sebaceous and/or apocrine glands in certain regions, to structures which have evolved elaborate anatomical or biochemical characteristics. In some cases also, structures have evolved which facilitate microbial activity and thus microbial scent production from primary skin products.

Specialized skin gland complexes occur in most mammalian species, many species possessing an array of different skin scent glands in different regions of their body surface. Many such glandular complexes have been described anatomically (Schaffer, 1940; Adams, 1980; Brown and Macdonald, 1984), and many scent marking behaviour patterns in which the products of these and other semiochemical sources are deposited in the environment have been the subject of detailed study (Ralls, 1971; Johnson, 1973; Thiessen and Rice, 1976; Sebeok, 1977). But, in spite of the biological importance attributed to the products of such structures (Mykytowycz, 1970; Eisenberg and Kleiman, 1972; Mykytowycz and Goodrich, 1974; Halpin, 1980) very little linked chemical research has yet been undertaken. Those compounds which have been observed commonly fall into categories discussed in the previous chapter, including:

(1) long chain fatty acids and related compounds;
(2) terpenoids derived from intermediates on the biosynthetic route to cholesterol;
(3) sterols;
(4) substances derived by microbial action.

4.2 ARTIODACTYLA

Chemical signalling plays a very important part in the lives of the hoofed mammals or ungulates of the orders Artiodactyla and Perissodactyla. Semi-

Figure 4.1   The gliding phalanger, *Petaurus breviceps*, a nocturnal, tree-dwelling marsupial, employs substances from a multiplicity of sources semiochemically. Individual, group and species odours have been noted and territorial, self- and partner-marking behaviours observed. Signals arise from frontal (forehead), sternal (chest) and perianal glands. The frontal gland, developed in the male only, is arrowed here. Male saliva is also odorous. No chemical studies have been reported. (Reproduced by permission of Springer Verlag from Schultze-Westrum, 1965)

ochemical sources include urine, faeces, and a very large range of specialized skin gland secretions. A good, brief introduction to the biological literature concerning chemical signalling in these species has been produced by Grau (1976). However, only a few of these skin gland secretions have been studied chemically and almost all these have concerned representatives of the Artiodactyla.

Odocoileus hemionus: subspecies *columbianus*—black-tailed deer
: subspecies *hemionus*—Rocky Mountain mule deer

Of these two subspecies, the former has been studied in considerably greater detail. Both possess closely similar scent glands; of these, the tarsal and metatarsal organs are most clearly involved in chemical communication (Quay and Müller-Schwarze, 1970, 1971). Other scent organs which have been described include the preorbital (or antorbital) sacs and the interdigital glands.

The **preorbital sac** is, in both sexes, a large, non-glandular, thin-walled pocket anterior to the forward corner of the eye, in which accumulate cornified epidermal cells from the sac wall, secretions from the orbit of the eye

Figure 4.2   A male black-tailed deer, *Odocoileus hemionus columbianus*, with its preorbital sacs held widely open as it rubs a post. (Photograph courtesy D. Müller-Schwarze)

and various particles of environmental debris. The secretion of the orbit of these and other species include Harderian gland secretion, a secretion known to have semiochemical function in at least one species of rodent (see section 4.5). The **interdigital glands** are pockets supporting enlarged sebaceous and apocrine secretory components and are present in both sexes being equally developed in fore- and hind-feet. Although the tail is used in this species primarily in visual signalling, it seems that a caudal odour is also released under certain circumstances. However, histological studies have failed to reveal anything but a slight enlargement of the sebaceous and apocrine components of the dorsal and lateral surface of the tail. Large **caudal glands** have been described in other species of deer (red deer, *Cervus elaphus*; musk-deer, *Moschus moschiferus*; reindeer, *Rangifer tarandus*).

There is also evidence that a social scent arises from the forehead, for it has been noted that these deer rub their foreheads on dry twigs and that these 'marked' twigs then becomes centres of social attention for other members of the deer group and for newcomers introduced into the group. Such behaviour is reported also in the male roe deer (*Capreolus capreolus*) which possesses a specialized forehead skin gland. In the black-tailed deer, forehead skin merely contains slightly enlarged apocrine glands.

The **metatarsal gland** consists in either sex of a highly developed patch of modified skin situated on the lateral or outer surface of the metatarsus of each hind leg, and comprises a central non-glandular keratinized ridge surrounded by hairy skin containing greatly enlarged apocrine glands and slightly enlarged sebaceous glands. When the deer encounters a 'fear-inducing' situation, the garlic-like odour of metatarsal scent is released. The odour is part of the complex of visual, auditory, and olfactory signals associated with alarm behaviour. When presented alone to captive animals, metatarsal scent tends to inhibit feeding. The chemical nature of this scent which appears to result from substances present in trace quantities, has not been determined.

Most chemical attention has, however, been given to the tarsal scent. The **tarsal organ**, situated on the inside of the hock in both sexes, consists of enlarged sebaceous and apocrine glands associated with a conspicuous hair tuft (Quay and Müller-Schwarze, 1970). Tarsal organs are present only in deer of the subfamily Odocoileinae, which includes the moose, reindeer and roe deer.

Black-tailed deer sniff and lick each other's tarsal organs in a variety of circumstances so that the gland's products clearly play a part in intraspecies chemical communication (Müller-Schwarze, 1971). The tarsal scent of each member of a group of deer is commonly inspected roughly hourly by another group member. However, at night or if a newcomer is introduced into the group, the tarsal scent inspection rate of all group members increases. In addition the tarsal scent of an intruder is sniffed as the possible prelude to attack and pursuit. In adult males, the erection of the tarsal hair tuft and the exposure of its scent are part of threat behaviour. In quite different circumstances, the tarsal scent also appears to play an important role in the female's recognition of her own fawns, and them of her. Further, it seems to convey information relating to the sex of an individual.

On chemical investigation a complex mixture of volatile compounds was found to be associated with the tarsal hair tuft lipids. The major volatile component of the male tarsal hair tuft (10–80 μg per hair tuft) was isolated by preparative GC and identified as (Brownlee *et al.*, 1969) the γ-lactone, (Z)-4-hydroxy-6-dodecenoic acid lactone, or (Z)-6-dodecen-4-olide.

In the female tarsal hair tuft, although present, this compound was not the major volatile component.

Employing a behavioural assay based on the sniffing, licking and following responses of other deer to individuals to whose hind legs test substances (commonly 2 μg of a single compound or of isolated natural components) had been applied, it was apparent that this γ-lactone elicited a behavioural response comparable with that elicited by the entire secretion (Müller-Schwarze, 1969). However, three other unspecified substances isolated from male tarsal scent and tested individually also evoked behavioural responses of a similar magnitude. Increased behavioural activity was noted as secretion fractions trapped by preparative GC were combined to yield a mixture more closely approaching the composition of the complete secretion. Studies with chemicals structurally related to (Z)-6-dodecen-4-olide showed that the configuration of the unsaturated double bond is essential to behavioural activity as determined in this way. Thus the (E)-6-dodecen-4-olide as well as the saturated γ-lactones dodecan-4-olide, undecan-4-olide and decan-4-olide all lacked activity (Müller-Schwarze, 1969; Müller-Schwarze *et al.*, 1976).

It now appears that the (Z)-6-dodecan-4-olide does not originate in the tarsal organ itself, but in urine. Indeed, it is a major contributor to the characteristic odour of this urine, which is quite different from that of, for example, male white-tailed deer which lacks elevated levels of this compound. It is transferred to the tarsal organ secretions presumably with other urine lipids, via a behavioural sequence known as **rub-urination** in which the deer rhythmically rubs its hocks together while slowly urinating over them. Rub-urination corresponds to a natural lipid extraction procedure of urine. This behaviour is exhibited by deer of both sexes and of all ages, but most commonly by mature males at rut, and is associated with threat behaviour in the male and with distress signalling in the fawn. Similar urine transfer is noted in the reindeer and in the red deer, although in the former the interdigital gland region is soaked in urine while in the latter the rutting stag drenches its belly hair with urine to create an olfactory and visual signal (red-brown stain).

Gas chromatographic evidence that the (Z)-6-dodecen-4-olide arises from urine in the black-tailed deer includes (Müller-Schwarze, 1977; Müller-Schwarze *et al.*, 1978a):

(1) the absence of this compound from secretion taken from directly within the secretory tissue of the tarsal organ;

(2) the presence of large quantities (1.2 mg/l) of this compound in the urine taken by catheterization from a sedated adult female;

(3) the observation that this and other lipids are partially extracted from urine passing through the tarsal hair tuft.

A detailed study has revealed that the tarsal hair tuft is specially adapted to extract lipids from urine (Müller-Schwarze *et al.*, 1977b). Electron microscopy reveals that the short stiff hairs at the centre of the tarsal tuft are modified to hold large quantities of lipid deriving from the sebaceous tissue of the tarsal organ. These have enlarged chambers between their cuticular scales while the scales themselves possess comb-like ridges on their surfaces. Such specialized hairs are termed osmetrichia, or odour hairs.

In rub-urination, just described, urine is rubbed into the tarsal hair tufts. Here it encounters the oily secretion supported on the osmetrichia and here urine lipids are extracted from the largely aqueous urine. Excess urine is then licked from the tuft and the tuft is closed by drawing together the outer, longer, unspecialized tuft hairs. In this way a profile of urine lipids is preserved for future display inside the tuft. It is significant that the surface temperature of the hock is the region of highest body surface temperature in this species, thus facilitating evaporation of the volatiles even at low ambient air temperatures.

It follows that chemical signals present in urine in this species can be broadcast in a variety of ways.

(1) directly by urine being deposited in the environment;
(2) indirectly via the tarsal organ
    (a) with the tarsal hair tuft closed. Very short range; approaching sniffing, licking among group members;
    (b) with the tarsal hair tuft open. Maximal exposure of tarsal scent; threat and alarm situations.

Very little work has been conducted on the occurrence of specialized scent hairs or **osmetrichia** in other species. A remarkable example of osmetrichia is provided by the rare crested rat *Lophiomys imhausi* (section 4.5) while the hair of the ventral gland of the mongolian gerbil (*Meriones unguiculatus*) also seems to be specialized. The metatarsal hair of the black-tailed deer, the tarsal hair of the white-tailed deer (*Odocoileus virginianus borealis*), the subauricular hair of the pronghorn, and the scalp, axillary and pubic hair of man all failed to reveal such adaptations.

A comparison of the γ-lactone, (Z)-6-dodecen-4-olide, present in black-tailed deer urine with optically pure synthetic enantiomers indicated that the enantiomeric composition of the natural compound is $89(R)$-$(-)/11(S)$-$(+)$. It appears that the deer may be able to distinguish between these enantiomers, exhibiting a slightly stronger response to the $(-)$- than to the $(+)$- enantiomer (Müller-Schwarze et al., 1978a).

A behavioural comparison of the tarsal scent of the black-tailed deer with that of the Rocky Moutain mule deer subspecies indicated a clear preference of the black-tailed deer for their own subspecies scent (Müller-Schwarze and Müller-Schwarze, 1975). Even to the human nose, the tarsal scents of these two subspecies differ.

Odocoileus dichotomus: the marsh deer
(Jacob and von Lehmann, 1976)

The lipid chemistry of the nasal gland secretion of this south American species has been reported. Nothing is known of the function of the secretion or of its volatile components, although the lipid is predominantly (47%) type 1 wax diesters with about 10% each of wax monoesters and cholesterol esters. The alcohols of the wax esters include a large proportion with unusual ω-9 monounsaturation.

Rangifer tarandus: subspecies *tarandus*—reindeer
subspecies *caribou*—caribou

Reindeer and caribou are greatly dependent on olfactory signals in their social relationships and possess specialized preorbital, interdigital, tarsal and caudal scent gland complexes. The histology of these glands have been described by Quay (1955). Like most cervids, they possess no metatarsal gland.

80

Figure 4.3   A male Maxwell's duiker, *Cephalophus maxwellii*, from the forests of West Africa, using the secretion of its highly developed maxillary glands to mark vegetation. (Aeschlimann, 1963. Reproduced by permission of *Acta Tropica*)

Preorbital sacs consist of pockets (2–3 cm long) situated anterior to the eye. In comparison with other glandular areas, the epidermis lining these pockets is thin and supports relatively few sebaceous and apocrine glands. Little secretion can be obtained on sampling these glands in reindeer except in the case of mature males in rut (September–October) when a massive increase in sac contents occurs. This accords both with behavioural data and with the histological observation that the apocrine component of the sac lining undergoes some development at the time of the rut, presumably in response to elevated circulating androgen levels, although considerable variations between individual males are noted. (Mossing and Källquist, 1981). However, the preorbital sacs are the only skin scent organs which in this species show histological evidence for some degree of seasonal sexual variation. The glandular sac also accumulates secretions draining from the orbit of the eye. The gland appears to function in threat behaviour and is opened and its contents displayed in agonistic encounters.

Of the mixture of volatile compounds associated with male rut preorbital pouch secretions, two ketones have been identified (Andersson, 1979):

Heptan-4-one

2-Methylheptan-4-one

These compounds have also been noted in reindeer urine.

Interdigital glands are highly developed in both sexes, comprise both apocrine and sebaceous components, and are located predominantly in the dorsal interdigital skin of the hind-feet. The glands of the hind-feet are highly developed, forming a deep glandular sac bearing stiff hairs to which secretion adheres. A cytochemical study (caribou; Quay, 1955) has shown that the apocrine secretory cells are characterized by the accumulation of lipofuscin pigment granules.

About 50 mg of sharp smelling lipid may be extracted in organic solvent

Table 4.1. Reindeer interdigital gland volatiles

Component		Mass isolated by preparative gas chromatography from 52 glands
6-Methylheptan-2-one		0.5 mg
7-Methyloctan-3-one		0.7 mg
7-Methyloct-1-en-3-one		0.8 mg
7-Methyloctan-3-ol		—
2-Phenylethanol		—
1-Hydroxy-7-methyloctan-3-one		1.1 mg
7-Methyloctan-3-ol acetate		—
p-Cresol		—

Andersson *et al.*, 1979.

from an excised reindeer interdigital gland (Brundin *et al.*, 1978). TLC and IR show this to consist of sterol ester (60% secretion), triglyceride, free fatty acid, free sterol and unidentified lipid fractions and (at this level of analysis) to be similar in many regards to lipids present elsewhere on the body surface. Free cholesterol, lanosterol and 24,25-dihydro-Δ^8-lanosterol have been identified by GCMS. The relative proportions of some of these compounds seem to vary with the season and with the sex, although further experiments are required to clarify this.

Free volatile fatty acids are also present at low levels (0.1–1 μg/gland) in the interdigital gland secretion (GC), principally isobutyric, isovaleric and isocaproic acids but also including acetic, propionic, butyric, 2-methylbutyric, valeric and caproic acids (Brundin *et al.*, 1978). Longer-chain (C_{12}–C_{20}) free fatty acids are also present.

Volatile fatty acids are present also in combination as sterol esters (but not as glycerides). Alkaline hydrolysis of the sterol ester fraction yielded a tenfold increase in the quantity of these volatile fatty acids per gland.

A Russian study (Sokolov *et al.*, 1977) also noted mixtures of cholesterol and lanosterol esters and triglycerides in skin secretions taken from a variety of locations on the reindeer body surface. However, free sterols and free fatty acids occurred in quantity only in certain areas such as preorbital and interdigital glands, suggesting that these are areas in which ester hydrolysis occurs.

The **tarsal organs** in an area of thickened glandular skin, bearing more developed apocrine and sebaceous secretory tissue and thickened, longer, stiffer hairs than the surrounding skin, and is situated on the medial side of the hock. It is larger in the female than the male and larger quantities of secretion appear to be associated with female than with male glands. Histologically it is the least developed skin gland complex in this species (Mossing and Källquist, 1981). Volatiles present in the secretion coating gland hair tufts include aldehydes, alcohols, and hydrocarbons, the major component being nonanal (1–10 μg/gland) (Anderson *et al.*, 1975). The compounds listed in Table 4.2 were noted. Valeric and isovaleric acids are major components of certain samples.

Although the **caudal gland** has only recently been described in the reindeer extending the length of the tail from base to tip, both ventrally and laterally, consisting predominantly of apocrine secretory tissue with some sebacous tissue and bearing thick long white hair. In this species, the tail brushes in a

Table 4.2. Reindeer tarsal hair volatiles

Aldehyde	Alcohol	Hydrocarbon
Heptanal	Dodecanol	C_{11} diene
Octanal		C_{12} diene
Nonanal	Tetradecanol	C_{12} diene
Decanal	Hexadecanol	Heptadecane

Andersson *et al.*, 1975.

Table 4.3. Reindeer caudal gland hair volatiles

Aldehyde	Fatty acid
Heptanal	Acetic
Octanal	Propionic
Nonanal	Isobutyric
Decanal	Butyric
	Isovaleric
	2-Methylbutyric

(Müller-Schwarze *et al.*, 1977a).

lateral motion over the rump patches of hair either side of the anogenital area thus creating a large 'odour field' dominated by the gland at the rear of the animal. The volatiles associated with the yellow odorous secretion adhering to the hair of the caudal gland include compounds also noted in the tarsal gland secretion, but in much larger quantity. Caudal gland components may be involved *inter alia* in olfactory recognition between mother and young (Källquist and Mossing, 1982).

Mixing the scents

However, the scents from such sources do not for long remain uncontaminated on the body surface of a living animal. In the female, caudal gland secretion adhering to the hair of the ventral surface of the tail and of the rump presumably becomes mixed with urine and vaginal secretion components, while in the rutting male, agonistic behaviour includes a rapid tramping of the hind legs together with intermittent urination so that the lower hind legs, hoofs and interdigital regions become soaked in the strongly smelling urine (Espmark, 1964; Mossing and Damber, 1981). Urine may also be transferred to the head as when a bull confronting another bull urinates and then rubs his nose in the urine patch for some minutes. A further behaviour pattern through which scents may be moved about the reindeer's body is the so-called hind leg–head contact behaviour (HHC) in which reindeer rub or tap their foreheads or the tips of their antlers with the interdigital region of their hind legs. Interdigital gland secretion may be transferred thence over the body and to the environment by subsequent head or antler rubbing. Thus, antler tips may be rubbed successively in the hind interdigital gland and in the groin. As Taylor (1966) has shown with goats, horns or antlers may be very well supplied with blood vessels and act as important heat exchangers with the environment. As a result, they may provide a particularly suitable site from which to broadcast volatile semiochemicals. The possible link between thermal regulation and chemical signalling is mentioned elsewhere in this text (section 4.5). HHC behaviour may occur throughout the year, but is maximal in May when new antlers are completing their growth. The velvet of growing antlers

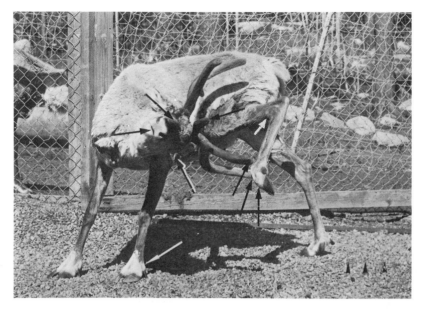

Figure 4.4 A male reindeer, *Rangifer tarandus*, performing hindleg–head contact behaviour (HHC), in which the forehead or antler tip is placed on the rear interdigital gland. Experiments marking the secretion with dye indicate that interdigital gland secretion is, as a result, transferred widely over the body surface (locations marked by arrows). (Photograph courtesy T. Mossing)

may also bear its own indigenous scent as it supports some apocrine and sebaceous glands.

Behavioural observations have shown that odour signals are very important in the life of reindeer. Reindeer produce and follow tracks during their migrations and it is assumed that the secretion of the highly developed interdigital glands is important here. Most deer have well developed interdigital glands and are potentially capable of leaving interdigital gland scent in their tracks. Reindeer will sniff the footprints of other reindeer in their absence, this 'tracking behaviour' being particularly evident to the observer in fresh snow. Reindeer in a test pen show considerable interest in interdigital gland secretion taken directly from the gland (Müller-Schwarze *et al.*, 1978b).

Even to the human nose, interdigital gland secretion odour can be observed to change over a period of 15 minutes, losing its sharp vinegar-like odour and assuming a more cheesy note, so that it would be quite possible for the reindeer to distinguish fresh from aged tracks. Reindeer exhibit significant interest in nanogram quantities of isovaleric and isobutyric acids (present *inter alia* in interdigital gland secretion) placed on filter papers in a test pen (Brundin *et al.*, 1978). In comparison, their reactions to pivalic acid, an isomer of isovaleric acid not found in nature, were indistinguishable from those associated with blank filter papers. However, in these behavioural tests we encounter a problem common to all such mammalian studies, namely, even if the immediate behavioural response is slight and ill-defined, this does not necessarily reflect a lack of potential importance of the signal for the animal. The overt response will depend critically on context, for in mammals the signal may often be 'informative' rather than 'releasing' and the information gained from the signal is integrated with other sensory information and past experience in determining the behavioural response. In the tests with isobutyric and isovaleric acids, for example, the criterion of interest was merely the length and frequency of a sniffing reaction. Reindeer seldom examined any filter more than once, presumably having extracted all the necessary information from it on their first visit.

Observations (Müller-Schwarze et al., 1977a) do suggest that the caudal gland is one of the more important sources of chemical signals in reindeer. During social encounters and particularly when individuals meet again after a period of separation, the tail region is sniffed more than any other body area (Müller-Schwarze et al., 1979). Much social sniffing, however, occurs at distances of up to several metres rather than at close contact. In contrast the rutting male sniffs the tail area of its female and frequently sniffs and licks both her urine and her anogenital region. The scent of the caudal gland also appears to function in the female's recognition of her young. A cow reindeer often sniffs the tail of her calf and will not enable it to suckle until she has done so. Finally, the caudal gland is possibly the source of alarm substances (compare the metatarsal gland of the black-tailed deer and the ischiadic gland of the pronghorn) for when alarmed, the reindeer raises its tail, spreading and thus exposing maximally the white hair on its ventral side.

Moschus moschiferus: the musk deer
(Lederer, 1950; Do et al., 1975)

This small, rare, solitary deer from the forests of the high mountains of central Asia has for centuries been sought after for the musk product of the adult male's scent gland or 'pod'. This comprises a cavity about the size of a walnut in the skin of the abdomen in which a thick secretion accumulates. As well as being valued as a source of perfume, this material is noted for its pharmacological properties (Seth et al., 1973) and for this reason has been used in traditional oriental medicine. Interestingly, male musk deer faeces are also said to have a musky odour. Secretion production is under hormonal control and may be increased by administering androgen. The gland emits its viscous, dark-brown, strong-smelling product particularly at the time of rut when secretion may be rubbed onto objects in the environment.

The attention of chemists was attracted to this secretion early this century when a musky ketone, muscone, was found to occur at levels of 0.5–2% in this substrate. This culminated in the structural elucidations in the 1920s by Ruzicka of the macrocyclic musks, muskone and civetone (from civet, section

5.6). Natural muscone is optically active. Subsequently a related macrocycle, muscopyridine, was identified in the same secretion.

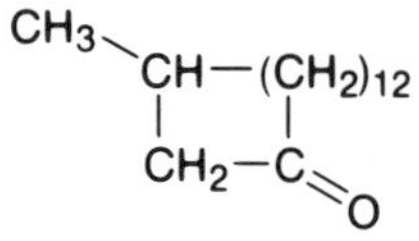

$$CH_3-CH-(CH_2)_{12}$$
$$\qquad\quad|\qquad|$$
$$\qquad\quad CH_2-C{=}O$$

Muscone

$$CH_3-CH------(CH_2)_8$$

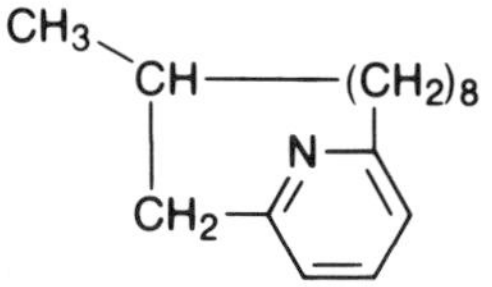

Muscopyridine

More recently, the secretion has been shown to contain a number of androstane steroids as well as cholesterol and cholest-4-en-3-one. Much of the secretion also consists of wax esters and cholesterol esters derived from C_{14}–C_{40} carboxylic acids, including a substantial contribution from branched chain acids.

Table 4.4 Androstanes identified in male musk deer secretion

5α- 5β- } Androstan-3,17-dione		
5α- 5β- 5α- } Androstan- { 3β- 3β- 3α- } ol-17-one		The androstane nucleus
5α- 5β- 5β- } Androstan- { 3β,17α- 3α,17β- 3α,17α- } diol		
Androst-4-en-3,17-dione Androst-5-en-3β-ol-17-one Androst-4,6-dien-3,17-dione		

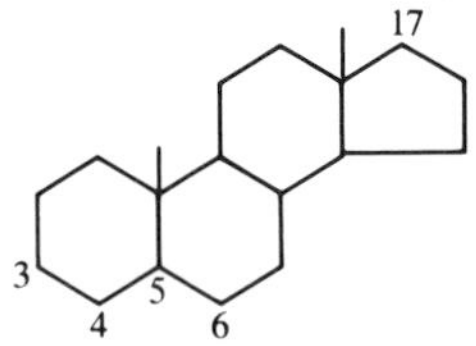

Do *et al.*, 1975.

Antilocapra americana: the pronghorn

Four scents are important in the social life of this animal (Müller-Schwarze and Müller-Schwarze, 1972); those of the paired subauricular (postmandibular or 'jaw patch') glands, of the paired ischiadic ('rump patch') glands, of the single dorsal gland which is located along the mid-dorsal line of the back just anterior to the tail, and of urine. The pronghorn also possesses interdigital glands on fore- and hind-feet. Subauricular and dorsal scents are produced only by males.

The **subauricular scent** is clearly apparent to the human nose and has been described as similar to the odour of burnt fried potatoes. When the male confronts another individual, he frequently approaches with head raised, stops, and throwing his head to one side, presents one of his subauricular glands directly to the face of the other animal. This behaviour pattern commonly occurs between male and female as part of courtship. It also occurs between males, when it acts as a threat display. Females sometimes lick the subauricular glands of males.

Subauricular gland secretion is also deposited on vegetation in a process by which the male marks its own territory, the scent marks formed being easily detected by the human nose. Males can distinguish their own scent marks from those of other animals. On encountering a strange scent mark, a male will commonly sniff, lick and thrash the mark with its horns, then overmark it with his own subauricular gland secretion. Females react in a similar way but lack subauricular glands to overmark.

Pentane extracts of the hair associated with pronghorn subauricular gland yielded the major volatile components shown in Table 4.5.

Tests in a pen showed that isovaleric acid presented alone in microgram quantities elicited the strongest behavioural reaction in male pronghorns (sniffing, licking, marking and thrashing) comparable with that elicited by whole subauricular gland secretion taken directly from a live animal.

The fatty secretion of the **dorsal gland** possesses an odour similar to the subauricular gland. Its function is unknown, although the female has been observed to lick this male gland. Males erect the hair around the dorsal gland

Table 4.5. Components of pronghorn subauricular scent

		Approx. quantity per gland
2-Methylbutyric acid	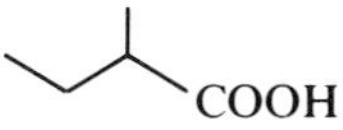	10 μg
Isovaleric acid	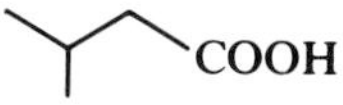	10 μg
13-Methyltetradecan-1-ol	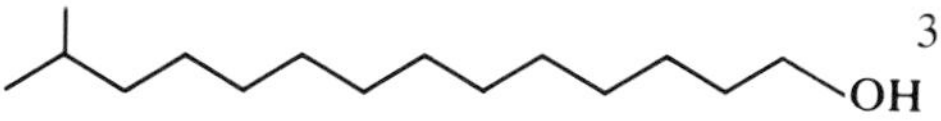	30 μg
12-Methyltetradecan-1-ol	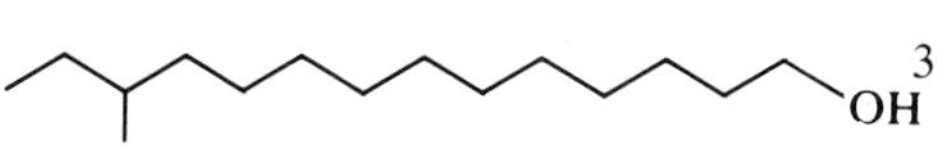	30 μg
plus the four possible esters of the above two acids and two alcohols		70–100 μg each component

Müller-Schwarze *et al.*, 1974.

(thus spreading its odour field), prior to and during copulation, and in certain other activities. No chemical studies have been reported.

The scent of the **ischiadic glands**, apocrine glands situated in the white hair patches of the rump, is discharged as a brief puff before and during flight and so perhaps acts as an alarm pheromone. The odour, which has not been studied chemically, is said to resemble 'buttered popcorn'.

Table 4.6 Compounds of the dorsal gland secretion of the springbok (Burger *et al.*, 1978; 1980; 1981a)

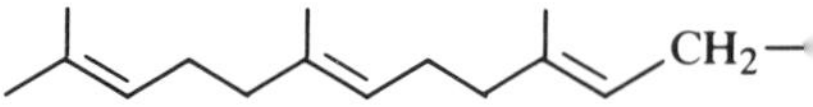

Hydrocarbons

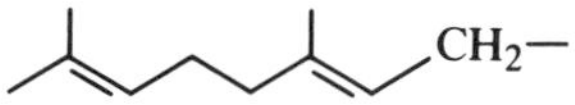

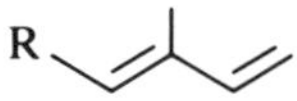

α-Farnesene

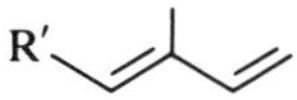

α-Springene

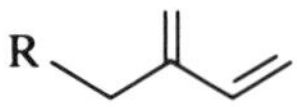

β-Farnesene

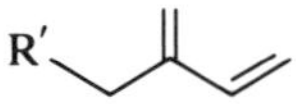

β-Springene

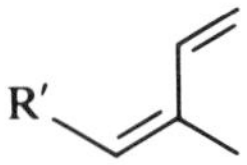

R′—R′ squalene

Aldehydes

$CH_3(CH_2)_nCHO$ where $n = 4, 5, 6, 7$

Ketones

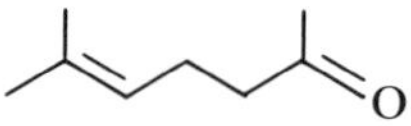

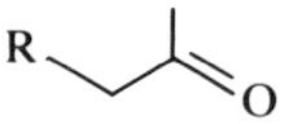

Geranylacetone

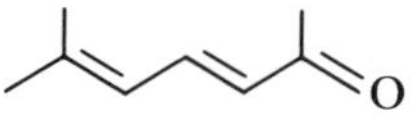

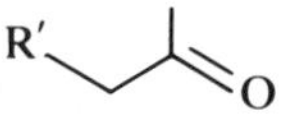

Farnesylacetone

$CH_3(CH_2)_nCOCH_3$ where $n = 10, 11, 12$

Diols/diol esters

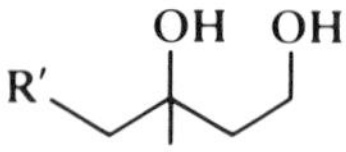

(?)

$CH_3(CH_2)_nCHOHCH_2OCO(CH_2)_mCH_3$ where $\begin{cases} n = 12, m = 1 \\ n = 13, m = 1 \\ n = 13, m = 2 \end{cases}$

$CH_3(CH_2)_4$  $(CH_2)_6CHOHCH_2OCOCH_2CH_3$

Young animals regularly urinate on their hind-feet which become stained and smell strongly as the result.

Antidorcas marsupialis: the springbok

The springbok is one of a number of South African antelopes which have been studied by a group at the University of Stellenbosch. When alarmed, this species opens up the white hair covering its **dorsal gland** and displays a very conspicuous visual signal. At the same time the products of the dorsal gland are also exposed and it is supposed that these could include alarm semiochemicals. Dorsal secretion taken up from some 300 trapped animals of both sexes yielded some 5 ml of yellow oil which was found to be unusually rich in isoprenoid and terpenoid hydrocarbons and ketones, some of which had not previously been noted in nature. For example, some 3 mg of 'β-springene' could be isolated by preparative GC. As with carotenoids, these compounds proved to be very susceptible to air oxidation when isolated. Preliminary GC indicated that considerable sex-related differences in profiles existed. The major component noted on GCMS was squalene.

Damaliscus dorcas dorcas: the bontebok

The yellow waxy secretion of the interdigital gland of this rare South African antelope was taken on swabs inserted into the interdigital cavity of wild

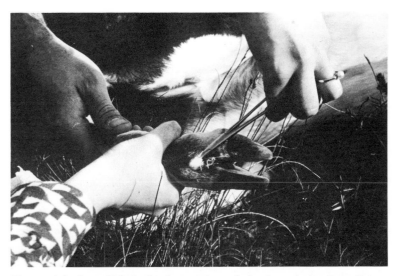

Figure 4.5 Interdigital gland secretion of the bontebok, *Damaliscus dorcas dorcas*, from South Africa being sampled by inserting gauze into the gland cavity. (Photograph courtesy B. V. Burger)

Table 4.7. Compounds of the interdigital gland of the bontebok

Heptan-2-one	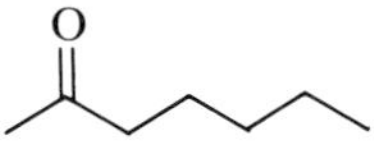
Nonan-2-one	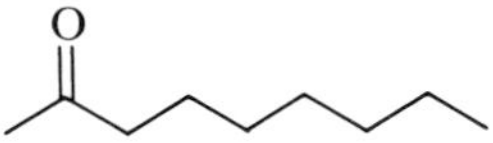
Undecan-2-one	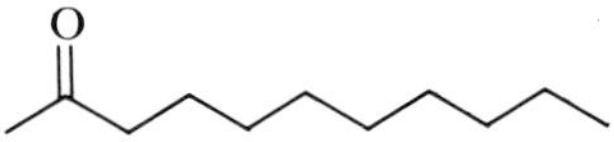
(Z)-5-Undecen-2-one	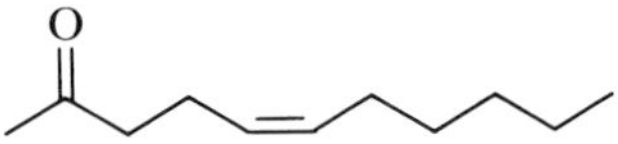
α-Terpineol	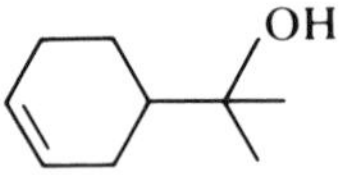
2-Heptylpyridine	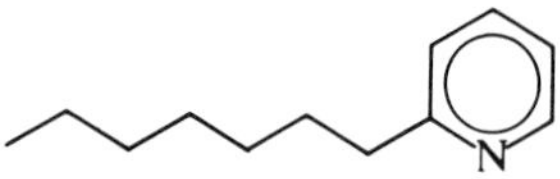
2,5-Undecanedione	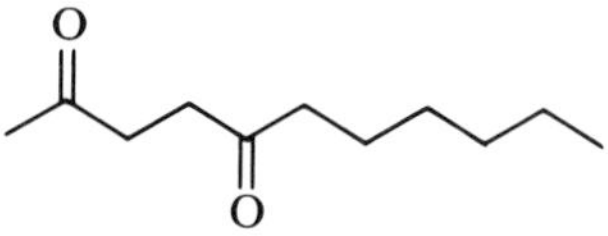
m-Cresol	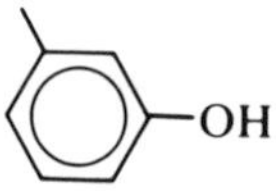
(Z)-6-Dodecen-4-olide	

animals of either sex, the swabs extracted and the extracts combined. Volatiles associated with this oil were examined by GCMS and the major components separated by preparative GC. Table 4.7 lists the components identified (Burger *et al.*, 1976, 1977). A large number of volatile components present at a lower level, however, remained unidentified.

The chemical composition of this secretion showed very little individual variation. The ketones were also noted in the interdigital gland secretion of the **blesbok *D. d. phillipsi.***

Preliminary studies involving presenting a captive male bontebok with (Z)-6-dodecen-4-olide and (Z)-5-undecen-2-one, the major components of this secretion, and with whole interdigital secretion failed to show any clear responses.

Raphicerus melanotis: **the grysbok**

This small solitary antelope occupies dense scrub and relies heavily on olfactory communication. Both sexes produce similar quantities of preorbital sac secretion and both secretions contain many of the same compounds, although in different proportions. The male marks extensively with this secretion. A chemical study on preorbital gland secretion (Bigalke *et al*., 1980; Burger *et al*., 1980, 1981b) revealed a large number of volatile components which were principally straight chain (saturated and unsaturated) alcohols, formates and carboxylic acids, as follows.

Alcohols (all primary alcohols):
saturated C_{11}, C_{12}, C_{13}, C_{14}, C_{15}.
monounsaturated (Z) $C_{13:1\Delta4}$, $C_{15:1\Delta6}$, $C_{17:1\Delta8}$, $C_{19:1\Delta10}$.
diunsaturated (Z) $C_{15:2\Delta6,9}$, $C_{17:2\Delta8,11}$.
formates derived from the following alcohols (all primary):
saturated C_{14}, C_{15}, C_{16}, C_{18}, C_{19}, C_{20}, C_{21}, C_{22}, C_{23}, C_{24}, C_{25}.
monounsaturated $C_{17:1\Delta8}$, $C_{21:1\Delta12}$, $C_{23:1\Delta14}$.
Aldehydes C_6.
Carboxylic acids C_1, C_9, C_{16}, C_{18}.
Ketones (all alkan-2-ones) C_7, C_8, C_9.

In many cases GCMS was insufficient to provide identifications and components were separated by preparative gas chromatography and examined by NMR. Unsaturated components were subjected to ozonolysis to locate the double bond positions.

Parasites in semiochemical systems: an example

The possibility that mammalian semiochemical systems might be utilized by parasites in locating their hosts seems to have been demonstrated in the case of a newly described African tick, *Ixodes matopi*, the adults of which specifically parasitize the **Klipspringer antelope**, *Oreotragus oreotragus* (Colborne *et al*., 1981; Spickett *et al*., 1981). Adults of this tick are attracted to and aggregate on the preorbital scent gland marks which the Klipspringer leaves on vegetation. In this way the tick stands an excellent chance of making contact with its host when the scent gland mark is revisited. Phenolic compounds are known to be attractants to some other Ixodid ticks (Wood *et al*., 1975) although whether such compounds play any part in this interaction has not been investigated.

Domestic bovids: **the ox, the goat and the sheep**

Bovine skin glands have already been mentioned (Chapter 3). A degree of interest in bovid semiochemistry has attached to the role of mammal scent in attracting insects. One of the most promising semiochemical approaches to the control of the African tsetse fly (*Glossina morsitans* and *G. pallidipes*), the vector of trypanosomiasis (sleeping sickness), has been provided by the recent discovery of its intense attraction to ox body odour components (Vale and Hargrove, 1975; Hargrove and Vale, 1979; Vale, 1979). These have yet to be identified although it is reported that carbon dioxide and certain volatile ketones seem to be contributors. Human body odour, in contrast, includes components which tend to repel these flies and it has been suggested that eccrine lactic acid plays a part here. Although lactic acid is relatively involatile, it is one of the components known to attract mosquitos to human skin (Acree *et al.*, 1968; Schreck *et al.*, 1982). Some studies on the chemistry of bovine sebum have been reported (Smith and Ahmed, 1976).

Just as the tsetse attractant seems to derive from the head and neck of the ox, so the characteristic rutting odour of the male goat appears to derive from sebaceous skin glands in the head and neck (Jenkinson *et al.*, 1967). A recent chemical study (Sugiyama *et al.*, 1981) has shown that the neck hair lipids from mature male goats contain unusual 4-ethyloctanoic, -decanoic, -dodecanoic and -tetradecanoic acids, both as free acids and as esters, and that these free acids, particularly the 4-ethyloctanoic acid, are responsible for the characteristic male goat odour. Other fatty acids of these carbon numbers are present only as minor constituents.

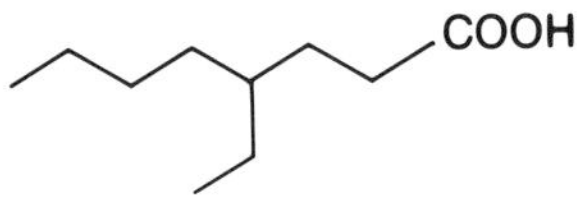

4-Ethyloctanoic acid

Sheep wool components have attracted some attention and something of the composition of wool wax has already been mentioned (see Section 3.5). It now appears that the wool of the ram emits unidentified semiochemical substances which stimulate ovulation in the ewe (primer pheromone; see Section 6.3). Urine is reported not to be a source of this substance (Knight and Lynch, 1980).

Camelus bactrianus: **Bactrian camel**
(Wemmer and Murtaugh, 1980; Ayorinde *et al.*, 1982)

Camelus dromedarius: **Arabian camel**
(Singh and Bharadwaj, 1978)

Male camels of both species possess similar **occipital or poll glands** located in the skin of the back of the head. In the Arabian camel these are a pair of oval tubuloalveolar structures some 10×5 cm in extent and 2 cm thick, divided by the midline of the neck and situated about 6 cm below the nuchal crest. These enlarged apocrine structures secrete into pilary canals above the sebaceous duct and appear to be under androgen control. They are present in the male from birth, enlarging with maturity and may reach a mass of 120 g, frequently with loss of the hair. They regress following castration and are absent in the female. In the intact male, their activity seems to be related to testicular function. They are particularly active at the time of the winter rut when they exude large quantities of a pungent, aqueous, dark brown secretion. This is a time of increased aggressive behaviour and scent marking as well as increased male libido.

In the Bactrian camel it is observed that at the height of the gland's activity, secretion, which is described as having a 'heavy, somewhat sweet aroma', will flow out, saturate and encrust the hair, including the long hair of the nape of the neck. Secretion is transferred by rubbing the gland area on objects in the environment leaving behind a darkly stained scent mark. This occurs particularly at places where male–male contact occurs and it seems the secretion may be involved in male–male rivalry. The secretion is also applied to the hump by leaning the head backwards.

Histological examination shows the secretion to be rich in mucopolysaccharide. A chemical study (GCMS) of the secretion of the Bactrian camel has revealed the presence of the odorous steroid 5α-androst-16-en-3-one (up to 100 μg for some animals) as well as dihydrocholesterol and other unidentified steroids closely related to cholesterol (but not cholesterol itself). Long-chain free acids (C_{15}–C_{25}) are present together with the more volatile acids, isovaleric, hexanoic and decenoic acids (isoC_5, and nC_6, $nC_{10\Delta}$ and the lactone, γ-dodecalactone). Very great differences in composition were noted between different samples and these are probably related, as is the quantity of the secretion, to seasonal factors.

4.3 LAGOMORPHA
(Bell, 1980, 1984)

Chemical communication is clearly important among the rock-dwelling Ochotonidae (pikas) for example (Harvey and Rosenberg, 1960; Svendsen, 1979). These species mark their environment by rubbing rocks with the secretions of their well developed, hormone-dependent apocrine cheek glands. The secretion which is said variously to be odourless to humans and to have a urinous odour is also transferred via their paws. However, no chemical studies have been reported on these or any other lagomorph save the European rabbit. The reason for the concentration of interest is the practical need to seek new, effective means of rabbit control in those areas where it has become a serious agricultural pest. Among the pioneers in this field have been Australian government scientists at C.S.I.R.O.

Oryctolagus cuniculus: **European rabbit**
(Mykytowycz, 1966; Hesterman and Mykytowycz, 1968; Goodrich and Mykytowycz, 1972; Hesterman *et al.*, 1976, 1981; Mykytowycz *et al.*, 1976; Schalken, 1976; Goodrich *et al.*, 1978, 1981a, 1981b)

Rabbits are gregarious territorial animals which live in small, relatively permanent breeding groups dominated by a breeding pair. These dominant animals are responsible for most of the reproduction within the group. The dominant male is also by far the most active scent-marking animal of the group and as a result his scent dominates the group environment. Chemical communication is clearly very important in the lives of these animals.

Among the sources of the chemical signals employed by the rabbit are the secretions of the apocrine anal, submandibular (chin) and inguinal (groin) skin gland complexes. The latter also includes a sebaceous component. All are androgen sensitive, being larger in the male than in the female, and larger in dominant, sexually active rabbit than in the subordinate, sexually inhibited animal. Behavioural observations show that rabbits are able to distinguish their own anal and chin gland secretion from those of other rabbits and also that the experimental introduction into a neutral territory of an animal's own odour, particularly as conveyed by its anal gland (faecal pellet) or chin gland secretion, so enhances its confidence that its chances of dominating or expelling a conspecific in that environment are very greatly increased. The possible role of the Harderian gland remains unknown, although it is clearly under hormonal control, exhibiting differences in size with sex, season and social status, and is discussed briefly elsewhere (see Section 4.5). Urine is also semiochemically important and epuresis, or enurination, in which one animal sprays a conspecific with a short jet of urine, is commonly observed in the male in its relation to the female (Southern, 1948; Bell, 1980). Here the dominant male sprays conspecifics with short jets of urine following a chase thereby impregnating them with its odour. This behaviour, which forms part of reproductive behaviour when a male sprays a female before mating, is also elicited in response to danger. Similar enurination has been observed in some other species. Nothing is known of the chemistry of rabbit urine semiochemicals.

It is not easy to distinguish different communicatory functions for the different scent sources. As Bell (1980) has indicated, a range of different scents may be applied by a rabbit in a given context and it is possible that similar messages may be transmitted by different scent sources. Equally, the message value of a particular scent is likely to be highly context dependent and to vary from situation to situation. It seems that marks formed by the secretions of the anal and of the submandibular glands are both particularly important in territoriality, increasing the confidence of the group within its own territory and advertising their territorial claim to strangers. The secretion of the inguinal gland seems not to function in this way. It is not involved in scent marking and may instead be involved in individual recognition.

The characteristic 'rabbity' odour seems to be associated with apocrine anal and inguinal gland components—particularly with the non-polar, possibly hydrocarbon, lipid fraction. In contrast the secretion of the submandibular gland is, to the human nose, odourless. A study (using GC, TLC and electrophoresis) of anal, submandibular and inguinal gland tissue homogenates (Goodrich and Mykytowycz, 1972) has shown that chemical differences exist both between the different glands in a given animal and also between identical glands in different animals. Variations associated both with sex and with individual identity have been noted. Such differences occur for example in the hydrocarbon fraction GC profiles (C_{11}–C_{26} alkane volatility range) of such extracts. Lipid extracts of gland homogenates contained fractions of TLC mobility indicative of hydrocarbons, wax esters, free fatty acids, sterols, and in the case of the submandibular glands and the sebaceous component of the inguinal glands, glycerides.

Electrophoresis of anal and submandibular gland secretion showed that both contained protein, concentrations being particularly high in male samples. Protein-bound carbohydrate was particularly evident in male submandibular secretion. A number of protein components not present in rabbit serum were noted.

The submandibular gland
(Mykytowycz, 1965)

The submandibular or chin gland is an apocrine skin gland complex which produces a copious secretion leading to a yellow encrustation of the chin hair in the male rabbit. The gland (~400 mg) of the dominant sexually active male is commonly more than twice the size of that of the subordinate sexually inhibited male and more than four times the size of that of the female. The glands are largest and most active in the breeding season. The European hare, *Lepus europaeus,* also possesses a similar gland although it is much smaller than in the rabbit (male ~75 mg; female ~15 mg). The comparatively small chin and anal glands in this species may be related to its non-territorial nature.

Secretion is transferred to objects in the environment—including other rabbits—using a chin-rub behaviour pattern known as 'chinning'. It is most frequent in the male, particularly in the dominant, sexually active male. It is transferred to the environment during territorial scent marking and is evoked by rabbit scents, particularly by the scent marks of strange rabbits. It is also employed following aggressive encounters between males. At copulation, the male may mark the female with his chin gland secretion. He may also mark juveniles to facilitate their acceptance by the group. Very little is known about the chemistry of this substrate. To the human nose it appears to be odourless when fresh.

The inguinal gland
(Montagna, 1950)

Very little is known about the chemistry of these glands. Located in each groin just under the skin, each complex consists of a white sebaceous structure and a brown, branched, tubular, apocrine gland each of which secrete through separate ducts into a pouch or a fold of hairless skin close to the genitals where a thick waxy secretion accumulates. The combined secretion coats the skin with a malodorous brown encrustation. This strong rabbity characteristic odour arises principally from the brown apocrine secretion. This apocrine gland is active from an early age and seems not to be very greatly influenced by gonadal hormones. Volatile fatty acids, especially acetic and isovaleric acids, are, however, present as are microorganisms which might be responsible for their formation and in fact for the inguinal odour (Merritt *et al.*, 1982). Such fatty acids cause changes in heart rate when presented to rabbits.

The European hare possesses much larger inguinal glands than the rabbit. In contrast with the other rabbit skin glands, the inguinal gland is larger in the female than in the male. The secretion is not dispersed in the environment and is not involved in territorial marking. It may be of value in individual or possibly group membership recognition. Information concerning sex identity is also conveyed. Thus members of *all male* groups of rabbits were misidentified and attacked if smeared with inguinal gland secretion from other rabbits of either sex, although this did not happen if they were smeared with chin gland secretion or urine from unfamiliar rabbits or with a commercial perfume. Identical behaviour was shown by *all female* groups of rabbits with the exception that they showed no aggression toward unfamiliar *male* inguinal gland odour (Hesterman and Mykytowycz, 1982a,b).

The anal gland

This paired apocrine structure secretes at the junction between the rectum and the anus. At one time, it was believed that the principal function of this gland was to provide a lubricant to facilitate defaecation. Surgical removal of the gland, however, has shown that this is not so. The hormone sensitivity of the gland also suggests another function.

In the rabbit, the anal glands increase greatly in size with the coming of sexual maturity particularly in the male. The rather solitary European hare also possesses similar anal glands, but in spite of its greater body size, they are only a fraction of the mass of the rabbit's glands. In the hare also, they show no marked sexual dimorphism. In both rabbit and hare anal gland weight and secretory activity are highest in the breeding season.

Rabbits produce two types of faecal pellet. One type is small, soft and mucus covered. These the rabbit reingests. The other, larger, harder, more fibrous, it does not. These latter pellets may be coated (marked) to varying degrees with anal gland secretion by the rabbit during defaecation and in this way, the secretion of the anal gland enters the environment. Behavioural tests also show that rabbits react differently to faecal pellets which are likely to be heavily marked with secretion (those deposited in a strange environment)

Table 4.8. Typical anal gland weights (Mykytowycz, 1966)

Rabbit	Male	sexually active, dominant	2200 mg
		sexually inhibited, subordinate	590 mg
	Female	sexually active, dominant	1500 mg
		sexually inhibited, subordinate	430 mg
Hare	Male and female		110 mg

than they do to those which are likely to be less heavily marked (those deposited in their home range). Also they react more strongly to fresh faecal pellets (<5 hours) than to old faecal pellets (5 days). Rabbits mark their faecal pellets in this way particularly at certain marking points. As in the cases of some other mammals, rabbits form particular 'dungheap' or 'latrine' sites about their territories (Mykytowycz and Gambale, 1969) which act as a kind of olfactory 'notice board' not only for the resident group but also for intruders. These latrine sites may be constantly inspected and added to for periods of years. They are maintained principally by male animals and are visited by dominant animals following aggressive encounters. Not only are faeces and with them anal gland secretions deposited, but chin gland secretion and urine may also be added to the site.

To the human nose, a distinctly 'rabbity' smell is associated with the rabbit's anal gland. This is detected on swabs taken from the anus/rectum interface and also in anal gland tissue homogenates. Some people can detect this odour even in tissue homogenate diluted by a factor of 10^9 in water and presumably the rabbit is equally sensitive to this scent! Anal gland odour intensity is also found to vary seasonally, being greatest in the breeding season, and also to be greater in samples from males than from females.

A chemical analysis has been undertaken of the volatiles associated with the rabbit anal gland (Goodrich *et al.*, 1978). Because volatiles are present at such low levels in the secretion itself and because so little secretion is obtainable by sampling, studies were conducted on the headspace volatiles associated with an aqueous homogenate of a pool of some 20 rabbit anal glands. The complex mixture of volatiles obtained was examined by high efficiency gas chromatography and GCMS. Fractions of possible biological importance were selected for further examination by constraining a rabbit to sample constantly the GC column effluent while monitoring its heart rate radiotelemetrically. Fractions which depressed the heart rate by more than 3% (up to 9%) were separated and subjected to more detailed chemical analysis. Components of interest identified among the large number of components present in the headspace volatiles are shown in Table 4.9. All are aldehydes. Other fractions (possibly including some C_{13} and C_{14} aldehydes) also possessed some activity. It is quite possible that components which do not elicit a heart-rate deceleration in this crude but convenient assay may yet be of semiochemical importance also.

Other major components identified in rabbit anal gland headspace volatiles but which did not produce heart-rate deceleration are listed in Table 4.10. These include other aldehydes together with some alcohols and a ketone. Numerous headspace components remained unidentified.

The composition of the anal gland volatiles is clearly very complicated. The component the odour of which is subjectively most characteristic of anal tissue is (Z)-undec-4-enal. The saturated aldehydes in the range $C_{10}-C_{12}$ have a similar odour.

A similar chemical study of the headspace volatiles associated with marked

Table 4.9. Components of rabbit anal gland headspace volatiles causing a heart-rate deceleration response. (Reproduced by permission of Plenum Publishing Corp. from Goodrich *et al.*, 1978)

	Saturated aldehydes			Unsaturated aldehydes	
		Branched			
	Unbranched	iso-	anteiso-	Unbranched	Branched iso-
C_8	Nonanal (b)		6-Methyloctanal (a)	(*E*)-Non-2-enal (b)	6-Methylhept-5-enal (a)
C_9	Decanal (b)		7-Methylnonanal (a)	(*E*)-Dec-2-enal (b)	
C_{10}	Undecanal (b)	8-Methylnonanal (a)		(*Z*)-Undec-4-enal (b, c)	
C_{11}	Dodecanal (b)			(*Z*)-Dodec-5-enal	
C_{12}				(*E*)-Dodec-5-enal (b, c, d)	

(a) GCMS, provisional identification.
(b) GCMS, comparison with authentic material.
(c) Hydrogenation/ozonolysis.
(d) (*Z*) : (*E*) ratio, 5 : 1.

Table 4.10. Additional compounds identified in rabbit anal gland headspace volatiles (not producing a heart-rate deceleration response) (Reproduced by permission of Plenum Publishing Corp. from Goodrich *et al.*, 1978)

	Alcohols	Aldehydes/ketones
C_5	2-Methylbutan-1-ol	3-Methylbut-2-enal
	3-Methylbutan-1-ol	
	2-Methylbut-2-en-1-ol	
	3-Methylbut-2-en-1-ol	
C_6	Hexan-1-ol	
C_7	Heptan-1-ol	
C_8	Octan-1-ol	Octanal
	Oct-1-en-3-ol	6-Methylhept-5-en-2-one
C_{10}	Linalool	Citronellal
C_{12}		9-Methylundecanal (a)
C_{13}		10-Methyldodecanal
		Tridec-5-enal (a)

(a) Provisional identification.

Table 4.11. Headspace volatile components from adult male rabbit faecal pellets which elicited heart-rate changes (Reproduced by permission of Plenum Publishing Corp. from Goodrich *et al.*, 1981a)

Hydrocarbons
Normal alkanes, C_8, C_{10}–C_{14}
Two monoterpenes
β-Gurjenene
$C_{15}H_{30}$, $C_{16}H_{32}$, $C_{16}H_{34}$

Alcohols
Normal primary alcohols C_4–C_9
2-Methylpropan-1-ol
3-Methylbutan-1-ol
3-Methylbut-2-en-1-ol
2-Phenylethanol

Ketones
Pentan-2,3-dione
Heptan-2-one
6-Methylheptan-2-one
6-Methylhept-5-en-3-one
Octan-3-one
2,2,6-Trimethylcyclohexanone
Nonan-2-one

Carboxylic acids
Normal saturated, C_2–C_4

Others
Dimethyl disulphide
Naphthalene

Esters
Methyl butanoate
Ethyl propanoate
Ethyl butanoate
Ethyl pentanoate
Ethyl hexanoate
Ethyl 2-phenylpropanoate
n-Propyl butanoate
n-Butyl pentanoate
n-Butyl hexanoate
n-Pentyl butanoate
n-Hexyl butanoate
3-Methylbutyl acetate

Phenols
Phenol
o-Cresol
p-Cresol
2,6-Di-tert-butyl-*p*-cresol

Heterocycles
Pyridine
Trimethylpyrazine
Tetramethylpyrazine
2-Ethyl-3,5,6-trimethylpyrazine
2-n-Pentylfuran

Figure 4.6   A plastic film 'balloon' chamber used to test the behavioural effects of odours and GC odour fractions on rabbits. The metal 'hood' contains two compartments by which two rabbits may be introduced into the chamber simultaneously. The hood also incorporates a ball valve through which volatiles from gas chromatograph traps can be introduced into the chamber. The ancillary equipment to the right controls the flow of gas used to introduce the volatiles. The apparatus has been used to study the effect of faecal pellet volatiles on interactions between pairs of rabbits.
(Photograph courtesy B. S. Goodrich)

faecal pellets from adult male rabbits revealed a complex mixture which included components which either increased or decreased by more than 15% heart rates of rabbits monitoring the GC effluent. These are listed in Table 4.11. The active aliphatic aldehydes reported in homogenized anal gland preparations were not noted but could have been present at low levels. The presence of these total headspace volatiles considerably increased the territorial confidence of the donor rabbit in a test environment. Thus observations of competitive encounters between pairs of adult male rabbits placed together in a controlled odour environment test chamber showed that the outcome was

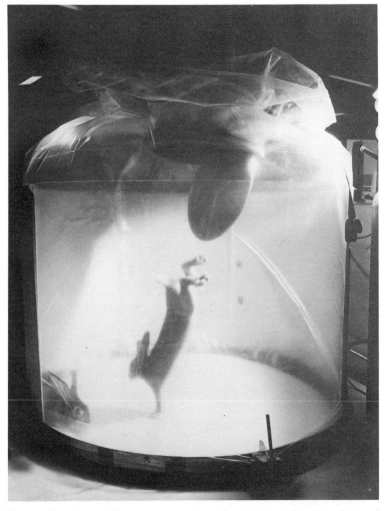

Figure 4.7 A rabbit is released into the balloon chamber from the compartment in the hood. (Photograph courtesy B. S. Goodrich)

substantially weighted in favour of the individual whose total faecal pellet volatiles had previously been introduced into the chamber. A technique for the GC fractionation and representation of these headspace volatiles has now been elaborated and successfully tested and offers scope for determining in some detail which components are of major significance in this behavioural assay (Goodrich *et al.*, 1981b).

4.4 PROBOSCIDEA

Loxodonta africana: **African elephant**
Elephas maximus: **Asiatic elephant**
Eisenberg *et al.*, 1971; Buss *et al.*, 1976; Estes and Buss, 1976; Adams *et al.*, 1978; Wheeler *et al.*, 1982).

Temporal glands are unique to elephants. They are modified apocrine structures deeply embedded in the subcutaneous tissue on each side of the head midway between the eye and the ear and may reach up to 1.5 kg in mass. They are functional in both sexes but the brown watery/lipoidal secretion is most noticeable in the male at musth (rut) when it may be observed running from the gland, down the side of the face to the mouth. The temporal gland secretion is of particular interest to the female at that time. Secretion also flows profusely in other situations whenever the elephant is stressed or excited. Secretion may be transferred to trees and other objects in the environment via the trunk or by head rubbing. Its precise functions are uncertain but a multiplicity of roles has been suggested including reproductive, alarm and individual recognition functions.

Figure 4.8   Secretion from the temporal gland flowing down the side of the head of a male Asiatic elephant, *Elephas maximus*. (Photograph courtesy I. O. Buss)

The odour of the secretion has been variously described as musky and phenolic, and chemical studies on the composition of African elephant temporal gland secretion have indeed revealed large quantities of p-cresol. Phenol and m-cresol have also been reported in certain samples. Other compounds of interest includes all-*trans*-farnesol, a farnesol hydrate and a farnesol dihydrate.

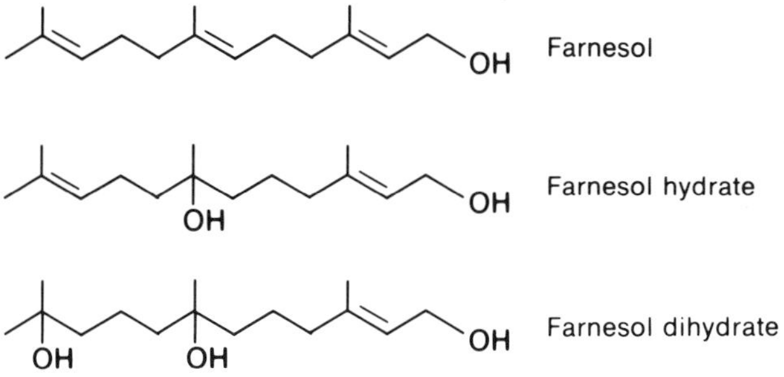

Farnesol

Farnesol hydrate

Farnesol dihydrate

Considerable variations of cholesterol (free and esterified), total protein and total urea levels were recorded in the secretion and some variations in individual lipoprotein and glycoproteins were noted.

4.5 RODENTIA, INCLUDING THE HARDERIAN GLAND

Although there are some 1600 different rodent species alive today, remarkably few have received any kind of semiochemical study. Of these, major attention has been directed to the function of urinary components (Chapters 6 and 7). Skin gland complexes have been described in a substantial number of species. Microtine rodents, for example, are commonly characterized by enlarged sebaceous glands in the caudal, rump, hip, flank or perineal areas (Quay, 1968; Doty and Kart, 1972; Stewart and Brooks, 1976) while anal glands are well represented in the hystricomorph rodents which include the porcupines, cavies, agoutis and the coypu, *Myocastor coypus,* an introduced pest species in Britain (Kleiman, 1974). In the case of the rare maned or crested rat, *Lophiomys imhausi,* from East Africa mentioned earlier remarkably highly developed osmetrichia (scent hairs) have been described (Stoddart, 1979, 1980). On being alarmed this provides both a visual signal, by raising the banded black and white hairs running along its back, and also an odour signal by raising the lipid laden osmetrichia which run along the side of its body above a large skin gland complex. For most of their length these short stiff hairs exhibit a unique open sponge-like appearance admirably adapted to holding large quantities of glandular secretion.

The possession of scent glands in the anal area is common in certain groups of rodents (Kleiman, 1974). The male mara, *Dolichotis patagonum,* a large hare-like rodent from South America, is reported to deliver secretion from anal

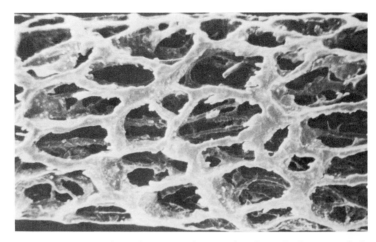

Figure 4.9 Scanning electron micrograph of a single scent hair (osmetrichium) from the lateral gland of the crested rat, *Lophiomys imhausi*. The polygonal spaces which pervade the hair are up to 0.08 mm across and enable the hair to retain large quantities of secretion. (Photograph courtesy D. M. Stoddart)

glands via teat-like papillae, thus presumably providing a continued slow release of secretion (Taber and Macdonald, 1984). One of the many sciuromorphs possessing trilobed perianal scent glands is the prairie dog, *Cynomys ludovicianus* (Jones and Plakke, 1981). These glands, which appear identical in either sex, incorporate both apocrine and sebaceous tissue secreting into reservoirs which connect with the anal region via ducts. In this species also, visible secretion was not observed to be discharged even when the animals were highly excited and the nipples which terminate the duct were fully extruded. The scent issuing from this region is, however, clearly semiochemically important and is involved in intraspecific identification behaviour. In this sequence, both animals crouch. One turns round, everts the three perianal gland duct nipples through the anal orifice and presents them to the second animal which then sniffs them. The sequence is repeated with the second animal.

Although such biological findings have been reported for a variety of rodents, in very few have any chemical studies been undertaken. Those that have include the guinea pig and capybara.

Ground squirrels (Sciuridae) possess well developed dorsal (apocrine) and anal (apocrine/sebaceous) gland complexes used in scent marking, the former being well placed to mark burrows and overhanging rocks, as well as an apocrine complex in the oral angle used in greeting behaviour (Kivett *et al.*, 1976; Kivett, 1978). The preputial secretions of the rodents are considered in Chapter 7.

Cavia porcellus: **domestic guinea pig**
Cavia aperea: **wild guinea pig**
(Beauchamp, 1974; Berüter *et al.*, 1974; Wellington *et al.*, 1979)

Guinea pigs possess a perineal gland, a closed pocket lined with sebaceous tissue located between the genitalia and the anus. For scent marking and in other situations related to aggressive encounters and reproduction, the odorous contents of the pocket are exposed. Scent marking is undertaken principally by dominant males and involves drawing the anal region across the surface to be marked (perineal drag) when secretion and also some urine may be deposited. Behavioural observations suggest that secretion contains individual and species recognition information.

Volatiles have been isolated from *C. porcellus* and *C. aperea* perineal secretions by a low temperature vacuum microdistillation (80 °C, 0.2 torr) and examined by GCMS. The compounds listed in Table 4.12 were identified. A number of unidentified components were also noted. The concentrations of individual compounds ranged from between 0.03 and 3.5 mg/g secretion and very great differences in compound profiles were noted between samples. Differences were also apparent between the wild and domestic species. However, far more striking differences between the species which were consistent were noted when room temperature headspace volatiles were compared. The headspace volatiles of *C. porcellus* were entirely volatile carboxylic acids plus phenol while these same volatile fatty acids were almost totally absent from *C. aperea* headspace volatiles which consisted entirely of neutrals and phenols. This could be explained by the presence of sufficient weak base, such as ammonia, in *C. aperea* secretion to retain the fatty acids as involatile salts thus excluding them from the headspace. Such salts would, however, dissociate on distillation or on direct GC injection.

There is now evidence that the odour of the secretion arises in large measure from bacterial action in the perineal pocket, analogous to the situation in

Table 4.12. Volatile components of guinea pig perineal secretion (Reproduced by permission of Plenum Publishing Corp. from Wellington *et al.*, 1979)

Saturated carboxylic acids
nC_2, nC_3, nC_4, nC_5, nC_6, $isoC_4$, $isoC_5$, $anteisoC_5$, $anteisoC_6$

Alcohols
Alkan-1-ols C_{10}, C_{15}
Alken-1-ols C_9, C_{10} (Δ^3, Δ^4, Δ^9), C_{11}, C_{12} (6 isomers)

Ketones
Alkan-2-ones C_{11}, C_{15}, C_{17}

Phenols
Phenol, *p*-cresol, *p*-ethylphenol

Other
Indole, 2-piperidone, methylnaphthalene

the carnivore anal sac (see Section 5.3). The attractiveness of the secretion for the guinea pig and its perceived odour increase with sebum residence time in the perineal pocket. A number of the components noted such as the volatile fatty acids are expected microbial products and their formation has been shown to be prevented when microorganisms are eliminated from the pocket.

The guinea pig also possesses an androgen-sensitive supracaudal sebaceous skin gland complex (Martan and Price, 1967) which has not been studied chemically.

The steroid chemistry of the androgen-dependent sebaceous chin gland of the closely related South American male cuis (*Galea musteloides*), the secretion of which appears to be involved in reproduction possibly by assisting in the induction of oestrus in the female, has been studied histochemically (hydroxysteroid dehydrogenase activity) (Holt and Tam, 1973). The *in vitro* steroid metabolizing activity of the tissue has been studied also using radiolabelled steroid substrates. The natural secretion has not, however, been analysed chemically.

Hydrochoerus hydrochaeris: the capybara
(Macdonald, 1981; Macdonald *et al.*, 1984)

This pig-sized species, the largest of all rodents, which lives beside the rivers of Central and South America is unique in that the adult male possesses a large, protuberant gland above its snout, a largely naked swelling of dark, sponge tissue some 7 cm long and 5 cm wide, which may stand out up to a height of 3 cm above the surrounding tissue. This nose gland consists of sebaceous tissue and is covered with a film of secretion which issues copiously as a creamy viscous oil from pores when the gland is rubbed against vegetation or other objects. Such scent marks seem to be of considerable social importance in this species. The female possesses a similar structure, but it is much less developed, is covered with hair and is but rarely used in marking.

As in many other caviomorph rodents males and females also possess well developed perineal scent glands. These are sebaceous and line the inner wall of the anal pocket, an area in which the skin is normally kept folded shut. The anal pocket is also associated with the openings of the anus and the urogenital tract. The interior of the pocket is commonly greasy with secretion and, while the perineal glands are most highly developed in the female, the male anal pocket is distinguished by possessing secretion-laden hairs (osmetrichia) which emerge from the slit to the exterior and which are extremely easily dislodged, thus providing a novel means of secretion dispersion. The hairs are found to be coated with accumulated annular rings of dark, crystalline, calcium-rich salts derived from the pocket's contents. Following nose gland marking, the male may subsequently overmark by dragging its anal pouch area and/or by urination. As the result of this behaviour, the inner thighs of the male are commonly wet with urine (see Section 6.2).

Such chemical studies as have been undertaken indicate that both the nose gland and the anal pocket secretions are highly complex containing numerous compound classes (by TLC). The two secretions differ from each other in composition, and secretions from different animals also differ, and it has been suggested that this might provide a basis for individual recognition through the scent marks. In both secretions sterols (notably cholesterol) occur and it is conjectured that terpenes might also occur. GC and MS studies show that the nose gland secretion contains esters of long-chain acids with long-chain alkan-2-ols and a major hydrocarbon component of formula $C_{30} H_{50}$.

Arvicola spp: **water voles**
Apodemus spp: **field mice**

The water vole possesses a sebaceous flank organ the function of which is hormonally controlled. It is most highly developed in the male, its activity being correlated with testis weight, both being maximal in the breeding season. Its function is not clear although secretion seems to be deposited in the animals' environment. Chemical (GCMS) analysis (Stoddart *et al.,* 1975; Stoddart, 1980) of lipid encrusting the glandular surface has revealed a complex mixture of monoester waxes in the mass range 340–480 amu. These waxes contain both straight and branched chain moieties. Their patterns of occurrence as revealed by GC seem to exhibit seasonal variation and also adult-juvenile and group (family and population-distinguishing) features, although no sex related differences in these components were noted. However, the study has been limited by the use of low resolution columns and by a lack of emphasis on the more volatile odorant components. How pertinent the observed differences are semiochemically remains to be answered.

Apodemus species (particularly *A. flavicollis*) have been examined by GC in a similar way. Compound identifications have not been reported although it has been suggested that wax monoesters are present. In these species, interest has been directed to the milky, odorous secretion of the sebaceous subcaudal gland. This gland is considerably more highly developed in the male than in the female. Attention has again been focused on the complex mixture of relatively involatile lipid components (Stoddart, 1973, 1977). Gas chromatograms showed that adult male secretions were more complex and richer in components than those obtained from the female or the closely similar juvenile male secretions both of which gave evidence of relatively few compounds. However, as already indicated, there is considerable danger in equating too simply GC chromatograms of relatively involatile lipids with semiochemically significant odorant profiles.

Meriones unguiculatus: **Mongolian gerbil**
(Thiessen *et al.,* 1969, 1974; Thiessen, 1973; Yahr, 1977a,b, 1980; Halpin, 1976, 1980)

As well as employing the Harderian gland, as a source of semiochemicals, this rodent possesses a large abdominal scent gland from which it rubs secretion onto distinctive low objects in its environment. In this way it marks its environment and also conspecifics. This gland is most highly developed in dominant mature males which also mark more frequently than subordinate males or females. Both scent marking and glandular development are under hormonal control, being stimulated by testosterone and diminishing following castration. The hair associated with the gland seems to be specialized (osmetrichia) and contains deep longitudinal grooves which may facilitate distribution of sebum from the gland. The scent mark itself seems to function in a variety of ways depending on context. Through scent marking, the dominant male impresses its odour on its territory. Maternal sebum is also employed in enabling the mother to recognize its pups.

A chemical study of male sebum components indicated that phenylacetic acid was a semiochemically active component of this secretion, although other unidentified components are likely also to be important. One compound could not, for example, for a basis for individual recognition. Phenylacetic acid has a powerful, animal-like odour.

Sebum was warmed in an air stream and the volatiles, which retained the characteristic odour, were fractionated by TLC. Each fraction was examined for odour and one, which was found by GCMS to contain phenylacetic acid, was identified as having an odour similar to that of whole sebum. Semiochemical activity was inferred by comparing the reactions of gerbils to sebum volatiles, sebum fractions, phenylacetic acid, and controls both in a behavioural conditioning experiment and in their investigative responses to odours introduced into a test chamber via an odour port.

Behavioural evidence suggests that ventral gland sebum odours are indeed complex, even being influenced to some degree by diet, for preweanling young prefer the odours of sebum taken from adults fed on the same diet as their own parents (Skeen and Thiessen, 1977).

Recent studies on the male abdominal scent gland of the related Israeli jird (*Meriones tristrami*) (Kagan *et al*., 1983) has revealed a complex lipid mixture including a range of alkyl and alkenyl acetates, C_{16} and C_{24} predominating. In this species male gland extracts were attractive to female jirds as were a number of the acetates tested, particularly the unsaturated acetates such as the $C_{14:1 \Delta 9}$, $C_{16:1 \Delta 9}$ and $C_{18:1 \Delta 9}$ compounds and the $C_{18:2 \Delta 9,12}$ compound. It is of interest to note the parallel with the female attracting acetates observed in rat and mouse preputial tissue and previously thought to be unique to these tissues (see Section 7.1). Although the lipid components of Mongolian gerbil abdominal sebum have been surveyed (Jacob and Green, 1977; Kagan *et al*., 1983), no such acetates have been detected in that species.

Mesocricetus auratus: the golden hamster
(Johnston, 1977)

Golden hamsters, like gerbils, employ Harderian gland secretion semio-chemically. In addition, they utilize a variety of other scent sources semi-ochemically including urine, and vaginal secretion (Murphy, 1980), and poss-ibly also preputial gland components, although these have not been studied. They do however possess androgen-dependent sebaceous flank glands which are employed in scent marking particularly with agonistic arousal, for golden hamsters are highly intolerant of each other. No chemical studies have been reported, although olfactory bulb removal experiments indicate that olfactory cues are important in eliciting aggressive behaviour.

The Harderian gland
(Rock, 1977)

Although this gland occurs in all non-primate terrestrial mammals, its semi-ochemical implications have been explored mainly for rodent species. It was first described in the stag by the Swiss naturalist, Johann Harder, in 1694. It is located behind the eyeball and its oily secretion acts, together with the pro-ducts of the lachrymal glands, as a source of lubrication for the eye. However, a number of other functions have also been postulated for this gland. Its secretion, together with the other secretions of the orbit contribute to the semi-ochemically active contents of the preorbital sacs where they are present. Although in higher mammals the Harderian gland is less developed relative to the lachrymal glands (and in man is present only in the foetus), in rodents it is well developed. Here it is typically a bilobed tubuloalveolar apocrine lipid-rich structure opening through a duct on to the posterior surface of the nictitating membrane, the rudimentary 'third eyelid' commonly noted at the inner corner of the eye. In a number of species it has been shown that large quantities of this secretion are released by way of the nares, mixed with saliva and employed in thermoregulatory self-grooming. Such species include the gerbils *Meriones unguiculatus, M. tristrami,* and *M. lybicus,* the hamster, *Mesocricetus auratus* and the rat, *Rattus norvegicus.* In at least two species, the mongolian gerbil and the golden hamster, recent studies have shown this secretion to have semiochemical significance.

In rodents the Harderian gland is particularly prominent and is the site of excess porphyrin accumulation which results in the tissue exhibiting a red fluorescence under ultraviolet irradiation. This phenomenon is not charac-teristic of the Harderian glands of mammals in general, however, such fluorescence being absent from the glands of the rabbit, dog, cat, guinea pig and many other species.

Meriones unguiculatus: the Mongolian gerbil

The Harderian gland of the Mongolian gerbil has received particular atten-tion from Thiessen and his associates (Thiessen *et al.,* 1976, 1977; Thiessen,

1977a,b; Thiessen and Kittrell, 1980) who have been interested to explore what might initially seem to be a somewhat unlikely general link between thermoregulation and chemical signal generation.

In the gerbil the gland, which is much larger than the eye itself, is highly vascularized, highly pigmented with protoporphyrin (see Table 4.14) and heavily innervated with sympathetic and parasympathetic fibres, contains two distinct types of secretory cell (lipid and pigment-producing) and yields an oily, porphyrin-rich secretion by an apocrine mechanism (as in the case of the hamster) similar to that which occurs in the mammary gland. No obvious differences in the glands are noted between the sexes. A lipid extract of the gland possesses a floral odour and shock-avoidance and taste-aversion conditioning experiments show that gerbils can both smell and taste this material.

Secretion is released under neural/hormonal influence into the conjunctival sac and thence both into the nasal corner of the eye and also via the nasolachrymal duct to exit at the nares (nostrils), where it is mixed with saliva and spread over the surface of the nose, chin, cheeks and paws (and also over the ventral scent gland area) during facial grooming behaviour. It is simple to monitor the distribution of the secretion over the body surface in this way, not only in this species but in a number of other rodents including the golden hamster, the laboratory rat and other species of gerbils, by utilizing porphyrin ultraviolet fluorescence.

Grooming behaviour thus spreads a saliva/Harderian secretion mixture over the face to lower body temperature by evaporation while also displaying Harderian secretion components. The stimulus which initiates this behaviour is physiological arousal (excitement and stress through exposure to novelty, social interaction, etc.), leading to increased metabolic rate and to increased body temperature. Heat stress also stimulates this behaviour and it seems that when placed in a thermal gradient, dominant animals both preferentially occupy the warmer regions and exhibit increased self-grooming behaviour. The same stimuli initiate salivation, grooming behaviour and Harderian secretion release. There are a number of instances known of social interactions between animals of other species leading to self-grooming and Thiessen speculates that this type of link between body temperature regulation and chemical signalling might have some generality. Particularly efficient for both temperature regulation and chemical signalling would be the display of secretion on those bared, vascular areas of the body surface which are regions of high heat flow.

This grooming behaviour pattern appears to be contagious among groups of gerbils as excitement spreads so that members of small groups are more likely than not to have comparable amounts of secretion. It appears that part of the signal is airborne so that during and shortly after facial grooming, other gerbils are attracted and show considerable interest (sniffing and licking) in the displayed saliva/Harderian gland mixture followed by self-grooming although after a minute or so the interest wanes considerably. It is fascinating to note that this decay of interest mirrors the decay of fluorescence on the gerbil's face. Although Harderian gland extracts maintain their fluorescence for long periods of time, this property is lost rapidly if the secretion is mixed either with human or with gerbil saliva. This is clearly an enzymic reaction for the ability of saliva to bring about this change is destroyed by heat treatment. So, not only is saliva a medium for the display of Harderian secretion, it also limits the life of a particular signal. The Harderian gland itself has a latency of some 5 minutes during which a second secretion discharge cannot be obtained. As a result, a second discharge occurs only if the stimulus is maintained to a time after the initial Harderian signal has decayed to zero. Dominant males more actively spread their Harderian secretion than others. Gerbils which are unable to participate in this behaviour pattern (by surgical removal of their Harderian gland) are social outcasts, losing all social status and, if males, being unable to reproduce as a result of being rejected by females even in the absence of other males.

Thiessen regards the Harderian signal as a generalized primer which presets the conditions for more specific social interactions and without which these interactions are inhibited.

Mesocricetus auratus: *the golden hamster*

There is evidence in some species, for example in certain strains of mice (Margolis, 1971), that Harderian gland porphyrin production may be sex related. However, the most striking example reported to date concerns the golden hamster in which clear links with reproductive function have been described. Although the significance of these variations in hamster glandular activity remain uncertain, they would provide a basis for a semiochemical system by providing external evidence of the physiological state of the animal in question.

In the golden hamster, Harderian gland histology and porphyrin levels are related to hormonal status, major differences being apparent between the sexes.

In the female, porphyrin accumulation occurs to the extent that solid porphyrin is observed within the glandular alveoli, while in constrast, in the intact male solid porphyrin inclusions are not observed and levels are minimal (Payne *et al.*, 1977). In the castrate, the situation is closer to that of the female, although porphyrin levels are reduced by administering androgen. In the female, porphyrin levels vary seasonally, and perhaps more significantly, over the oestrus cycle, being maximal at oestrus (Payne *et al.*, 1979). They are even higher in pregnancy. Although very clear effects are noted statistically, very large individual variations in porphyrin levels are reported between different females.

In view of the sexual dimorphism of the Harderian glands of the golden hamster, it has been suggested that one function of its secretion could be to provide information concerning the sex and hormone status of the emitting

112

Table 4.13. Sexual dimorphism of the golden hamster Harderian gland (Payne *et al.*, 1977, 1979)

	Male	Female
Gland weight/unit body wt.	Heavier	Lighter
Histology	Two secretory cell types	One secretory cell type
Porphyrin level (nmol/g)	Low ~4	High 1200 (at oestrus)

animal. As the aggression between males is greater than the aggression shown by a male towards a female hamster, the frequency of agonistic behaviour between males one of which had been treated about its eyes with an homogenate of female Harderian gland tissue was investigated (Payne, 1977). It was noted that although threat behaviour was not reduced, actual attacks were reduced by this treatment in comparison with controls (male treated with water or with female muscle tissue homogenate) so that this female chemical cue can reduce aggresson even where the other factors of the interaction remain masculine.

By placing gland tissue homogenates in a test pen (Payne, 1979) and noting both the number of times test animals visited these targets and the length of time they spent investigating (sniffing/licking), it was observed that both male and female gland homogenates are attractive to male animals (relative to controls of muscle tissue homogenate) and that female gland homogenates are considerably more attractive than male gland homogenates. However, no correlation was noted between the attractancy of female glandular material and either the stage of the oestrus cycle at which it was obtained or the concentration of protoporphyrin present. Protoporphyrin IX itself was found not to be attractive to male hamsters. This tends to accord with behavioural observations that males will attempt to mount females at any stage of the oestrus cycle but are deterred by female aggression when the female is unreceptive, that is, that they gauge female receptivity by behavioural rather than by chemical cues. In this context, other workers have shown that in this species males cannot distinguish between vaginal discharges of females at different stages of the cycle. It also suggests that protoporphyrin IX itself is not semiochemically active. However in assessing this, the severe limitations of the artificial testing procedure must be borne in mind.

The rat, the mouse and the rabbit

It has long been known that as in the case of the golden hamster, the principal porphyrin of the Harderian gland of the rat and of the mouse is protoporphyrin IX, and that enzymes are present in this tissue for the total biosynthesis of this substance from glycine and succinate (Tomio and Grinstein, 1968; Margolis, 1971). A more detailed study of rat Harderian gland tissue revealed the presence of three porphyrins (Kennedy, 1970; Kennedy *et al.*, 1970) (Table 4.14).

Table 4.14. The porphyrins of the Harderian gland

porphyrin	R^1	R^2	% (1)
Protoporphyrin IX	$-CH=CH_2$	$-CH=CH_2$	64
Harderoporphyrin	$-CH=CH_2$	$-CH_2CH_2COOH$	29
Coproporphyrin	$-CH_2CH_2COOH$	$-CH_2CH_2COOH$	9

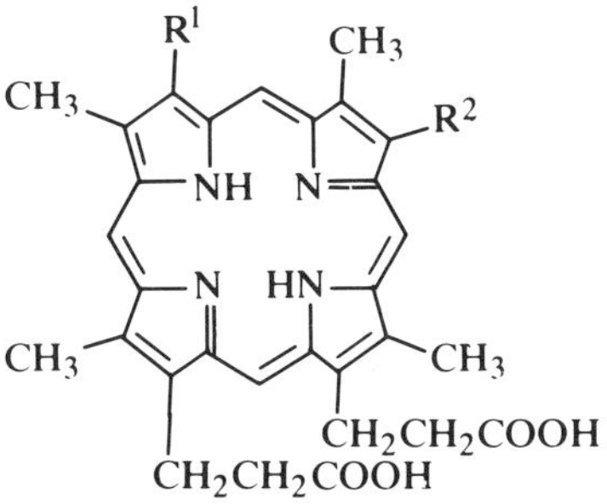

(1) Based of ~5 mg porphyrin isolated from the Harderian glands of 700 rats. Considerable variations in the proportions of the various porphyrins have been reported (Kennedy, 1970; Kennedy et al., 1970).

The relatively few studies which have been conducted on the other lipids of the Harderian gland have also revealed some unusual features. In the laboratory rat (Murawski and Jost, 1974) some 66% of the Harderian gland neutral lipids are esters which are notable in that both the fatty acid and the fatty alcohol moieties are monounsaturated principally at the unusual ω-7 position. Although a range of chain lengths are present, the principal fatty acid is $nC_{20:1}$ and the principal fatty alcohol, $nC_{24:1}$. The high biosynthetic activity of this tissue is indicated by macroautoradiographs showing that, following administration of 1-^{14}C-acetate to the intact rat, the highest radioactivity concentration of any body tissue occurs here, principally incorporated into lipid (Hais et al., 1968).

The dependence of the Harderian gland on the endocrine state of the mammal, and particularly on pituitary function, is illustrated by experiments on the rat. Rat lacrymal gland weights show a similar dependence. Although the weights of the Harderian and of the lacrymal glands are largely unaffected by castration or by the treatment of castrates with testosterone, gland weights are greatly reduced by hypophysectomy-castration. The weights are only partially restored by testosterone treatment and in the case of the lacrymal gland, the restoration is minimal. Pituitary growth hormone alone only partially restores Harderian gland weights and is ineffective, either independently or synergistically with testosterone, in restoring the lacrymal gland. Pituitary α-MSH treatment has no independent and only a slight synergistic effect with testosterone in restoring the weights of either gland.

A function of the Harderian gland in the *rat* is also suggested by another experiment (Brooksbank *et al.*, 1973) in which radiolabelled androstadienone (androsta-4,16-dien-3-one) on injection is found to accumulate very rapidly in a limited number of organs, including the Harderian gland. A similar experiment with testosterone showed a different pattern of accumulation. Since the only known biological function of androstadienone is as the immediate precursor of the odorous steroid 5α-androst-16-en-3α-ol which acts as a sex semiochemical in the pig, this has led to speculation that such a steroid might also be present in Harderian gland secretion. This has yet to be confirmed by direct chemical studies.

In the *mouse*, the normal Harderian gland is reported to be rich in alkyl-diacylglycerols, although most research here has been conducted on Harderian gland tumour lipids (Kasama *et al.*, 1970) (mouse preputial gland, Section 7.1)

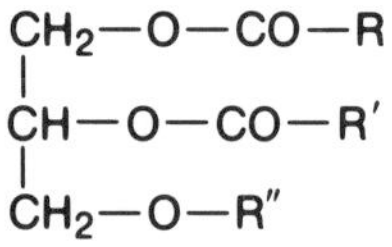

$$\begin{array}{l} CH_2-O-CO-R \\ | \\ CH-O-CO-R' \\ | \\ CH_2-O-R'' \end{array}$$

Rabbit Harderian gland, the size of which is hormone dependent, consists of two distinct tissue types, described conveniently as 'white' and 'pink' (Kühnel, 1971). Both tissues are rich in neutral lipid. In the *white* tissue, these are principally triglycerides and type 1 wax diesters (Rock *et al.*, 1976)

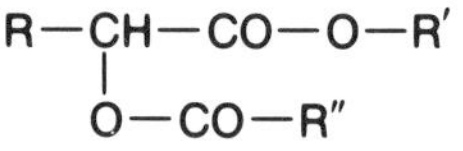

$$\begin{array}{l} R-CH-CO-O-R' \\ | \\ O-CO-R'' \end{array}$$

where R, R' and R'' are all saturated and

R is C_{12}, C_{13}, C_{14}
R' is C_{20}, C_{21}, C_{22}
R'' is C_{13} to C_{21}, principally C_{15}.

The pink tissue lipids are, however, quite different and include compounds which appear to be unique to this tissue. Some 60% of the total lipid were alkyldiacylglycerols (Jost, 1974) including:

(1) fully acylated hydroxyalkylglycerols (47% of tissue dry weight) (Kasama *et al.*, 1973)

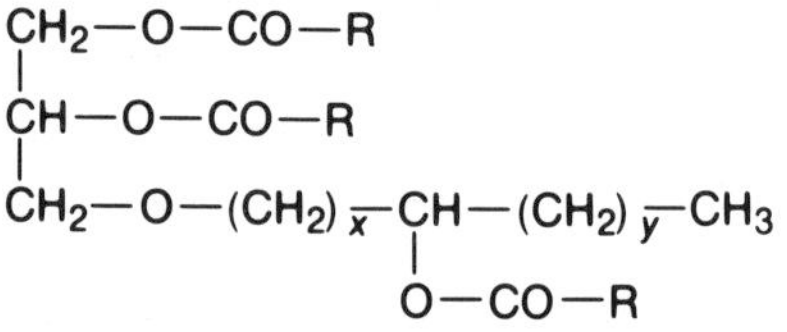

$$\begin{array}{l} CH_2-O-CO-R \\ | \\ CH-O-CO-R \\ | \\ CH_2-O-(CH_2)_{\overline{x}}CH-(CH_2)_{\overline{y}}CH_3 \\ \qquad\qquad | \\ \qquad\qquad O-CO-R \end{array}$$

$x = 9, y = 5$
$x = 10, y = 4, 6$
$x = 11, y = 5$

where R is saturated nC_{15} and nC_{17} plus up to 10% of unusual branched chain C_{16}. Certain components of this type also have isovaleric acid residues esterified at the three position of glycerol.

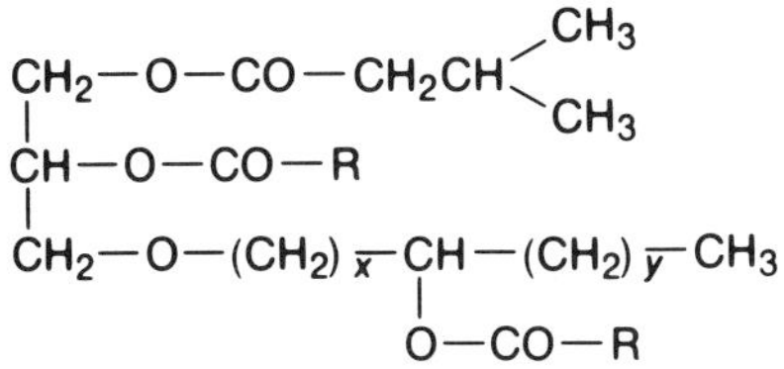

$$CH_2-O-CO-CH_2CH\begin{smallmatrix}CH_3\\CH_3\end{smallmatrix}$$
$$CH-O-CO-R$$
$$CH_2-O-(CH_2)\overline{x}CH-(CH_2)\overline{y}CH_3$$
$$O-CO-R$$

Minor quantities of the related deacylated monohydroxy-and dihydroxy-compounds were also present.

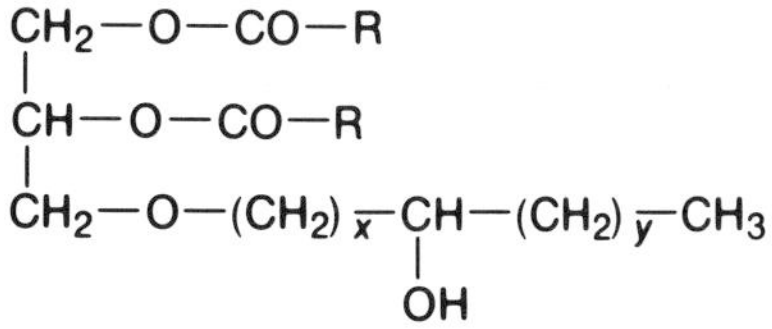

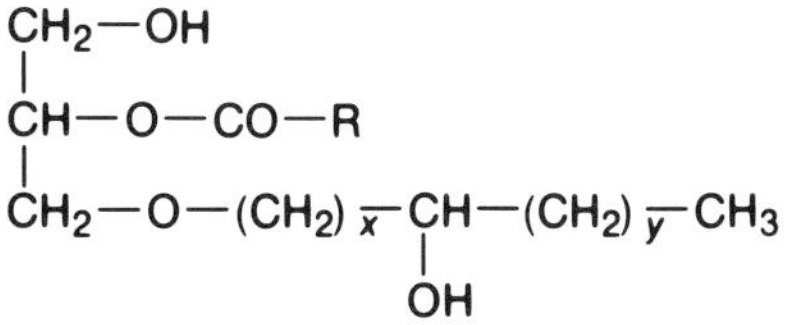

$$CH_2-O-CO-R$$
$$CH-O-CO-R$$
$$CH_2-O-(CH_2)\overline{x}CH-(CH_2)\overline{y}CH_3$$
$$OH$$

$$CH_2-OH$$
$$CH-O-CO-R$$
$$CH_2-O-(CH_2)\overline{x}CH-(CH_2)\overline{y}CH_3$$
$$OH$$

(2) alkyldiacylglycerols containing the isovaleric acid moiety (Blank *et al.*, 1972)

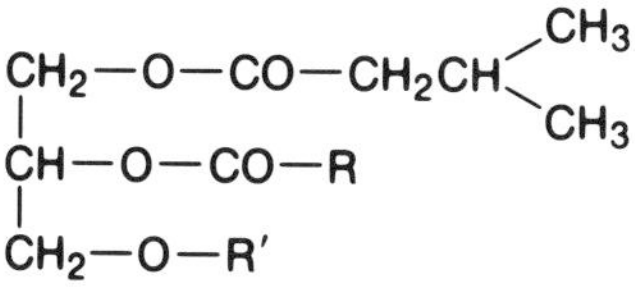

$$CH_2-O-CO-CH_2CH\begin{smallmatrix}CH_3\\CH_3\end{smallmatrix}$$
$$CH-O-CO-R$$
$$CH_2-O-R'$$

with R mainly $nC_{15:0}$ and $nC_{17:0}$; R' mainly $nC_{16:0}$ and $nC_{18:0}$ with 30% unusual branched chain $C_{15:0}$ and $C_{17:0}$.

Other examples of isovaleric acid esters as animal lipid components include the subauricular gland of the pronghorn and the interdigital gland of the reindeer, in both of which secretions free volatile fatty acids are known. They also occur in the milk lipids of ruminants (Patton and Jensen, 1975; Christie, 1978) and in the acoustic tissues of the porpoise.

Whether or not rabbit Harderian gland secretion has a semiochemical function remains uncertain, although Bell (1984) notes that caged rabbits devote appreciable time to sniffing and licking around the eye.

4.6 CARNIVORA

Carnivores possess an array of specialized scent glands (Ewer, 1973; Gorman, 1980). Important among these are the anal sacs and anal pockets which are considered in detail in Chapter 5. In addition, they possess a variety of other specialized skin glands, although virtually nothing is known of their semiochemistry.

Domestic cats, for example, are believed to possess a semiochemically active skin gland complex above and between the eye and the ear, and it is thought that much of the 'affectionate' head-rubbing behaviour seen in pet cats is an example of 'conspecific' scent-marking behaviour using these

Figure 4.10   One cat scent marking another by rubbing with its forehead
and transferring secretions from its forehead skin

glands. Other skin glands of semiochemical importance occur at the margins
of the lips, under the chin and on the tail.

Although nothing is known of the chemistry of these glands, it is interesting
to speculate on the peculiar behavioural responses of domestic cats and certain other Felidae to certain plant products. The essential oil of the catnip or
catmint plant, *Nepeta cataria,* contains a very large proportion of the mono-
terpene, *cis-trans*-nepetalactone (Regnier *et al.,* 1968; Eisenbraun *et al.,*
1980), the odour of which, even at exceedingly low levels (1 part in 10^9–10^{11})
can evoke in these species a remarkable behavioural response (the catnip
response) resembling behavioural patterns associated with sexual and inges-
tive behaviour. Sensitivity appears to be genetically inherited, to be highest in
animals of reproductive age, to be manifest in either sex and to vary considerably between felid species. The usual sequence involves seeking the scent
source, sniffing, licking and chewing with head shaking; chin- and cheek-
rubbing; and, finally, rolling over and body rubbing. The oriental vine,
Actinidia polygama, also contains terpenes (Sakai *et al.,* 1980) which elicit
similar remarkable effects on cats and it seems possible that these plants may
contain compounds closely resembling or identical with as yet undetected
felid semiochemicals (Todd, 1962, 1963; Hill *et al.,* 1976; Harney *et al.,*
1978).

Weight is added to such a possibility following studies on another carnivore
skin scent gland, the **supracaudal** (dorsal tail) **gland** of the **red fox**, *Vulpes
vulpes,* which appears to generate a complex mixture of related compounds
(Albone, 1975; Albone and Flood, 1976). At one time this gland was known
as the violet gland on account of its scent. In the red fox it consists of some

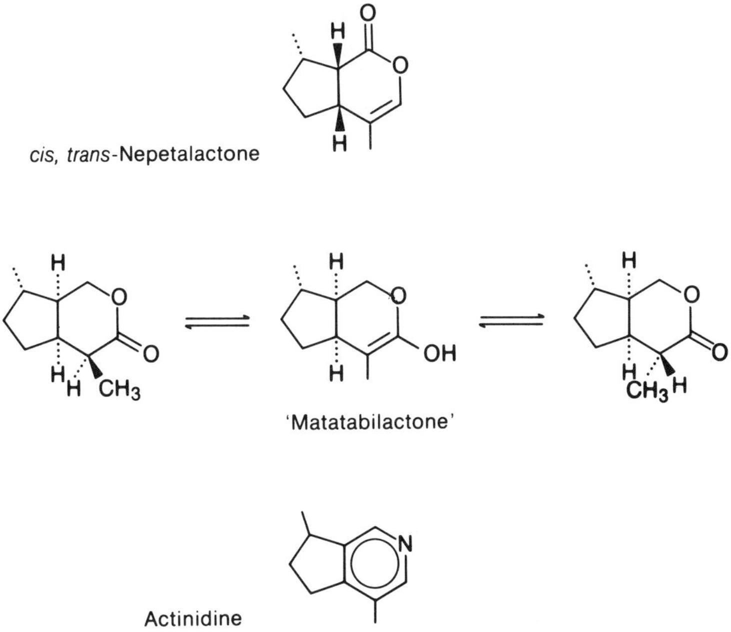

117

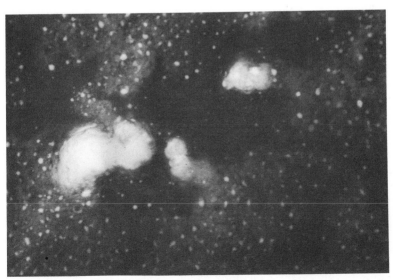

cis, trans-Nepetalactone

'Matatabilactone'

Actinidine

Figure 4.11   Some related compounds eliciting a catnip (catmint) response

Figure 4.12   Photomicrograph of a section of red fox supracaudal sebaceous gland tissue illuminated with ultraviolet light. Lipid droplets and lipid accumulated in the duct system show an unusual, short-lived photolabile fluorescence. (Photograph courtesy P. F. Flood)

200 mg of secretory tissue, massively developed sebaceous glands interspersed with some large apocrine glands, occupying a small area on the top surface of the tail a third of the way from its base. The stiff guard hairs of this area arise from simple follicles which lack the multiplicity of fine secondary hairs arising from compound follicles of the general body surface (Lovell and Getty, 1957) and these with the skin, commonly bear yellowish waxy particles of stale secretion. Most canids possess such a gland (Hildebrand, 1952), although in the domestic dog it appears not to be functional while in the grey fox (*Urocyon*) it runs 20 cm or so along the tail. It is reported to be absent in the hunting dog, *Lycaon pictus* (van Heerden, 1981). Very little is known about the significance of this gland in the life of the fox but it does appear to elicit interest in conspecifics, particularly in the less gregarious social species. The distinctive steroid biochemistry of this tissue, as revealed by HSD histochemical studies, has already been mentioned (see Section 3.4).

Attention was directed to this gland on noting an unusual photolabile yellow-green UV fluorescence which appeared immediately after an incision was made in the gland. Subsequently this fluorescence was found to occur to a lesser extent in other red fox sebaceous tissue, and to arise from lipid droplet in the sebaceous cells and in the ducts. TLC, UV, fluorescence spectroscopy and field desorption mass spectrometry showed that a number of fluorescent components were present, that these were probably carotenoid and that the major component involved a conjugated carbonyl function. A conjugated polyene hydrocarbon was also present as a minor component. The fluorescence observed in the stale secretion adhering to the skin surface was bluer and exhibited a different fluorescence spectrum. GCMS of the volatiles present in stale secretion revealed a complex mixture including many substances clearly related to the catnip active plant components already discussed. However, very few of the components could be identified definitively on the basis of low resolution GCMS data alone. It is probable that the occurrence of volatile terpenes in this secretion is linked with the presence of the sebum carotenoids. For example, dihydroactinidiolide, one of the few components of this secretion for which GCMS provided a relatively definitive identification, which occurs also in *Actinidia polygama,* and which has been shown to have behavioural effects on cats, may be obtained from β-carotene by photoxidation.

Clearly the chemical and behavioural study of such carnivore skin gland has much to offer to future research. It remains to be seen what components are present in freshly expelled secretion which are absent from the stale secretion adhering to the hair so far studied. Other canid skin gland secretions, such as those of the sebaceous perianal or circumanal skin glands, have received no chemical attention as yet.

4.7 PRIMATES AND MAN
(Epple, 1974, 1976)

We know remarkably little of the nature of the chemical signals employed by primates even though a considerable body of evidence points to the importance of chemical communication in this order and even though the order has the special interest of including man among its species.

It is particularly among the prosimians and the marmosets and tamarins that we encounter a multitude of specialized skin scent glands and ritualized patterns of scent marking behaviour. Here also the importance of olfaction is emphasized anatomically by macrosmatic nasal structures and by the relative dominance of olfactory structures in the brain. However, chemical communication is not confined to these species. Most primates show a marked olfactory and even gustatory interest in the chemical products of the bodies of conspecifics and this includes the old world monkeys and apes which are largely devoid of specialized skin scent structures as well as the new world monkeys in which these structures are relatively common (see urine; vaginal secretions, Chapters 6 and 7). Concentrations of apocrine glands seem to be commonly related with the generation of musky odours associated with many primates.

Sebaceous and apocrine skin glands occur throughout the order (Montagna, 1972). Eccrine glands are confined to the friction surfaces in all but the Cercopithecoids (old world monkeys) and the Hominoidea (apes and man), and only in man are eccrine glands the dominant sweat glands of the general body surface. The precise nature of the eccrine glands also depends considerably on the taxonomy of the species concerned. Aggregates of apocrine glands producing musky odours commonly occur in some area of the body surface, while sebaceous glands add lipid carrier for this apocrine scent, whether or not sebaceous semiochemicals are also present.

First consider the prosimians (Schilling, 1979). Apart from some studies on the bushbaby, nothing is known of the chemistry of communication in these primitive primates. However, the biological importance of chemical communication in the suborder is so great that there can be little doubt that any chemical studies here will be richly rewarded. It is worth briefly mentioning something of the untouched wealth which awaits investigation.

Tupaiidae: tree shrews
(Martin, 1968; Montagna *et al.*, 1962)

Chemical communication is important in the tree shrews, for, in addition to the numerous apocrine glands secreting via the pilary ducts in the general skin, they possess specialized glandular fields in the gular/sternal and in the abdominal regions. The gular field of *Tupaia belangeri* includes highly developed sebaceous and apocrine components particularly in the mature male yielding an 'oily and sticky' secretion. The active regions of both gular and abdominal gland fields are marked by bare skin. Not only are the glands apparently under androgen control but so are the specific behaviour patterns by which secretion from either gland is applied to objects in the environment ('chinning' and 'sledging'). Both behaviour patterns are manifest principally by mature males although sledging is considerably less common than chinning. The male may also chin the back of its female partner in a process which appears to strengthen pair bonding. It also marks objects in its environment, particularly when encountering the scent of a strange mature male. The products of these glands appear to contain information concerning the sex, sexual status and individual identity of the marking animal, although nothing is known of its chemistry.

In addition, tree shrews employ urine and faeces, and possibly also saliva in scent marking. The Tupaiidae are in many ways anomalous and it is a matter of debate whether these primitive creatures are properly primates at all. They alone among prosimians have eccrine as well as apocrine glands in the general hairy skin, even though these eccrine glands are somewhat primitive. They are abundant in the skin of the rhinarium and genitals as well as friction surfaces. The hair follicles also are arranged in groups which are more typical of insectivores than of primates.

Lorisidae: *Lorisinae*: lorises

(Manley, 1974; Montagna and Ellis, 1959, 1960; Montagna *et al.*, 1961; Montagna and Yun, 1962c)

Chemical communication is very important in the lives of these nocturnal tree dwellers. In the general skin surface, sebaceous glands are poorly developed and one or two apocrine glands secrete directly to the skin surface at the periphery of each hair follicle cluster.

The potto, *Perodicticus potto,* from the forests of tropical Africa, is characterized by a musky odour emanating from the extremely rich fields of apocrine glands which are present in the male in the skin of the scrotum and, in the female in that of the pseudoscrotum. Their secretions seem to have social functions. Not only do these regions appear to be of considerable general interest to these animals, but a specialized behaviour pattern involving them (and specific to the potto) has been described as a frequent component of the commonly occurring sequence of mutual grooming. This specialized behaviour pattern has been called 'genital-scratching grooming' (GSG) and results in the transfer of secretions from these gland fields to the fur of the animal being groomed. GSG does not appear to be a component of sexual behaviour. Typical prosimian grooming is performed with tongue and teeth so that there also exists the possibility that saliva-borne semiochemicals are being distributed.

At times of stress and fear, these same glandular areas may secrete profusely and elicit great interest in conspecifics, so that they may also function as a source of alarm semiochemicals.

The closely related angwantibo, *Arctocebus calabarensis,* also possesses similar although less extensive glandular fields in the scrotal/labial area which produce a greasy secretion (Montagna *et al.,* 1966b). The male marks the female's back fur with these glands in a distinctive 'passing over' behaviour pattern so that when a strange female is introduced, major attention is devoted initially to sniffing her back fur, presumably to note any male sent marks. An alarm or fear scent may be produced both in males and females, as in the case of the potto.

Both the mutual grooming behaviour of the potto in which secretion is transferred to the partner's fur and the passing over behaviour of the angwantibo are examples of the exchange and mixing of animal scents. In this species, urine scent may also be shared in partner grooming, for the grooming potto will occasionally take a drop of his urine in his hand and rub it in his partner's fur (Epps, 1974).

The female potto also possesses a pair of flask-shaped glands which secretes into the vulva just inside the orifice and from which can be expelled a malodorous, viscous, soft, semi-solid keratinous material. Its function is unknown.

In contrast, the oriental lorises, the slender loris, *Loris tardigradus,* and the slow loris, *Nycticebus coucang,* each derive their characteristic scent from large, apocrine glands present in both sexes on the medial surfaces of the upper arms. These so-called branchial organs, for which a defensive function has been suggested, are not to be confused with the quite different brachial organ of *Lemur catta.* The brachial secretion of the slender loris has been described as a watery, yellow fluid of unpleasant, musky odour, while that of the slow loris is said to be a yellow viscous substance of foetid yet citrous, fruity odour. Histochemical peculiarities of the glands and their innervations have been noted in both species.

Lorisidae: *Galaginae*: bushbabies

Bushbabies scent mark in a variety of ways, the most important of which involve chest-rubbing, anogenital rubbing and urine-washing (see Section 6.2) (Bearder and Doyle, 1974; Tandy, 1976). In these species, sebaceous and apocrine glands are generally poorly developed (Yasuda *et al.,* 1961). However, in the great bushbaby, *Galago crassicaudatus,* histochemically distinct sebaceous glands of the lips, perinasal, perioral and perianal areas, the prepuce and scrotum are large and multiacinar. Apocrine glands are also more numerous in the skin of the lip, chin and face than on the general body surface. It has been suggested that the apocrine glands act principally as scent producers and contribute to the unpleasant body odour characteristic of this species, which is particularly associated with the rich field of large sebaceous and large apocrine glands situated behind the scrotum.

Table 4.15. Chest gland components from the male great bushbaby G. crassicaudatus (Reproduced by permission of Plenum Publishing Corp. from Crewe et al., 1979)

Compound	% of whole secretion
Benzyl cyanide	8
2(p-Hydroxyphenyl)ethanol	17
p-Hydroxybenzyl cyanide	56

Female secretion contained the same components although the first component was present in relatively lower concentration. A number of unidentified minor components were also noted.

Although the degree of development varies very considerably from individual to individual, the great bushbaby also possesses highly developed chest glands. This skin gland complex, which extends from the base of the throat in both sexes, may be highly developed and is frequently moist with secretion which is transferred to objects in the environment by a characteristic chest-rub behaviour sequence. The bushbaby employs this secretion extensively for scent marking and may convey information concerning sex, age and identity to conspecifics (Clark, 1982a,b). There is evidence of discrimination between the scent marks of subspecies. Although the chemical investigation of this secretion has only begun, some very interesting results have already been obtained (Wheeler et al., 1977; Crewe et al., 1979; Katsir and Crewe, 1980). Some 3–4 μl of secretion were collected in capillaries from each bush-baby and on examination by GCMS, unusual skin gland components were identified (Table 4.15). On presenting a sample of 2(p-hydroxyphenyl)-ethanol in the form of an artificial mark to a G. crassicaudatus an unusual response was noted. The animal slowly and deliberately placed its open mouth over the mark as if to press the inside of its mouth and/or the back of its tongue against the source. This response itself suggests that this substance could have biological importance.

Since the benzyl cyanide is very considerably more volatile than the other two components it is possible that short- (~ hour) and long- (~ some days) term messages may be associated with this scent depending on the presence or absence of this component. Following field studies it seems that the scent mark may function on a short-term basis for territorial marking purposes. The signals appear to be very short range and are effective only if placed on objects on regular routes used by these animals. Message value seems to be associated with mixtures rather than with individual components.

Other species of bushbaby possess similar but less developed chest glands.

Tarsiidae: tarsiers

The tarsiers are not lacking in scent glands, although none has been studied chemically. Thus, the male Philippine tarsier, Tarsius syrichta, possesses a large epigastric sebaceous (with some apocrine components) gland field which yields a waxy brownish red secretion (Arao and Perkins, 1969). This species also possesses a gland on its upper lip (labial gland) consisting of gigantic sebaceous glands, a large circumanal glandular area (sebaceous and apocrine) and very large meibomian glands (Montagna and Machida, 1966).

Lemuridae: lemurs

(Montagna, 1962; Montagna and Yun, 1962a, 1963b; Montagna *et al.*, 1967; Yun and Montagna, 1964; Sisson and Fahrenbach, 1967; Evans and Goy, 1968; Harrington, 1974; Schilling, 1974; Chandler, 1975)

Although not nocturnal, these highly social prosimians show a considerable biological investment in chemical modes of communication. It is very likely that the performances of many of the specialized scent dispersal behaviour patterns employed by these animals may simultaneously serve as visual signals. An example is tail flicking behaviour of the male ring-tailed lemur, *Lemur catta,* which simultaneously displays both tail-borne scent and also the tail's black and white banding pattern.

The skins of lemurs are exceptionally rich in sebaceous glands so that the dense fur is typically greasy. Depending on the species and the body region, the sebum may be dark brown or black due to the presence of abundant melanin granules arising from active melanocytes associated with the sebaceous tissue itself. It is not clear what if any function this pigment has.

Highly developed circumanal glands are characteristic of lemurs. These consist of massively developed multiacinar sebaceous glands together with a number of large apocrine glands. As with other sebaceous and apocrine tissues, these structures are larger in the male than in the female. The highly developed glandular complex may render the circumanal skin hairless and cover the genitalia and perianal skin with greasy, malodorous secretion. These substances are employed in scent marking. The male *L. catta,* for example, is noted to squat and positioning itself over the object to be marked, transfer secretion by a sideways rubbing movement. The female may also adopt a similar stance in scent marking. Indeed in this species genital marking is more common in the female (particularly in dominant females) than in the male and is especially common during and just after oestrus. In the female the labia are widely reflected and the inner surfaces pressed against the object to be marked. Secretion plus mucus, and possibly including some urine, are smeared to form an odour streak which retains its attractancy for some days. Studies with *L. fulvus,* the brown lemur, demonstrate that scent marks contain information relating to the individual identity and the sex of the marking animal. The reproductive condition may also be communicated. For example a distinct must odour emanates from the vulva of the female mouse lemur, *Microcebus murinus,* at oestrus and could have considerable importance in signalling this condition in this nocturnal species (Martin, 1972).

The mixing of male and female scents also commonly occurs. Not only are female scent marks overmarked by the male, but the bodies of the partners may also be scent marked. Thus, *L. fulvus* and *L. macaco* males and females commonly scent mark each other with anogenital scent. Such partner marking and overmarking is common in certain other primates.

Concentrations of large multiacinar sebaceous and large apocrine glands thus occur mainly in the anogenital and scrotal areas of lemurs. They also occur on the face and elsewhere. The male sifaka, *Propithecus verreauxi*, for example, marks with its throat gland as well as with its perineal area (Richard, 1974). *Lemur catta* is remarkable in that in addition to the scent structures possessed by other lemurs, it is distinguished by the possession of two highly specialized scent organs, the brachial (arm) and antebrachial (forearm) organs. This is a characteristic it shares with the grey gentle lemur, *Hapalemur griseus*. The following remarks refer to *L. catta*.

The **brachial organs** occur each as a walnut-sized swelling on the medial surface of the upper arm above the axilla. They are small reservoirs lined with sebaceous tissue which accumulate sebum and cellular debris. On being squeezed, an unpleasantly smelling, viscous liquid exudes from a central orifice.

The **antebrachial organs** consist each of a round cushion of smooth, soft black tissue on the medial surface of the forearm situated about one-third of the distance along the forearm from the hand and at the terminus of a narrow strip of palmar skin which extends along the forearm from the palm. On being squeezed, whitish droplets of apocrine secretion exude from the tissue. In addition to atypical apocrine cells, the organ consists of clusters of distinctive 'interstitial' cells histologically and histochemically suggestive of steroidogenic cells of an endocrine organ. In close association with this organ and on its ulnar side is the **antebrachial spur,** a small conical horn unique to this species. The spur is traversed by the ducts of underlying sweat glands of a type present in the palmar skin.

The brachial organ is present only in the male while in the female the

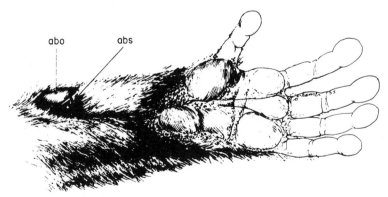

Figure 4.13   The antebrachial complex of the ring-tailed lemur, *Lemur catta*, showing the antebrachial organ (abo) and the antebrachial spur (abs). (Reproduced with permission of The Zoological Society of London from Evans and Goy, 1968)

antebrachial organ is non-functional and the antebrachial spur is often absent. Both organs are under androgen control. These brachial and antebrachial organs clearly contribute in important ways to chemical communication within this species. A male pays particular attention to his antebrachial organs, sniffing and licking them frequently and using them to mark branches and other objects, including his own tail. In all such marking behaviour, it seems that brachial secretion may also be transferred, for each sequence is preceded by 'shoulder-rub' in which the two glandular areas come into contact. When branches are marked, they are simultaneously scarred by the antebrachial spur and secretions from the active palmar sweat glands underlying the spur may also be deposited. Marked branches are preferred to unmarked branches, such markings remaining attractive for several days. There is evidence that individual identity is encoded in the mark and that a lemur can distinguish his own mark from those of others. Marking of this and all other kinds occur at territorial boundaries and particularly at sites of territorial disputes between troops. Such behaviour may be associated with maintaining the status and dominance both of the individual lemur and of the lemur troop in question.

The male will also mark its own tail with antebrachial secretion, gripping it and rubbing it in such a way that brachial secretion is also likely to be transferred to the tail and also, via the tail, between the two glandular areas in question. Such behaviour is a common prelude to aggressive 'stink fights' between males in which dominance is tested and may itself be sufficient to cause the opponent to flee. If not, the male then stands before his opponent with tail arched and flicking rapidly above his back.

Tail flicking also occurs as part of the response of the male to the female particularly with the approach of the brief period of receptivity in late oestrus. This chemical and visual display might have some influence on ovulation in the female. The male for its part receives chemical signals from the female by direct contact. As oestrus approaches the male shows increased interest in the female, spending increasing time licking her anogenital area. At this time the female produces a profuse vaginal secretion. It has been suggested that the peaking of the chemical signalling which occurs between lemurs at the breeding season, and of which this is just one example, could have a 'priming effect' in coordinating physiological events in the two sexes. This would be particularly valuable in species possessing a very short breeding season (see Section 6.3).

Callithricidae: marmosets and tamarins
(Wislocki, 1930; Epple and Lorenz, 1967; Epple, 1970, 1978, 1979, 1980; Richard, 1974; Epple *et al.*, 1979)

All evidence indicates that in this family also olfaction plays a central role in regulating their complex social structures. Here, however, the first steps towards a chemical elucidation of the bases of these interactions are being taken.

Marmosets and tamarins exhibit a similar array of specialized skin gland structures, principally in the sternal and circumanal areas but also in other regions (Perkins, 1966, 1968). These provide examples of skin gland complexes which are commonly more highly developed in the female than in the male, as evidenced, for example, by the suprapubic sebaceous/apocrine gland field in such species as cottonheaded tarmarin, *Saguinus oedipus* (Perkins 1969a) where the gland seems to have a sex signalling function. The female Geoffroy's tamarin, *Saguinus geoffroyi* possesses a highly developed, black cushion of sebaceous and apocrine tissue covering the labia majora, as well as the pubic, perineal and circumanal skin. In the male, the glandular tissue is less highly developed and less extensive. Before mating, the female will commonly impregnate her tail with circumgenital secretion and urine and the male will sniff and lick the tail intensively. Similar behaviour has been noted in the little-known new world monkey, *Callimico goeldii,* which is closely related taxonomically with the Callithricidae (Perkins, 1969b). This animal possesses extensive gland fields in the sternal, genitopubic and perianal regions consisting of highly developed apocrine and sebaceous tissue histochemically and histologically distinct from similar tissue of the general skin.

Most chemical work has, however, concerned *Saguinus fuscicollis,* the saddle-back tamarin. In common with most Callithricidae this species lives in extended territorial family groups dominated by a single adult breeding pair, both males and females possessing extensive and specialized skin scent glands in the circumgenital–suprapubic (sebaceous/apocrine) and sternal (largely apocrine) regions. These they rub on objects in their surroundings—including other tamarins—during scent marking. Scent marking with the hairless circumgenital–suprapubic glandular pad, which is usually moist with secretion, is much commoner than with the sternal gland. The resulting mark contains an admixture of urine and vaginal secretion components in addition to sebaceous and apocrine skin gland secretions. The circumgenital–suprapubic glands develop fully with the coming of puberty as does the expression of marking behaviour itself. The female marks rather more frequently than the male. Castration prior to puberty considerably retards the development of the male glands while it appears to have relatively little effect on marking behaviour.

Scent marking is involved in the regulation of a variety of sexual and social behaviours and seems to communicate a wealth of information concerning the species, subspecies, identity, social status and reproductive state of the marking animal. In some species, males are attracted to female odours throughout the year, although most particularly at oestrus. However, females may also be attracted to male odours, as in *S. fuscicollis*. During sexual encounters, extensive licking and sniffing of the female's body by the male, especially of the circumgenital area, and also of its scent marks is common, as is scent overmarking by the male. Partner marking which results from rubbing the circumgenital glands against the bodies of a mate leads to the mixing of body odours and seems to be involved in the maintenance of the bond between a mated pair. This behaviour occurs particularly if a strange male or female is introduced. Also, the male, both of this species, and of *Saguinus oedipus,* is much attracted by the odour of his pregnant mate as she approaches term. At this time the female scent marks frequently and this could be connected with the fact that the presence and the assistance of the male at and after birth greatly enhances the chance of survival of the young.

Marmosets and tamarins are highly aggressive, particularly towards conspecific members of other troops and of the same sex. Such aggressive encounters greatly stimulate scent marking and it appears that these scent marks function to advertise aggressive intent and territorial claims. In the common marmoset, *Callithrix jacchus,* such scent marking is undertaken predominantly by the dominant male and female so that the pervasion of the living space by the scent of these animals could well function also in maintaining dominance relationships within the troop. 'Genital presenting' is another form of aggressive behaviour in species of the genera Callithrix and Cebuella. Here secretions present on the circumgenital skin of a dominant animal are displayed to conspecifics.

Behavioural tests on the ability of *Saguinus fuscicollis* to discriminate between samples show that individual identity is conveyed principally by the skin gland rather than by urinary components. Preliminary chemical studies now reveal a parallel although not necessarily a causal relationship between one aspect of the lipid composition of the scent-mark material and this phenomenon. When tamarins of either sex are allowed to mark clean glass plates, the light yellow oil ($\sim$ 0.2mg/mark) extracted from these marks continues to elicit a behavioural response. Analytical and preparative GC show that more than 95% of the relatively low molecular weight components of this secretion consists of a series of unbranched saturated and unsaturated butyrate esters plus squalene, and that the GC profile of each sample is characteristic of the individual marking animal (Table 4.16).

GC of the hydrolysate of the lipid extract revealed only two acids, n-butyric acid and isobutyric acid (ratio 98:2). Neither acid could be detected in free state in the scent mark.

Table 4.16. *Saguinus fuscicollis* scent-mark lipid components (Yarger *et al.*, 1977; Epple *et al.*, 1979; Golob *et al.*, 1979)

(a) Squalene
(b) Butyrates of the following straight alcohols

Alkanol	Alkenol		Alkadienol
C_{16}			
C_{18}	$C_{18:1\ \Delta9}$	$C_{18:1\ \Delta11}$	
C_{20}	$C_{20:1\ \Delta11}$	$C_{20:1\ \Delta13}$	$C_{20:2\ \Delta11,\ 14}$
C_{22}	$C_{22:1\ \Delta13}$	$C_{22:1\ \Delta15}$	$C_{22:2\ \Delta13,\ 16}$
	$C_{24:1\ \Delta15}$	$C_{24:1\ \Delta17}$	$C_{24:2\ \Delta15,\ 18}$

Gas chromatograms of the scent marks of three tamarin species (including two subspecies of *S. fuscicollis*) revealed profile similarities and differences between species and subspecies in this range of compounds. However, the identity of these compounds was confirmed by GCMS only in the case of *S. fuscicollis*. Differences between the profiles obtained from different individuals of a given species or subspecies were also noted although the individual profiles were always close to the species or subspecies norm, and were remarkably constant over a considerable period of time (~ year).

Cebidae: new world monkeys

(Hanson and Montagna, 1962; Perkins, 1975; Oppenheimer, 1977).

As with the Callithricidae, the Cebidae, although relying to a great extent on visual modes of communication, are almost all replete with specialized skin scent glands, including large and highly developed gular, sternal, epigastric and anogenital complexes. In these regards the new world monkeys differ markedly from the old world species. Distinct behaviour patterns in which these glandular regions are rubbed and secretion distributed are common, and sometimes an admixture of urine or saliva is employed. The precise distribution of skin glands depends on the species concerned but as a general rule it is found that these organs are most fully developed in the mature dominant males. For example, in the owl monkey, *Aotus trivirgatus,* which although lacking an epigastric glandular field, possesses a large apocrine/sebaceous subcaudal glandular complex in an area of naked skin at the base of the tail, as well as a large aggregate of sternal apocrine glands in the centre of the chest which causes the hair in the region to matt with encrusted secretion. However, not all Cebidae are so well endowed with scent glands. A case here is the squirrel monkey (Machida *et al.,* 1967) which uses urine as a major means of chemical communication (see Section 6.2).

The *Cercopithecoidea* and the *Hominoidea*: the old world monkeys, apes and man

The Cercopithecoidea (old world monkeys) and the Homonoidea (apes and man) are relatively devoid of specialized skin scent glands and ritualized scent-marking behaviour patterns. However, olfactory investigation, including mutual investigation, is very common in most species (Gautier and Gautier, 1977). This involves both sniffing and direct physical contact (nuzzling) so that a tactile component is implicated. The genitals provide a major focus of attention (vaginal odour, urine) but other areas can also be important as in muzzle–muzzle contact. Odours seem to play a part in individual recognition and fostering group cohesion, as well as having a sexual role. Even so, skin gland complexes do occur in some species. Many species possess concentrations of apocrine glands on the surface of the chest and the axillary organ of man and certain great apes is well described. Man is remarkable in possessing more sebaceous glands in his skin than any other primate with the exception of the lemurs (Montagna, 1972), although they seem to fulfil no vital function. A sternal gland has been reported in the drill, *Mandrillus leucophaeus* (Hill, 1944) while an unusual, highly developed eccrine/apocrine gland field is reported in the forehead and scalp skin of the stump-tailed macaque, *Macaca speciosa* (Montagna *et al.*, 1966a). This animal is characterized by the musky odour often associated with apocrine secretions, and like man also possesses very large sebaceous glands in the skin of its face and forehead. In contrast, the rhesus monkey, *Macaca mulatta*, in common with many other Cercopithecoidea, possesses no notable specialized skin gland structures (Montagna and Yun, 1962b; Machida and Montagna, 1964; Machida *et al.*, 1964; Montagna *et al.*, 1964; Perkins *et al.*, 1968).

The skin of the gorilla, *Gorilla gorilla* (Ellis and Montagna, 1962), and of the chimpanzee, *Pan satyricus* (Montagna and Yun, 1963a) show many human similarities. Eccrine sweat glands, histologically and histochemically similar to those of man, occur independently of the hair follicles over the general body surface, and are most concentrated on the friction surfaces. Apocrine glands although less numerous than eccrine glands also occur over the general body surface. Unlike human apocrine glands, they secrete directly to the skin surface.

In all three species the distribution of skin glands varies considerably over the body surface. In the chimpanzee, high concentrations of eccrine glands occur in the face, forehead and scalp, while sebaceous glands are widely distributed but poorly developed and apocrine glands are large and numerous in the skin of the face, throat, external auditory canal, mons and nipples. In the gorilla, apocrine glands are larger in the chest (especially in the areola), cheek and anogenital area while very large sebaceous glands are present in the skin of the cheek, brow and upper lip. In man apocrine glands are similarly restricted while sebaceous glands are highly developed particularly in the skin of the scalp and forehead (see Section 3.1). All three species are distinguished by the possession of an axillary (arm-pit) scent organ, which is most highly developed in man. In constrast, the closely related gibbons, such as *Hylobates hoolock*, lack such organs (Parakkal *et al.*, 1962). However, the orang-utan, *Pongo pygmaeus*, possesses a sternal gland, an area of large sebaceous glands located in a pit in the middle of the chest (Schultz, 1921; Wislocki and Schultz, 1925). There has, of course, been much speculation concerning the possible importance of communication by chemical signals in man (e.g. Wiener, 1966, 1967; Comfort, 1971; Doty, 1977).

The axillary organ
(Labows *et al.*, 1982)

The axillary or arm-pit scent gland complex is unique to man, the gorilla and chimpanzee and is most highly developed in man. It has been suggested that the evolution of this structure was associated with the adoption of an erect bipedal posture by our ancestors which reduces the proximity of the urogenital semiochemical source. Certainly, the odour generated by this organ, a dominant component of human body odour, does seem to have considerable sexual significance, although evidence in this regard has tended to have arisen tangentially in such diverse areas as psychiatry and anthropology (Ellis, 1905; Brody, 1975). The direct scientific study of the biological significance of human axillary scent is in its infancy. Even so, it does provide a channel by which the sex of the odour producer can be recognized and even, to some degree, his identity, for humans can distinguish significantly often their own body odour and that of their partner from those of other people. (Russell, 1976; Schleidt, 1980; Schleidt et al., 1981). However, not all experimental test results in these regards are very impressive so that the information transfer may not be very reliable. Reasons for this have been discussed (Doty et al., 1978). Certainly, this propensity which man shares with other mammals is culturally suppressed in a variety of ways. There is also a difference between cultures. In tests in Japan, a 'non-contact' culture characterized by a relatively low sensory contact between individuals, women tended to classify male odour as unpleasant, while in Italy, a 'contact' culture, this was not so.

In passing it should also be noted that there is evidence that human infants can distinguish the odour of their mother (and possibly their breast odour) from that of other women (Russell, 1976). Odour cues are known to fulfil an important function in developing an attachment of an infant to its mother. This has been demonstrated in the squirrel monkey (*Saimiri sciureus*) using artificial surrogates (Kaplan and Russell, 1974), in the rabbit, in the rat and in other species.

Nowhere else on the human body or that of any other mammal is there such a close association of very highly developed apocrine and eccrine sweat glands as exist in the axillae (Montagna, 1964). Using micromethods on 20–50 μg tissues, human axillary apocrine glands were found to exhibit the highest testosterone 5α-reductase activity of any site on the human body (Takayasu et al., 1980). Sebaceous glands are also present together with a growth of hair and a moist environment, which constitutes a haven for microbial growth.

The characteristic axillary odour (axillary osmidrosis) appears only after puberty and is related to apocrine gland function, although freshly secreted human axillary apocrine secretion is odourless. The formation of axillary odour varies with race, the scent being less developed among oriental peoples.

In the human axillary organ, small quantities of viscous apocrine secretion, a secretion rich in free cholesterol (see Section 3.2), are produced intermittently particularly at times of emotional excitement and stress, and this is then distributed over the skin and hair surfaces by the watery eccrine sweat. In the warm, moist, sheltered conditions of the arm-pit, bacteria thrive and it is as a result of their action that the characteristic odour is generated, for if axillary apocrine secretion is collected under sterile conditions, no odour is formed even after 2 weeks' storage at room temperature. However, if the same secretion is collected making no provision to exclude microorganisms, an acrid axillary odour becomes detectable after 6 hours and is very strong within only one day (Shelley *et al.*, 1953). In the absence of apocrine secretion, the characteristic odour is not produced although a different milder odour can be detected as the result of bacterial action on eccrine sweat and other materials of the skin surface. The presence of axillary hair does seem to increase axillary odour although the hair itself supports few microorganisms so that it seems likely that it functions primarily as an agent to facilitate the air diffusion of the scent materials after they have been formed.

A very thorough survey of the microbiology of axillary scent production has been carried out by Leyden *et al.* (1981) who examined the axillary microflora of some 229 subjects whom they classified as 'pungent body odour' or 'faint acid odour' producers. They found that regardless of sex the microflora of the axilla is stable with time and consisted of a mixed population of Micrococcaceae (principally *Staphylococcus epidermidis*, followed by *S. saprophyticus*, *S. aureus* and various *Micrococcus* species) and aerobic diphtheroids (all corynebacteria) with some propionibacteria (*P. acnes*, *P. granulosum* and *P. avidum*) and some Gram-positive rods, but, more significantly, that the two classes of odour producers possessed very different relative proportions of these bacterial types. The formation of 'pungent body odour' was associated with not only a rather higher axillary microbial population (mean $1.3 \times 10^6/cm^2$ vs $0.48 \times 10^6/cm^2$ for the two groups) but with the dominance of the aerobic diphtheroids (mean $0.81 \times 10^6/cm^2$ vs $0.053 \times 10^6/cm^2$ for the two groups) while micrococci dominated the 'faint acid odour' group.

Subsequent inoculation of sterile apocrine secretion showed that the aerobic diphtheroids alone of the bacterial types present are responsible for the characteristic 'body odour' of axillary origin. Micrococci generated a different odour typical of volatile fatty acids. Propionibacteria were, however, not tested as these microorganisms reside principally deep in the skin within the sebaceous ducts.

It is conjectured that the axillary odour could be due to the formation by these bacteria of odorous steroids similar to the musky 5α-androst-16-en-3α-ol (androstenol) and the related urinous ketone, 5α-androst-16-en-3-one (androstenone) already identified as a semiochemical in the pig (section 8.2). A preliminary GCMS analysis of volatiles present in the human axilla (headspace analysis using Tenax and also extraction of cotton pads worn in the area) revealed only materials of probable exogenous origin (atmospheric pollutants concentrated in the axilla, secreted substances of probable dietary origin, fragrance materials from soaps and perfumes) (Labows *et al.*, 1979a). A subsequent GCMS study (Labows *et al.*, 1979b) of uncontaminated human axillary apocrine secretion taken in capillaries directly from the apocrine ducts following intradermal adrenaline stimulation has shown that, although the freshly collected secretion is without odour, exposure to the heat of the GCMS (250 °C injector port) led to the production of pyrolysis products

132

having a strong axillary smell. Further investigation showed that these substances arose from the breakdown of two odourless androgen steroids in the freshly collected secretion, namely dehydroepiandrosterone (DHA) sulphate (0.8 μg/μl secretion) and androsterone sulphate (0.4 μg/μl secretion).

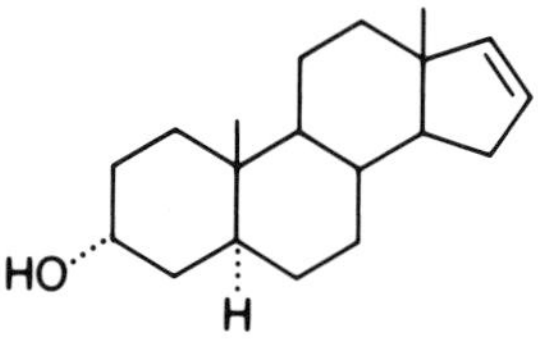

Androstenol

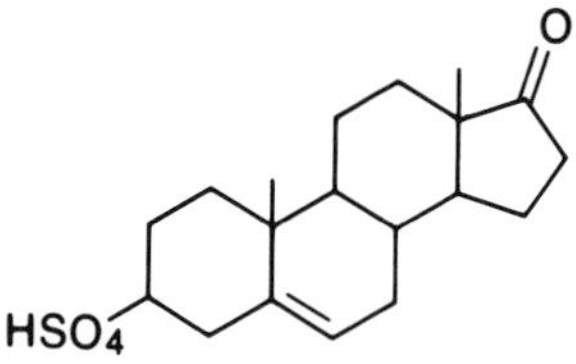

DHA sulphate

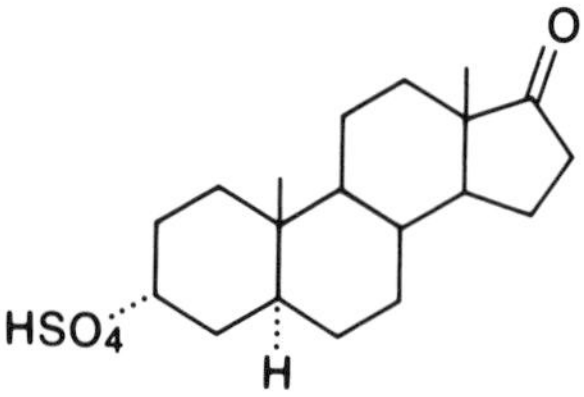

Androsterone sulphate

Both sterols had previously been found free in the axillary skin surface lipid. However, GCMS yielded no evidence of other steroids (save for cholesterol, 1000 μg/μl secretion) and no androstenol or androstenone was detected. This was not surprising since fresh secretion is odourless and the nose is a considerably more sensitive detector of these materials than is GCMS, the human odour threshold being extremely low (androstenone 0.001–0.005 ng/μl).

Their absence from fresh axillary secretion was confirmed in a further study (Brooksbank, 1970) involving the intravenous administration to a man of radiolabelled ^{14}C-progesterone and ^{3}H-pregnenolone, both of which are biochemical precursors of the steroids of interest. Interestingly, although radioactivity was, preferentially, dermally excreted in the axillae, TLC and GC showed that this was associated with fractions corresponding with androstenedione, dehydroepiandrosterone, pregnenolone and a number of minor components, but none of these corresponded with the odorous steroids androstenone and androstenol.

However, these odorous steroids *have* been detected in accumulated axillary sweat. Androstenol was identified by GCMS in material collected in axillary pads worn by men for periods of up to a week (4 μg per 600 mg lipid collected) (Brooksbank et al., 1974) and although androstenone was not observed on this occasion, it was reported in a similar study where only one male subject was involved (Gower, 1972). However, this ketone steroid has been detected in human axillary sweat using a radioimmunoassay technique sensitive to some 50 pg of steroid (Claus and Alsing, 1976a; Bird and Gower, 1980). It was estimated that androstenone formed in an axilla at the rate of about 14 ng/h.

The remarkable human olfactory sensitivity to androstenone and androstenol and their involvement in semiochemical systems in at least one other species have led them to be considered possible natural human semiochemicals.

Some preliminary experiments (Kirk-Smith *et al.*, 1978) seem to indicate that androstenol odour can influence human attitudes, although whether this is an innate response to a vestigial human pheromone or merely an association with other social situations in which musk odours are encountered remains unclear. In a carefully designed experimental situation odour was presented to or withheld from subjects who were presented with a random series of photographs about which they were asked to make judgements. The participants were given no indication that odour cues were a part of the experimental design. It did seem that the presence of androstenol made photographed women appear sexually more attractive and warmer in the judgement of both men and women. A similar study (Cowley *et al.*, 1977) on the effect of androstenol odour on human judgements of hypothetical job applicants described in interview reports does suggest that the presence of this substance modifies attitudes to some extent, particularly of women's attitudes to male applicant descriptions. However, such studies are clearly in their early stages.

Meibomian gland secretions
(Montagna and Ford, 1969; McFadden *et al.*, 1979; Nicolaides *et al.*, 1981)

The meibomian glands are large, modified sebaceous glands opening in a series of ducts along the margins of the eyelids. In some species, such as the

Table 4.17. Composition of meibomian gland lipid (% total lipid) (Nicolaides *et al.*, 1981)

	Bovine	Human
Hydrocarbon	0.06 (1)	0 (2)
Sterol ester	31.7	29.5
Wax ester	31.2 (3)	35.0
Wax diester (4)	11.4	8.4
Triglyceride	1.6	4.0
Free sterol	3.0	1.8
Free fatty acid	5.1	2.1
Polar lipid	13.3	16.0
Other lipid	2.8	3.2
Yield (5)	47 mg	3.7 mg

(1) Principally $nC_{18}-nC_{32}$ hydrocarbons.
(2) Exogenous hydrocarbons only detected.
(3) Combined liquid chromatography–mass spectrometry indicated a mass range of 500–700 amu ($C_{34}-C_{46}$).
(4) By TLC only; not definitely identified.
(5) Yield per pair of eyelids.

prosimian Tarsiidae and in the angwantibo (*Arctocebus calabarensis*) (Montagna *et al.*, 1966b) they may reach gigantic proportions. They are abundantly innervated and, like the Harderian glands of rodents and lagomorphs, seem to resemble a sebaceous gland modified for neural control. As yet no semiochemical function has been ascribed to their products, although they do seem to have important functions in reducing tear evaporation, contributing to the optical properties of the tear film, and preventing tear overflow.

Detailed chemical studies have so far been conducted on the meibomian lipids of man and the steer, and although human and bovine skin sebaceous lipid differ markedly from each other chemically, their meibomian gland lipids show many similarities between the species. In each case, meibomian lipid differs markedly from skin sebaceous lipid, which in man is principally hydrocarbon (squalene), 12%, wax monoester, 23% and triglyceride, 60% (Table 4.17). In the detailed compositions of the fatty alcohol and fatty acid contents of the various esters, meibomian lipid also differs from ordinary sebum.

Chapter 5
Microorganisms in Mammalian Semiochemistry

5.1 THE ROLE OF MICROORGANISMS

The mammalian body surface provides a variety of habitats within which a relatively restricted range of resident microbial species can flourish, utilizing and transforming the mammal's own secretions and excretions by their own metabolic activity. In the process, mammalian semiochemicals may be generated. It is immaterial that these signalling substances may be formed not by the mammal itself but by the microorganisms which its body supports. It is only necessary that the chemical source so formed should be sufficiently stable with time and sufficiently responsive to some significant mammalian biological parameter to be potentially usable by the mammal for communication. Microbially produced chemical changes can be of this kind. That bacteria can respond to their host's physiological state is exemplified by the striking colour change from white to red which occurs in the dense bacterial population inhabiting the accessory nidamental gland of the female squid, *Loligo pealei,* when this animal reaches sexual maturity (Bloodworth, 1977). In a mammalian context, the genital microflora of the rat is sensitive to levels of circulating oestrogen (Larsen *et al.,* 1977).

The possible involvement of microorganisms in the generation of mammalian semiochemicals is increasingly being discussed (Albone, 1977; Brown, 1979). Bacteria have also been implicated in some insect semiochemical systems (Brand *et al.,* 1975). Many important odorants identified in mammalian chemical communication systems are known microbial products, although only in relatively few cases have their microbial origins been proved in a specifically mammalian semiochemical context.

Also, many areas of the mammalian body surface are well suited for the formation of microbially generated odours. Pouches, sacs, invaginations and crevices which are moist, warm and hinder access of air are particularly conducive to this process, for odour production often results from the incomplete substrate oxidation associated with anaerobic processes. Strict anaerobes (Morris, 1977) are often important odour producers, and although certain aerobes are themselves also scent generators (Collins, 1976; Labows *et al.,* 1980), in this context aerobes contribute importantly in rapidly removing any

135

oxygen entering the microenvironment, thus maintaining reducing conditions. Unfortunately for any detailed understanding of this situation, the difficulties in comprehensively surveying the strictly anaerobic microflora of a mammalian scent organ can be considerable.

Important areas where there is some evidence that the contribution of microbially derived odours could be semiochemically significant include:

(1) vaginal odours (see Section 7.2),
(2) oral odours (see Section 8.1),
(3) skin odours (see Section 5.2),
(4) axillary odours (see Section 4.7),
(5) anal sacs, related structures and related scents (see Section 5.3),
(6) faecal odours (see Section 5.9).

5.2 MICROBIAL ACTIVITY IN SKIN

The skin surface is colonized by a variety of microorganisms and it is probable that these contribute to the general body odour which typifies many species. Most work has been conducted in the context of human dermatology (Marples, 1965; Skinner and Carr, 1974; Noble, 1981). Healthy human skin supports a relatively limited range of resident microorganisms, although within that range a large number of different microbial strains exist. The major bacterial contributors are members of the Micrococcaceae and various aerobic diphtheroids. These inhabit the outer layers of the stratum corneum and are dispersed in large numbers aerially when the skin fragments (squames) to which they adhere are detached from the skin surface. Anaerobic propionibacteria (*P. acnes*) inhabit the deep recesses of the sebaceous follicles but although these organisms are normally inhibited by oxygen, substances present in the skin surface appear to enable them to grow aerobically and they are also found on the skin surface (Evans and Mattern, 1979).

Fungi are also present in the form of the lipophilic yeasts, *Pityrosporum ovale* and *P. orbiculare*.

It appears that each person is unique in the details of the microbial ecology of his or her skin, for not only are there variations in the strains of bacteria present but the distribution of microorganisms over the skin surface differs from person to person (and is not even laterally symmetrical) as the result of competition between the bacterial strains for occupancy of the body surface 'terrain'. This patterning has been documented in the form of skin microflora maps (Bibel and Lovell, 1976). High population densities of skin microorganisms occur in the scalp, axillae, groin and perineum and high levels of human axillary odour have been found to be associated with that group of people in whom skin diphtheroids predominate over the Micrococcaceae.

As indicated, yeasts of the genus *Pityrosporum* inhabit the human skin, *P. ovale* being a major scalp microorganism, so that it is of some interest to note that the odorous γ-lactones which contribute to the odour of unwashed human hair are formed when cultures of microorganisms of this genus are incubated with human sebum constituents. Compounds noted by GCMS as incubation products are (Labows *et al.*, 1979a; Labows, 1981):

acids: volatile fatty acids

γ-lactones: (structure: γ-butyrolactone ring)—$(CH_2)_nCH_3$; n = 1 to 6

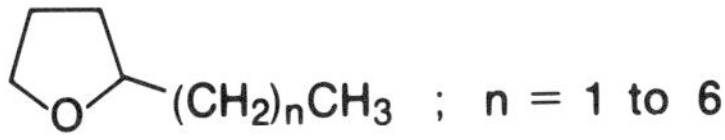

alcohols: (benzene ring)—CH_2OH (benzene ring)—CH_2CH_2OH

(branched alkyl chain)—OH   (straight alkyl chain)—OH

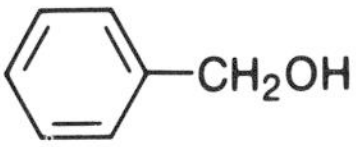

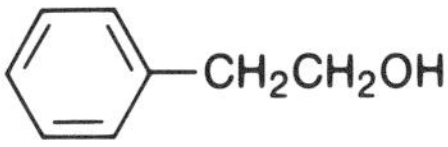

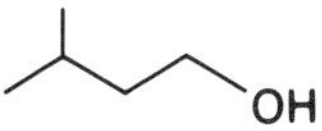

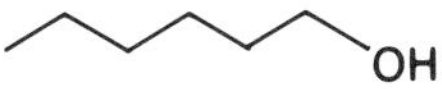

The formation of these γ-lactones appears to be a characteristic of *Pityrosporum* species.

As is discussed elsewhere, human skin surface film is characterized by high levels of long-chain free fatty acids. These are derived from triglycerides, the occurrence of which in large quantities is a characteristic of human sebum, by microbial hydrolysis (Marples, 1974; Shalita, 1974; Cove *et al.*, 1980). Studies using selective antibiotics show that *P. acnes* accounts in large measure for this hydrolysis. The activity of the microbial esterase responsible is maximal at pH 7.0 and, although the human skin surface pH is acidic, we have no data on the pH inside the follicle itself (Freinkel and Shen, 1969). The presence of these long-chain fatty acids itself has implications for the growth of other microorganisms many of which are inhibited by these compounds (Pillsbury and Rebell, 1952). Although the seba of all mammalian species examined contain esters of some kind, these are generally not triglycerides and are not readily hydrolysed microbially.

Although few semiochemically related studies have been conducted on the general skin microflora of non-human species, some interest has been directed to a study of the inguinal skin of the rabbit which forms a brown, malodorous encrustation containing volatile fatty acids, principally acetic and isovaleric acids (Merritt *et al.*, 1982). Such volatile fatty acids formed microbially on human skin are responsible, for example, for human foot odour (Tachibana, 1976).

5.3 ANAL SACS; RELATED STRUCTURES AND RELATED SCENTS

The anal sacs (Ewer, 1973) are odour producing organs common to most carnivores, and comprise a pair of reservoirs situated laterally to the anus, each communicating with the external environment through a short duct. The anal sacs of dogs are of some veterinary importance (Baker, 1962; Halnan, 1973). Anatomically they are simple invaginations of the skin. Here secretion

138

produced by the glands present in the sac walls, together with desquamated cells from the cornified sac epithelium, accumulates and is acted upon by sac microorganisms before being expelled into the environment by the mammal. Secretion is voided in a variety of situations, such as during territorial marking, sometimes although not always in association with defaecation, in aggressive encounters and at times of stress. In some species, as in the case of the skunk, anal sac secretion is used for defence. Here the sac is relatively large and is well supplied with muscles so that secretion may be ejected at an adversary over a considerable distance. The ability to expel anal sac secretion in jet form, although not necessarily for defence, has been noted in other species also, for example in the lion (Albone and Grönneberg, 1977).

The detailed structure of the anal sacs varies from species to species. In the red fox, for example, they form reservoirs of about one millilitre capacity, are located between the internal and external anal sphincter muscles, and open through short, narrow ducts on to the inner cutaneous anal region. A watery secretion may be readily obtained for analysis from these sacs using external digital pressure. In contrast with the situation in the red fox and in the domesticated dog alike, where the sac walls are well supplied with large coiled, tubular, apocrine glands (any few sebaceous glands being confined to the walls of the sac duct) (Montagna and Parks, 1948; Spannhof, 1969), the anal sac epithelia of the cat, the lion and the tiger contain discrete plaques of

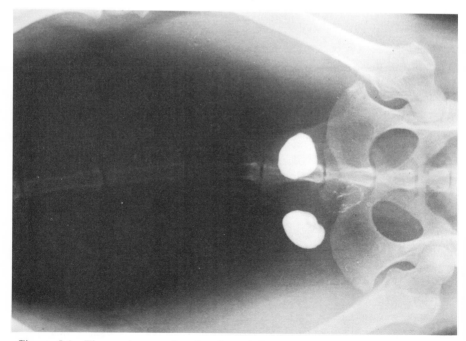

Figure 5.1 The anal sacs of a female red fox, *Vulpes vulpes*, revealed in this radiograph, had been filled with radio-opaque material. (Albone, 1977)

sebaceous tissue (Hashimoto *et al.*, 1963; Greer and Calhoun, 1966; Albone and Grönneberg, 1977) with the result that their lipid-rich anal sac secretion may include a mobile oil phase of volume comparable to that of the aqueous phase.

Although present in most carnivores, the anal sacs are considerably reduced in size in some species and are even reported to be absent for example in certain bears, in the sea otter, *Enhydra lutris,* in the kinkajou, *Potos flavus,* and in the coatis, although these possess a series of glandular pouches in this area. In the mongooses (Herpestinae) the anal sacs secrete into the anal pouch, a hairless infolding of the skin, which itself may contain enlarged glandular complexes. In this pouch a composite secretion accumulates which the mammal may apply to objects in the environment following pouch eversion. Anal sacs are also present in some rodents (Section 5.7) and members of other mammalian orders.

The anal sac and microorganisms

The anal sac as a mammalian fermentative scent source

(Albone *et al.*, 1977, 1978)

A further fact about the anal sac is that it provides a moist, warm, enclosed environment which supports an abundant microflora. This means that all inputs to the sac from its walls are subjected to the activity of a microbial ecosystem which substantially influences the nature and odour of the secretion finally voided by the mammal. In this it resembles many other similar microbial scent sources on the mammalian body surface.

The most detailed bacteriological studies have been conducted on the anal sac microflora of the red fox (Albone *et al.*, 1974, 1978; Gosden and Ware, 1976; Ware and Gosden, 1980). Here only certain types of microorganisms have been detected in substantial numbers, and this together with the degree of uniformity at the genus level of the aerobic microflora present in samples from different foxes suggests a structured anal sac environment. Preliminary measurements indicate that the microflora may generate highly reducing conditions in the enclosed sac with redox potentials extending down to -400 mV. In comparison, a redox potential of -200 mV has been reported in the caecum of the mouse (Celesk *et al.*, 1976). In the anal sac it appears that aerobes (facultative anaerobic organisms) create and maintain an anaerobic environment in which strict anaerobes, frequently active odour producers, can grow (Smith and Holdeman, 1968; Morris, 1975). Indeed a number of very oxygen-sensitive anaerobes have been detected in the fox anal sac.

Aerobes The range of microbial species which inhabit the fox anal sac is limited and is remarkably uniform. Table 5.1 demonstrates that the aerobic sac microflora is dominated by streptococci and *Proteus* species while other faecal organisms, such as coliforms and gut lactobacilli are largely absent even

140

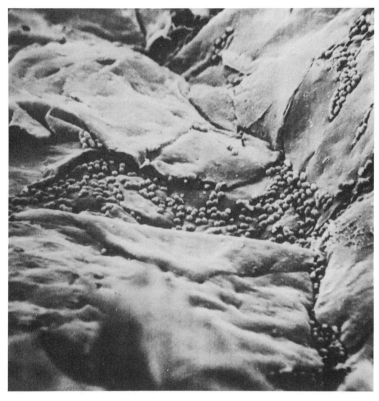

Figure 5.2   Scanning electronmicrograph showing bacteria present on the desquamating epidermis lining the anal sac of a red fox. (Photograph courtesy G. C. Ware)

Table 5.1. Occurrence of aerobic bacteria in red fox anal sac secretions. Adapted with permission from Albone *et al*. (1978). Copyright 1978 American Chemical Society

Organism	96 samples from 28 foxes examined Abundant (no. samples) (1)
Streptococcus faecalis	60
Streptococcus faecium	6
Streptococcus uberis	1
Streptococcus mitis	2
Proteus vulgaris	1
Proteus mirabilis	71
Proteus rettgeri	2
Escherichia coli	1

(1) Abundant means $>10^6$ organisms/ml. The following organisms, although commonly detected were abundant in no samples and were presumably transient contaminants: *Klebestella* spp. and other coliforms, *Serratia* spp., *Micrococcus* spp., *Staphylococcus* spp., *Neisseria* spp., *Bacillus* spp., *Diplococcus* spp., *Lactobacillus* spp.

though the sac is sufficiently close to the anus to permit passage of such microorganisms into the sac. Staphylococci, common skin microorganisms in many mammals, are also absent.

The selective nature of the anal sac microenvironment, itself presumably of chemical origin, is revealed by the failure of a number of attempts to establish *Escherichia coli* isolated from fox faeces as a resident species in an anal sac. For example, *E. coli* injected into a fox anal sac to give an initial population of 8×10^9 organisms/ml were reduced to 12% of this level in 2 days, to 3% in 4 days and were undetectable after 15 days, while streptococci and *Proteus* spp. in the same sample maintained populations in the ranges 10^8-10^{10} and 10^6-10^9 organisms/ml respectively. Similarly, a population of *E. coli* (1.8×10^6 organisms in 50 μl) incubated anaerobically *in vitro* in 250 μl unsterilized anal sac secretion was eliminated after 24 hours. This effect is also reflected by a comparison of the above data with the composition of faecal microflora. Of 18 faecal samples taken from 9 foxes, streptococci were detected in 17 and were abundant ($>10^6$ organisms/ml) in 12, while *Proteus* spp. were observed in only 7 samples and were abundant in one and, conversely, *E. coli* were observed in 14 samples and were abundant in 14.

Strict anaerobes (Morris, 1977)   The populations of strict anaerobes present in the fox anal sac have also been examined (Table 5.2) and it is clear that their description is as yet far from being complete. Fox anal sac secretion samples were examined anaerobically taking rigid precautions to avoid exposure to the air. Anaerobes were classified by obtaining API 20A biochemical profiles and employing single linkage cluster analysis to group

Table 5.2. Occurrence of strict anaerobes in red fox anal sac secretions. Adapted with permission from Albone *et al.* (1978). Copyright 1978 American Chemical Society

Reference organism (1)	A (2)	B (3)
Clostridium perfringens	50	37
Clostridium ramosum	16	0
Clostridium sporogenes	27	4
Clostridium bifermentans	21	1
Eubacterium lentum	13	4
Bifidobacterium eriksonii	0	2
Fusobacterium nucleatum	0	6
Fusobacterium negrogenes	0	17
Bacteroides fragilis fragilis	0	20
Peptostreptococcus anaerobius	0	7
Unclustered	0	9

(1) Isolates clustering at similarity coefficient 0.75 on the basis of API biochemical profiles, morphology and Gram stain reaction.
(2) Number of isolates form 66 secretions from 18 foxes, employing anaerobic techniques.
(3) Number of isolates from 37 secretions from 6 foxes, employing more rigorous anaerobic techniques, including an air-free sampling environment.

142

similar organisms with selected API 20A reference species. Anaerobic isolates from fox secretions yielded predominantly clostridia. However, an examination of further isolates from fox secretions, collected under the most rigorous anaerobic conditions using pre-reduced, anaerobically sterilized media and employing a specially designed container to exclude air and maintain a carbon dioxide environment around the hind-quarters of the sedated fox during sample collection, revealed the presence of other, more oxygen-sensitive anaerobes. More than half the isolates exhibited unique biochemical profiles matching no other profile exactly. No correlation was noted between particular foxes and their anaerobic anal sac microflora. Different foxes sometimes yielded identical strains while the same fox frequently yielded different organisms on different occasions. However, the isolation of strict anaerobes is surrounded with many uncertainties and the system cannot yet be considered to be fully elucidated.

Anaerobic population determinations were subject to error because of the uncertainty of percentage recoverability in primary culture. From microscopic counts, strict anaerobic populations were estimated to be comparable with those of aerobes, frequently in the range of 10^9-10^{10} organisms/ml secretion.

Volatile fatty acids

Chemical studies undertaken on the anal sac secretions of a variety of species have shown that many of the low molecular weight components present are, as expected, common products of microbial activity. Take again the example of the anal sac secretion of the red fox. This is a watery, turbid, straw-coloured, proteinaceous, odorous liquid having a variable pH (6.5–9.4) presumably reflecting differences in fermentation time, an increase in pH commonly being observed as proteinaceous substrates are fermented. The two anal sacs of a particular fox may differ not only in size, but their contents may differ both in appearance and *in composition*, although commonly those differences are not marked. Such differences as do occur presumably reflect differences in the extent to which fermentation has proceeded, for it is not

Table 5.3. Volatile fatty acids identified (GCMS) in red fox anal sac secretions (Albone and Fox, 1971)

Number of carbon atoms	Normal acids $CH_3(CH_2)_nCOOH$	iso-Acids $CH_3{\diagdown}CH(CH_2)_nCOOH$ $CH_3{\diagup}$	anteiso-Acids $C_2H_5{\diagdown}CH(CH_2)_nCOOH$ $CH_3{\diagup}$
2	Acetic	—	—
3	Propionic	—	—
4	Butyric	Isobutyric	—
5	Valeric	Isovaleric	2-Methylbutyric
6	—	Isocaproic	—

necessarily the case that the two sacs will be evacuated simultaneously or to the same extent during marking behaviour.

A range of volatile fatty acids has been observed in red fox anal sac secretion (Table 5.3), their total concentration commonly being in the range 5–175 mM. The secretion is, however, usually slightly alkaline (pH range 6.5–9.4) as the result of the presence of large quantities of ammonia, so that the volatile fatty acids (pK_a ~4.8), being principally in salt form, contribute less to the secretion odour than would otherwise be expected. They are, however, readily observed by gas chromatography owing to the heat lability of the ammonium salts and are available for release when the secretion is deposited in the environment. A study of the profiles of volatile fatty acids present in secretions taken from each anal sac of six captive foxes at approximately fortnightly intervals over ten weeks revealed that:

(1) the profiles obtained from the secretions of each of the two anal sacs of a given animal at a given time are usually but not invariably similar. Significant differences in composition between the secretions of the two sacs can occur.

(2) the profiles obtained from secretions from a particular animal may remain relatively unchanged on consecutive sampling dates, or they may change substantially.

(3) the short-term variability in the profile from a given animal may be greater than the profile differences noted between samples from different animals.

These volatile fatty acids are clearly of microbial origin. Not only are volatile fatty acids well known products of the anaerobic metabolism of many microorganisms, including clostridia of the type observed in the anal sac, but their microbial origin has been specifically demonstrated in this case. In the case of fox anal sac secretion, an inoculum of fresh anal sac secretion collected anaerobically and incubated anaerobically in pre-reduced Robertson's cooked meat medium at 37 °C for 48 hours yielded volatile fatty acids at concentrations (acetic acid, 170 mM; total C_3–C_6 acids, 95 mM, by gas chromatography) comparable with, or in excess of, those encountered in actual anal sac secretions. One hundred and fifty-two strictly anaerobic bacterial isolates from 78 anal sac secretions from 19 foxes were examined for volatile fatty acid production under similar conditions. Anaerobes of all genera produced some or all the volatile fatty acids noted in anal sac secretion, certain clostridia and eubacteria being the most active producers, although great variations in production were noted between different isolates from the same genus.

In the sac itself, microbial production of volatile fatty acids was confirmed by irrigating the sac with 1% saline sodium hypochlorite, followed with physiological saline and then by filling the sac with antibiotic (10% ampicillin/0.5% tetracycline in physiological saline). Neither bacteria nor volatile fatty acids were detected in the sac for in excess of 3 days subsequently.

These fatty acids could arise microbially from protein, various amino acid residues yielding the various volatile fatty acids (e.g. leucine yields isovaleric acid), and it is notable that the anal sac secretion is proteinaceous (mean 21 g/l; range 2–56 g/l protein).

Volatile fatty acids of microbial origin are doubtless present in all anal sac secretions. They have been reported in the anal sac secretions of

the domestic dog, *Canis familiaris* (Preti *et al.*, 1976);

the coyote, *Canis latrans* (Preti *et al.*, 1976);

the maned wolf, *Chrysocyon brachyurus* (Albone and Perry, 1976);

the bush dog, *Speothos venaticus* (Albone and Perry, 1976);

the lion, *Panthera leo,* (Albone *et al.*, 1974);

the tiger, *Panthera tigris* (Albone and Perry, 1976);

the domestic cat, *Felis catus* (Michael *et al.*, 1972);

the mink, *Mustela vison* (Sokolov *et al.*, 1980);

the Indian mongoose, *Herpestes auropunctatus* (Gorman *et al.*, 1974).

Anal sac secretion: semiochemical function

In most cases the precise semiochemical function of anal sac secretion remains obscure, although some clues may perhaps be obtained by noting seasonal variations in the secretory activity of the glands supplying the sac (Spannhof, 1969). Even so, the signalling function of anal sac secretion in the life of the fox in its natural environment remains unclear, reflecting the paucity of information which exists on the chemical ecology of wild mammalian species. As with urine, it probably conveys a variety of messages in a variety of circumstances (Macdonald, 1977). Field observations indicate that the secretion may be deposited with faeces or that it may be emitted without defaecation when the fox is alarmed. Anal sac evacuation has been observed at territorial boundaries following eviction of an intruder. However, an exhaustive analysis of the signalling role of this secretion has yet to be undertaken.

Laboratory studies, such as those of Doty and Dunbar (1974) on the attractancy to conspecifics of dog anal sac secretion in comparison with other semiochemical substrates (vaginal secretion; urine) are by their artificiality, of limited value particularly in studies on wild species such as the fox, although they do represent an advance on previously anecdotal observations (Donovan, 1967, 1969). Ideally one would investigate the detailed effect of chemical signals on the entire behavioural repertoire of wild animals. Field studies on the significance of scent marks in the life of wild mammals, however, remain few. Notable examples are those of Peters and Mech (1975) on the wolf and of Henry on the red fox (1977).

The attractancy of putrefying sources for carnivores   The products of the microbial decomposition of animal materials evokes a wide variety of responses from carnivores. In the canids (Albone *et al.*, 1978; Bullard, 1982) fermentative scent sources present in the mammal itself exhibit some similarities to environmental signals to which they respond. It has been noted, for example, that chemical and microbiological similarities do exist in the red fox between the anal sac and an important environmental scent source for this species, namely carrion. This is not to say that these scents are not distinguishable by the fox. Fundamental differences remain to the extent that the anal sac fermentation is controlled and limited by the living mammal. The observation of structure in the anal sac microecosystem seems to suggest the operation of selective pressures and to imply the existence of specific biological functions of the anal sac of adaptive value for the fox.

Carrion, which may be detected at least in part by odour cues, possesses considerable attraction for carnivores, whether as a food source or as a site for scent marking. Microbial putrefaction arises in the carcass of a dead animal largely as the result of bacterial penetration of the gut after residual tissue antimicrobial activity has been lost and when anoxic conditions have been established as the result of continued tissue metabolism in the absence of oxygen. Studies in this area are few and have been undertaken principally in the context of forensic science. A valuable review of this work is provided by Corry (1978) who reports that, although many different anaerobes reside in the intestine, only a few groups have been implicated so far as major colonizers of corpses during putrefaction. The most important of these is *Clostridium perfringens,* a vigorous saccharolytic, lipolytic and proteolytic organism which is also commonly resident in the fox anal sac. In addition, volatiles of the type identified in anal sac secretions are commonly produced when gut microorganisms are incubated anaerobically in protein-rich media. Thus, the low molecular weight metabolites produced by clostridia and other microorganisms include 5-aminovaleric acid, the volatile fatty acids and the amines identified in anal sac secretions (Moore *et al.,* 1966; Lewis *et al.,* 1967; Brooks and Moore, 1969; Moss *et al.*, 1970; Mead, 1971; Anema *et al.*, 1973; Hyatt and Hayes, 1975).

Putrefied animal matter has formed the basis for canid attractants of possible value in pest control programmes (Bullard, 1982). Thus, a putrefied fish formulation has been used as a coyote lure and, more recently, attention has been directed to a fermented aqueous suspension of chicken whole-egg powder, developed initially as an attractant for flies. The odour components of this material have been subjected to detailed chemical analysis by Bullard *et al.* (1978a) and are reported to include volatile fatty acids (77% total; 13 acids identified), bases (13% total, mainly trimethylamine, 9 amines identified), and headspace volatiles, including esters, aldehydes, ketones, alcohols, alkyl aromatics, terpenes and sulphur compounds (10% total, 76 compounds identified). Based on these data, a synthetic mixture, 'synthetic fermented egg' was formulated, composed largely of a mixture of 10 volatile fatty acids (81%), together with a diverse range of amines and other compounds. This mixture was found to be almost as attractive to coyotes as the fermented preparation itself. The volatile fatty acid component was found to be of considerable importance and to exhibit substantial coyote attractancy also (Shumake, 1977; Bullard *et al.*, 1978b; Roughton, 1982). It appears that among the volatile fatty acids (C_3 to C_{10}), the longer chain fatty acids are more attractive than the shorter, and studies at the US Department of Agriculture Laboratories at Berkeley show that a particularly effective coyote attractant in the salt trimethylammonium decanoate, a source both of trimethylamine and of decanoic acid (Fagre *et al.*, 1982).

As a further example of the attractancy of microbial products, an unfortunate case is quoted in the medical literature (Liddell, 1976) of a woman suffering from a skin condition which, as the result of secondary infection, emitted obnoxious odours which caused her to be molested by dogs. In a different context, Amoore and Forrester (1976) quoted Linnaeus' observation that the domestic dog is attracted by the odour of the plant *Chenopodium vulvaria,* the leaves of which are rich (400 ppm) in trimethylamine (see below).

Basing their methods on those of Linhart and Knowlton (1975) who investigated the use of a putrefied egg product as a coyote attractant, Albone *et al.* (1978) presented fermented products related in some measure to fox anal sac secretion to these animals in their natural environment. Linhart and Knowlton used a putrefied egg product in which chicken whole-egg powder had been exposed in aqueous suspension to the air at room temperature for 1–2 weeks. The nutrient was colonized by air microorganisms and these brought about putrefaction. As well as being uncontrolled, this method of inoculation excluded strict anaerobes which are among the most effective odour producers. In Albone *et al.*'s work, a fox tissue extract was incubated anaerobically with an inoculum of fresh, anaerobically collected anal sac secretion. This substance had the advantage of being readily prepared in large quantity and also mirrored, probably more closely than other readily available fermentation products, features of fox anal sac secretion. Samples were then presented in a small deciduous woodland where the movements of many of the resident foxes had earlier been established by radiotracking. Samples were placed at test sites in the evening and the following morning signs of fox activity in the vicinity of the various test sites were noted and compared. The experiment was run on a number of nights and the final statistical analysis showed that the attractancy of the fox tissue fermentation was very high, higher than that of fox urine samples which in turn was higher than that of controls of unincubated fox tissue extract. Unfortunately these interesting preliminary findings were not followed up. They do, however, indicate that this fox tissue extract anaerobic fermentation product is very attractive to foxes in the wild and that the attraction is dependent on microbial putrefaction. Preliminary experiments with similar incubation products derived from other media (Robertson's meat broth, casein acid hydrolysate and egg yolk medium) revealed relatively low attractancy compared with that obtained from the fox tissue extract.

Fermentation hypothesis of chemical recognition   Superimposed on the general attractancy which putrefying scent sources exert for canids, the mammal can convey more specific information to the degree that the microbial scent source responds to the physiological and social circumstances of the mammal.

Although the sensitivity of mammals to many of the microbial components of canid anal sac secretion has not been investigated, it is known that the domestic dog is sensitive to volatile fatty acids (Moulton *et al.*, 1960) and that the Indian mongoose can distinguish between different mixtures of these acids by olfaction (Gorman, 1976). Questions arise concerning the ways in which such odorous substances, the microbial formation of which appears to depend little on sex, sexual status or even on species, might acquire semiochemical significance.

A fermentation hypothesis of chemical recognition offers one possibility. By process of cross-infection, a group of animals living together would be expected to come to share a common microflora characterizing that group. The transfer of symbiotic microorganisms between animals can convey important benefits of other kinds. For example, characteristic gut microorganisms acquired from mature herbivores can be important in establishing effective digestion of plant materials in the young (Troyer, 1982). However, in the area of fermentative chemical recognition, a major problem surrounds the difficulty of obtaining a comprehensive survey of the species and strains of strict anaerobes, important odour producers, which may be present, for not only is their taxonomy complex but they can be extremely oxygen sensitive. Even when the species of microorganisms are narrowly defined, the possibility of many biochemical strains provides sufficient group-distinguishing potential. The hypothesis argues that if these microorganisms produce substances detectable by the mammal in question, the odours of individuals from a particular group would process certain recognizable common features characteristic of that group and its shared microflora (Albone *et al.*, 1974). It is known that characteristically different volatile metabolite profiles are produced by different strains of microorganism incubated under standard conditions (Lewis *et al.*, 1967). The situation is complicated by observations that incubation (residence) time can affect the profile (Moore *et al.*, 1966) and that the profile is also determined substantially by the detailed composition of the primary substrate which the microorganisms metabolize. It is clear that a detailed analysis of the relative contributions to profile form of microflora and substrate is likely to be intricate, for each must be important.

The possibility of microbial metabolic products present in anal sac/anal pouch secretion having individual or group recognition function has been discussed in relation both to the red fox and to the Indian mongoose, *Herpestes auropunctatus,* although the experimental evidence in support of such a fermentative hypothesis of chemical recognition is at present lacking. In his fermentation hypothesis, Gorman (1976) suggested that the Indian mongoose recognizes conspecifics as individuals on the basis of odour differences between individually characteristic profiles of anal sac volatile fatty acids. In 1974, Gorman *et al.* had clearly shown that these substances were of microbial origin, for they have noted that the anal sacs support a dense bacterial population and that volatile fatty acids were not produced in the presence of antibiotic (penicillin). A number of isolates capable of producing these substances in culture were obtained and identified as *Peptococcus grigoroffi*(?), *Peptostreptococcus plagarumbelli, Bacillus cereus mycoides* and *Eubacterium* and *Catenabacterium* species. A full bacteriological survey was, however, not conducted. Gorman also clearly demonstrated that the mongoose can distinguish between different mixtures of these volatile fatty acids and also between single secretion samples taken from a different mongoose, but he did not show that the volatile fatty acid profiles obtained at different times from the secretions of a particular animal were sufficiently alike to enable them to provide an effective basis for the animal to associate a particular profile with a particular animal. In view of the above discussion of fox anal sac secretion volatile fatty acid profiles, this must remain an open question for the present.

Anal sac secretion: further microbial products

Studies on lipid-rich lion anal sac secretions revealed the presence of aromatic acids, including the odorous compound phenylacetic acid, and it is possible that these may be present in fox anal sac secretion at low levels also. A TLC analysis of lion secretion showed the presence of a major lipid ester fraction and many of the low molecular weight lipids of lion anal sac secretion include substances expected as hydrolysis products of sebaceous esters (Albone *et al.,* 1974; Albone and Grönneberg, 1977). Cholesterol is the major sterol of lion and red fox anal sac secretion. In the red fox, mean levels of 0.93 mg/g were noted (% free sterol 56%) while in the domestic dog, these were 2.49 mg/g (65.8%).

Amino acids (Albone *et al.,* 1976)   An examination of the free amino acid composition of red fox anal sac secretion also revealed an anomalous profile with few amino acids present and with the non-protein amino acid 5-aminovaleric acid predominating (range 0.1–4.9 mM). The levels of this amino acid correlated positively with the levels of volatile fatty acids so that it is likely that both have in common a microbial origin, the levels of each depending on the length of time that anal sac fermentation has been in progress. Clostridia, for example, are known to produce this substance from proline or ornithine.   A peak corresponding to 2-piperidone has been observed on GC analysis of samples of anal sac secretion of the dog, the coyote and the mink.

2-Piperidone

Table 5.4. Lower molecular weight lipids of lion anal sac secretion (1)

Aromatic acids (2)

$\langle\!\!\!\!\;\rangle$—CH_2COOH $\langle\!\!\!\!\;\rangle$—CH_2CH_2COOH Microbial degradation products of phenylalanine and tyrosine

HO—$\langle\!\!\!\!\;\rangle$—CH_2COOH HO—$\langle\!\!\!\!\;\rangle$—CH_2CH_2COOH

Alkanoic acids (straight and some branched chain)   Present as esters in a
$C_nH_{2n+1}COOH$ $(n = 11-23)$ variety of sebaceous lipid components

Alkenoic acids
$C_nH_{2n-1}COOH$ $(n = 15-18)$

2-Hydroxyalkanoic acids   2-Hydroxyalkanoic
$C_nH_{2n+1}CHOHCOOH$ $(n = 12-22)$ acids present esterified in cat sebum as type 1 diester waxes

2-Hydroxyalkenoic acids
$C_nH_{2n-1}CHOHCOOH$ $(n = 12-20)$

1-Alkylglycerols (straight chain and branched chain)   Present as diesters in
$C_nH_{2n+1}OCH_2$—$CHOHCH_2OH$ $(n = 12-22)$ mouse preputial sebaceous lipid

Cholesterol

(1) Identified by GCMS. The composition of individual lion secretions is highly variable (Grönneberg and Albone, 1976; Albone and Grönneberg, 1977).
(2) A series of ω-phenylalkanoic acids also occurs in the malodorous secretion discharged by the 'stinkpot turtle', *Sternotherus odoratus*, when disturbed (Eisner *et al.*, 1977).

However, as 5-aminovaleric acid is a major component of the anal sac secretion, and as this substance readily loses water to form 2-piperidone in the gas chromatograph, the origin of this peak must remain in some doubt in these cases.

Amines The principal amine of red fox anal sac secretion is ammonia, the concentration of which is correlated with that of the volatile fatty acids. However, formation of fluorescent dansyl derivatives of anal sac secretion amines followed by TLC revealed the presence of other amines. In the case of the red fox these were identified (MS) as putrescine (1,4-diaminobutane) (0.5–30 mM) and cadaverine (1,5-diaminopentane) (cad/put ratio in range 0.01 to 3.5) (Albone and Perry, 1976). Both diamines were also noted in the anal sac secretion of the lion while mink anal sac secretion was found to yield a more complex pattern of fluorescent zones and again putrescine was identified (Sokolov *et al.*, 1980). Indole and trimethylamine are present in red fox anal sac secretion (Albone and Fox, 1971; Albone and Grönneberg, 1977) and indole has also been observed in the anal sac secretion of the mink, the ferret and the stoat. Trimethylamine has also been noted in the anal sac secretion of the coyote and the domestic dog (Preti *et al.*, 1976).

All these compounds are commonly encountered microbial products. Clostridia for example are known to produce putrescine from arginine or ornithine and cadaverine from lysine, while indole derives similarly from tryptophan.

The odorous fluid emitted by the tiger contains 2-phenylethylamine. Although a known microbial product, this substance is reported to be absent from similar secretions of the cat. It is uncertain from the published accounts whether anal sac secretion, urine, or a mixture of the two, was analysed (Brahmachary and Dutta, 1979). The aromatic amine, 2-methylquinoline, has been reported in the anal sac secretion of the skunk (Aldrich and Jones, 1897).

5.4 THE HYAENIDAE

(Fox, 1971; Kruuk, 1972, 1976; Wheeler *et al.*, 1975; Mills *et al.*, 1980).

In the hyaenas, the anal sacs open into a glandular anal pouch similar to those found in the Herpestinae and in some species, for example in the **striped hyaena,** *Hyaena hyaena*, these pouch glands are aggregated to form a pair of auxiliary glandular masses. In this species also, the secretion accumulating in the pouch may be presented to a conspecific as a part of an appeasement display, by evertion of the anal pouch following rectal protrusion. This is brought about readily by muscular action and is also employed in scent-marking behaviour when this scent is 'pasted' onto grass tufts at the height of 30–60 cm. A similar more strongly smelling ('like cheap soap boiling or burning') secretion is pasted by the **spotted hyaena,** *Crocuta crocuta*, to mark clan boundaries as well as in aggressive scent displays. The **brown hyaena,** *Hyaena brunnea*, is unusual in that it deposits two quite different scent marks, one white, the other dark brown, 2 cm apart with each pasting. This is accomplished since when the animal everts its anal pouch, a central sebaceous region producing the white secretion touches the vegetation. As the anal pouch is then retracted and the sebaceous area comes away from the vegetation, apocrine-rich areas of the pouch skin, which lie beside the sebaceous area and which produce the black secretion, make contact. The function of the double mark is not known precisely but it probably has a territorial significance. GC shows that the volatile profiles of each of the two marks are different and to the human nose the strong odour of the white mark (which darkens with age) lingers for more than a month while that of the black mark fades relatively rapidly. It is possible then that the double mark provides the hyaena with valuable information concerning the time of deposition of the mark as well as information concerning the identity of the marking animals. A group of brown hyaenas will saturate their territory with their anal pouch odour and it is estimated that in one year a group of five animals produced 145,000 pastings.

The unusual organosulphur volatile, 5-thiomethylpentane-2,3-dione has been identified as a component (~1% level) of male and female striped hyaena anal sac secretion.

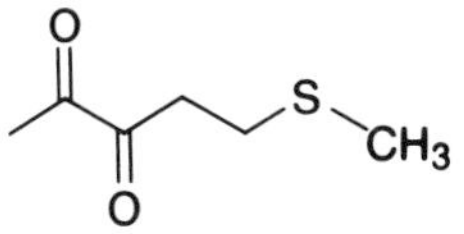

$$CH_3-CO-CO-CH_2-CH_2-S-CH_3$$

Another unidentified volatile sulphur-free component (M = 224) is also present.

Striped hyaena anal sac/anal pouch secretion scent marks (~500 mg secretion) collected on being 'pasted' were found to be yellow, odorous, viscous substances which dissolved almost completely in n-hexane to give a pale yellow solution. TLC followed by MS and other techniques showed that the secretion lipids were predominantly triglyceride (mainly of palmitic, stearic and oleic acid), while UV spectroscopy revealed a typical carotenoid spectrum corresponding to lutein at ~100 ppm concentration. Lutein is the major carotenoid of herbage and presumably derives from herbivore flesh. While lacking odour itself this is a possible contributory source of hyaena anal pouch scent as it is known that odours are formed when carotenoids are degraded (Sanderson et al., 1971; Jüttner, 1979). After purification, the lutein identity was confirmed by comparing its TLC mobility in two different systems (silica gel G, EtOAc/CH$_2$Cl$_2$ 50:50 and alumina G, EtOAc/CH$_2$Cl$_2$ 50:50) with that of authentic lutein (Albone and Watts, unpublished).

5.5 THE MUSTELIDAE

The **badger, *Meles meles***, possesses similar anatomical structures to those discussed in the previous section. Here the anal sacs also secrete into a glandular pouch, but this species also possesses an additional pocket (the subcaudal pouch) formed in the skin adjacent to and just above the anal pouch. The lining of the subcaudal pouch is hairy and glandular and within it a copious secretion accumulates which the badger applies to objects in its environment as well as to other badgers. Badgers discriminate between secretions obtained from different individuals, secretion identity being maintained in quite old samples.

Within the Mustelidae generally, anal sac secretion tissues commonly include both apocrine and sebaceous tissues and so yield lipid-rich secretions. Anatomical data have been used to suggest an evolutionary elaboration passing from the simpler '*Meles*-' and '*Lutra*-type' to the '*Martes*-' and '*Mustela*-type' anal sacs within this family (Stubbe, 1970).

Histological examination of **mink (*Mustela vison*)** anal sac tissue has revealed the presence of glandular tissue of two distinct types approximating to two concentric collars arranged around the neck of the sac. The inner, smaller collar which completely surrounds the sac duct consists of sebaceous

152

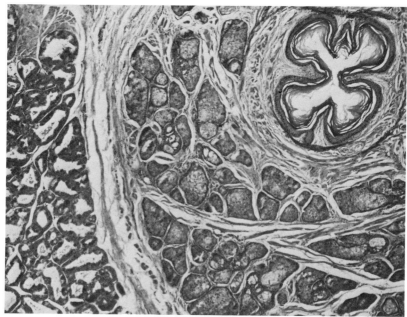

Figure 5.3 Transverse section of the duct and adjacent glandular tissue encircling the duct of the anal sac of a mink, *Mustela vison*. The duct is in the upper right, and is surrounded by sebaceous tissue. A part of the outer collar of apocrine tissue is on the far left. (Reproduced by permission of Plenum Publishing Corp. from Sokolov *et al.*, 1980)

glandular tissue. This in turn is almost completely surrounded by a bulkier outer collar of apocrine secretory tissue. The sebaceous glands appear to discharge into the duct and thence into the sac. while the apocrine glands open into the most caudal region of the sac itself. Both sac and duct are lined with stratified squamous epithelium. More detailed examination employing electron microscopy revealed electron-dense granules, up to 2 μm in diameter, in the luminal regions of the apocrine secretory cells and X-ray energy probe microanalysis showed these to contain substantially higher levels of sulphur than the adjacent cytoplasm. Histochemical staining reactions indicate that these granules consist principally of glycoprotein (Sokolov *et al.*, 1980).

Anatomical studies on the **otter (*Lutra lutra*)** show that both apocrine and sebaceous secretory tissues are present in an arrangement not unlike that in the mink. Quite apart from territorial marking, male and female otters deposit faeces, urine and the secretions of their anal sacs at established 'latrine sites' with a periodicity which mirrors the female's oestrus cycle, so that it appears likely that the chemicals present have reproductive function. As in otter species, the gelatinous anal sac secretions may be deposited either in association with or independently of defaecation. The secretions are commonly yellow, brown or green, and have a distinctive otter odour. Histochemi-

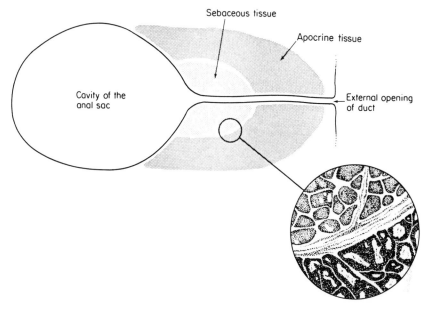

Figure 5.4   A section showing the arrangement of tissues in the mink anal sac.
(Reproduced by permission of Plenum Publishing Corp. from Sokolov *et al.*,
1980)

cally it was found that the secretion consists principally of protein and
mucopolysaccharide of apocrine origin with lipid droplets at least partly of
sebaceous origin also being present. The pigment is also of apocrine origin
and appears to be associated with histochemically demonstrable iron.
Acrylamide gel electrophoresis of fresh secretion reveals the presence of
protein patterns which can be used to distinguish secretions from individual
animals. Although some gas chromatography has been undertaken, the typical otter odour components have not been identified (Gorman *et al.*, 1978).

Chemical studies: the Mephitinae

(Aldrich, 1896; Blackman, 1911; Andersen and Bernstein, 1975)

The anal sac secretions of the Mustelidae are frequently highly offensive
substances, remarkable for their abundance of organosulphur compounds. In
some species, as in the skunks, the use of these substances has evolved into a
most effective means of chemical defence.

Thus, the anal sacs of the **striped skunk,** *Mephitis mephitis*, are embedded in
powerful muscles which are continuous with those employed to elevate the
tail, so that on raising the tail the sacs are tensed while the papillae through
which the anal sac ducts emerge are exposed. In this species, the noxious
secretion may be directed at an adversary at distances up to 3 metres.

For many years, the major odorant of skunk anal sac secretion was thought to be butane-1-thiol. However, this is not present in more than trace quantities. Rather, the principal organosulphur volatiles were found to be:

	Ratio
3-methylbutane-1-thiol	4
trans-2-butene-1-thiol	3
trans-2-butenyl methyl disulphide	3

identified by NMR, IR, derivative formation and comparison with authentic synthetic samples.

Striped skunk anal sac secretion consists of both aqueous and lipid phases. Of the latter, approximately half is volatile at 40 °C/0.2 torr and most of this comprises these three compounds.

Organosulphur compounds similar to those present in striped skunk secretion appear to be present in the anal sac secretions of closely related species, although the reports are rather old and further research is clearly required.

A number of less volatile lipids have been noted by TLC in skunk anal sac secretion, although many of these appear to be unstable. An old report also mentioned 2-methylquinoline (Aldrich and Jones, 1897). Perhaps mercifully, at high atmospheric concentrations the secretion volatiles are reported to induce anaesthesia.

Subsequent studies using capillary GCMS (Andersen *et al.*, 1982) to examine the nature of the volatiles more thoroughly revealed some 160 components of which 150 contained sulphur. One of the major components previously identified, *trans*-2-butenyl methyl disulphide was not detected on this occasion. No particular differences were noted with regard to the main chromatogram components between samples from two male and three female animals. The following compounds were identified by GCMS (percentage abundance by GC peak integration). Major components were thiols, sulphides and disulphides particularly of C_4 and C_5 alkyl and alkenyl groups.

Table 5.5. Striped skunk anal sac secretion components

Major components	Some minor components (<1%)

Thiols

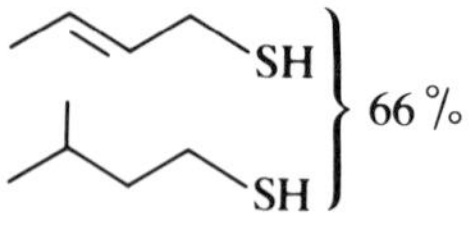

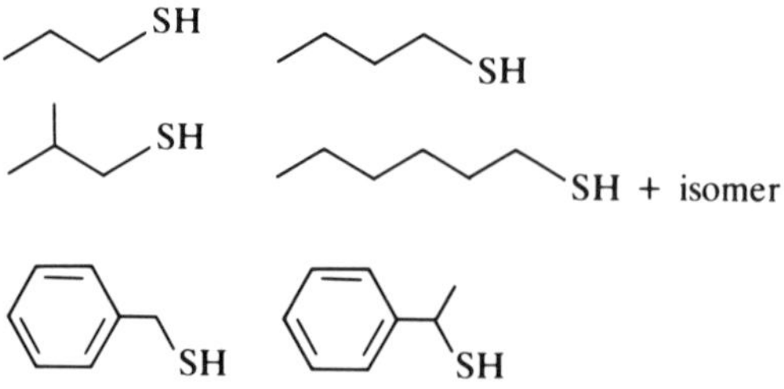

Table 5.5—*continued*

Major components	Some minor components (<1%)

Sulphides

7%

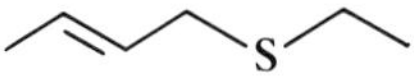

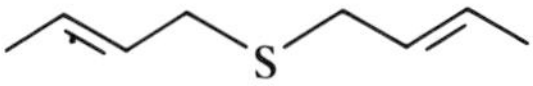

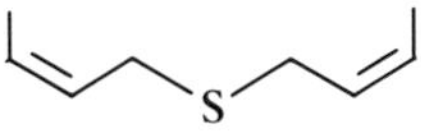

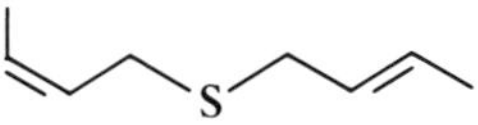

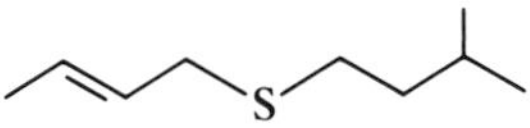

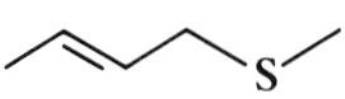

$C_7H_{16}S$

(isomeric C_4H_7)

Disulphides (possibly in part formed by air oxidation of related thiols)

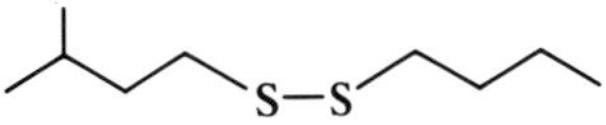

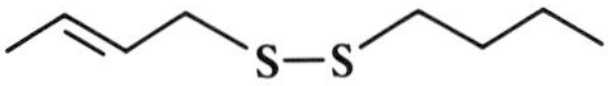

}7%

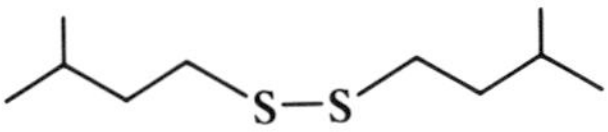

Isomeric $C_5H_{11}SSC_4H_9$ 3.6%

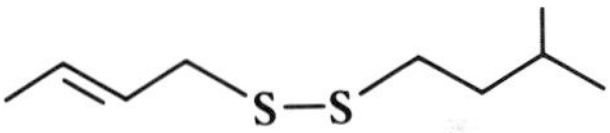

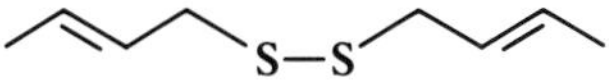

$C_4H_7-S-S-C_5H_9$

Others

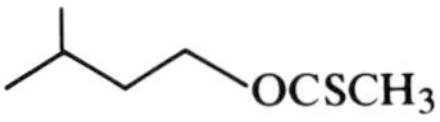

1.4%

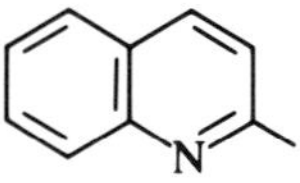

Isoamyl thioacetate

2-Methylquinoline

Unknown   7%

Aldehydes CH_3CHO
C_2H_5CHO

Numerous unknown compounds

No major contribution due to thietanes, observed in certain other mustelid secretions, was noted, although two trace components having m/e 102 and thus possibly isomeric with these dimethyl-thietanes occurred. (Andersen *et al.*, 1982).

Table 5.6. Sulphur heterocycles in mustelid anal sac secretions

Compound type	Compound	Mink (*Mustela vison*)	Stoat (*M. erminea*)	Ferret (*M. putorius furo*)
Thietane	2,2-Dimethylthietane	√*N	—	√
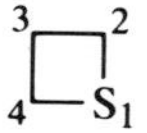	trans-2,3-Dimethylthietane ⎫	? two isomeric thietanes	—	√*
	cis-2,3-Dimethylthietane ⎬			√
	2-Ethylthietane	√	♀	—
	2-Propylthietane	—	√*	√
	2-Pentylthietane	—	√	√
1,2-Dithiolane	3,3-Dimethyl-1,2-dithiolane	√N	—	√
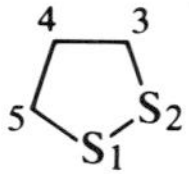	trans-3,4-Dimethyl-1,2-dithiolane	—	—	√
	cis-3,4-Dimethyl-1,2-dithiolane	—	—	—
	3-Ethyl-1,2-dithiolane	—	♀	—
	3-Propyl-1,2-dithiolane	—	√	√

*, major volatile organosulphur component of secretion.
♀, noted only in female.
N, structure confirmed by NMR.
Crump, 1980a, b; Sokolov *et al.*, 1980.

Chemical studies: the Mustelidae

(Wilz, 1967; Schildknecht *et al.*, 1976; Brinck *et al.*, 1978, 1983; Crump, 1980a,b; Sokolov *et al.*, 1980; Erlinge *et al.*, 1982).

The first chemical study of mustelid anal sac secretion in recent times was reported by Wilz (1967) in her doctoral thesis on mink. She identified the principal volatile organosulphur compounds present and showed that they contained the thietane and the 1,2-dithiolane ring systems. Today, this research has been extended by a number of workers.

Mink, stoat and ferret anal sac secretions are all yellow, turbid, odorous, lipid-rich, aqueous fluids containing related organosulphur odorants (Table 5.6). Studies on these and on other mustelid anal sac secretions suggest that these fluids are likely to exhibit species-distinctive variations on a common compositional theme.

In male stoat secretion, the organosulphur volatile profile is said to be relatively invariant from sample to sample, the principal component, 2-propylthietane, being present to the extent of about 100 μg (range 10–400 μg) per animal. In this species, female samples were more variable and, in addition, contained two organosulphur volatiles not noted in male secretion. Unidentified sulphur-free volatiles were also noted together with indole. However, Erlinge *et al.* (1982) noted differences in volatile profiles between male samples also and suggested that this could be related to the ability which stoats show in being able to distinguish between secretions of different individuals. Anal sac secretion seems to be involved in territorial marking, and as expected in a species in which males and females define their own territories separately, each territory excluding other members of the same sex, both males and females marked equally frequently, dominant members of either sex marking more frequently than subordinate members.

However, in ferret secretion, the principal volatile organosulphur component is *trans*-2,3-dimethylthietane and male and female samples appear to be indistinguishable and relatively invariant, with the major exception that at oestrus the female secretion lost its yellow colour (becoming a dirty brown) and much of its characteristic odour while chemical analysis showed that organosulphur volatiles were present only at a very low level. Indole and quinoline were also present.

Mink secretion is a similar fluid although here the dominant volatile organosulphur compound is 2,2-dimethylthietane. There appear to be no features in the volatile profile which distinguish male from female samples. Mink secretion (some 150 μl/sac) consists of mobile, immiscible lipid ($\sim$45% secretion by volume) and aqueous phases containing a large quantity of suspended solid. TLC analysis suggests that the oil consists principally of wax monoesters and elemental analysis shows that it contains 1.7% by weight of sulphur. The detailed GC profile of the less volatile lipid components appears to be characteristic of a particular animal over a period of many months and it has been suggested that in this way the secretion may convey information

related to individual identity, provided of course that the recipient is equipped to read this information. The straw coloured centrifuged aqueous liquic
contained ~0.7% sulphur. The aqueous phase (pH 8.0–8.3) appears tc
resemble closely the anal sac secretion of the fox, containing volatile fatty
acids, ammonia and amines such as 1,4-diaminobutane. It is interesting tc
note that totally different gas chromatograms are obtained depending or
whether headspace volatiles or crude aqueous secretions are injected onto the
gas chromatograph.

Of interest is the observation that among the trace volatile organosulphur
components detected was a substance tentatively identified on the basis of its
mass spectrum as isopentenyl methyl sulphide, A, and distinguishable from
the related compound, B, recently identified in hop oil (Moir *et al.,* 1980).
Just as the thietanes have only been observed in mustelid anal sac secretion,
so isopentenyl methyl sulphide had been thought to be a major odorant
unique to red fox urine.

A: $\diagup\!\!\!\diagdown\!\!\!\diagdown\!\!\!\diagup_{SCH_3}$ B: $\diagup\!\!\!\diagdown\!\!\!\diagdown\!\!\!\diagup_{SCH_3}$

Its occurrence in mink anal sac secretion, even if as a trace constituent,
suggests a biosynthetic relationship between the characteristic species odour
components of these two rather different species. The chemical similarity
between these two substances is immediately obvious and it is not difficult to
postulate mechanisms by which they might be biochemically interconverted,
for example following *S*-methylation of the thietane by *S*-adenosylmethio-
nine.

$$\boxed{\quad}-S \longrightarrow \left[\boxed{\quad}-S^+_{\diagdown} \longrightarrow {}_+\boxed{\quad}S_{\diagdown}\right] \longrightarrow \diagup\!\!\!\diagdown S_{\diagdown}$$

Other compounds identified in mink anal sac secretion include dimethyl
disulphide, dibutyl disulphide, butyl 3-methylbutyl disulphide, *bis*(3-methyl-
butyl) disulphide and indole.

As indicated, microscopic examination revealed not only the presence of
both apocrine and sebaceous secretory tissue, as expected from the two phase
nature of mink secretion, but also that an important input of sulphur into the
sac is derived from the glycoprotein secretory granules present in the apical
portions of the apocrine cells. The presence of sulphur in the lipid granules of
the sebaceous tissue remains to be investigated. But although X-ray energy
probe microanalysis provides a means of locating concentrations of sulphur
(state of combination unspecified) in the secretory tissue, the biosynthetic
origin of the organosulphur volatiles identified remains uncertain. It is not
clear what, if any, part the microflora of the sac play in their formation. A
number of these compounds are clearly isoprenoid and may be derived, as
suggested by Kjaer (1977) in his survey of low molecular weight organosulphur compounds of biological origin, perhaps from 3,3-dimethylallylthiol.
However, others are not isoprenoid and Crump has suggested another poss-

ible origin by recalling that α-lipoic acid is formed from octanoic acid by *Escherichia coli* (Parry, 1977).

5.6 THE VIVERRIDAE

In addition to anal sacs, many viverrids (civets and genets of the Viverrinae (Viverrini), Paradoxurinae and Hemigalinae) possess well developed perineal glands, skin glands secreting on to the relatively bare skin surface located between the anus and the genitalia, which skin is commonly infolded to form a pouch (Ewer, 1973). In the **African civet**, *Viverra civetta*, these 'perfume glands' are larger in the male than in the female and may be everted so that their contents can be rubbed onto objects in the environment when the animal scent marks (Ewer and Wemmer, 1974). 'Civet', the contents of the pocket, are subjectively unpleasant and presumably include microbial products, although this aspect has not been studied. However, components of the material have been used in perfumery for several hundred years, although synthetic materials are now commonly used. Ruzicka in 1926 identified about 3% of the musky macrocyclic ketone, civetone, in this secretion.

Civetone

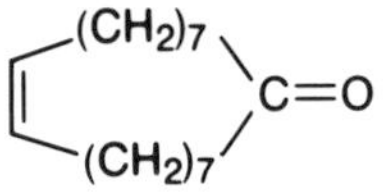

More recently, a complex mixture of related macrocyclic ketones, together with long-chain fatty acid esters of the related macrocyclic alcohols, has been noted (Van Dorp *et al.*, 1973).

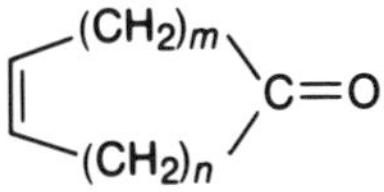

%of cyclic ketones		% of cyclic ketones	
$m = 4, n = 10$	3%	$p = 15$	1%
$m = 7, n = 7$	80%	$p = 16$	10%
$m = 7, n = 9$	6%		

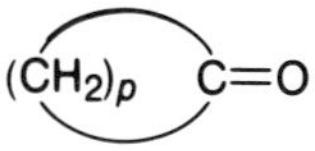

Triglycerides and cholesteryl esters also seem to be present (TLC). Much of the early work on this and other animal scent and perfume materials has been summarized by Lederer (1950).

5.7 RODENTIA

Anal sacs are not confined exclusively to the Carnivora. Similar structures have been described in certain rodents and in other orders also. Thus, a detailed study employing electron microscopy of the anal scent gland of the

160

Table 5.7. Castoreum components (Lederer, 1946, 1949). Reproduced by permission of the Royal Society of Chemistry

Alcohols
Benzyl alcohol
Cholesterol
1-Cholestanol
Mannitol
cis-5-Hydroxytetrahydroionol
Borneol

Phenols
p-Ethylphenol
p-Propylphenol
Pyrocatechol
Quinol
Quinol monomethyl ether
Chavicol (p-allylphenol)
2-Hydroxy-5-ethylanisole
4-Methyl-pyrocatechol
4-Ethyl-pyrocatechol
Betuligenol
2,4'-Dihydroxydiphenylmethane
2',3''-Dihydroxydibenz-2-pyrone
4,4'-Dihydroxydiphenic acid dilactone
A phenolic ether

Aldehydes
Salicylaldehyde

Ketones
Acetophenone
p-Hydroxyacetophenone
p-Methoxyacetophenone
Aromatic ketone
2 Isomeric hydroxy ketones

Aromatic acids
Benzoic
2-Phenylpropionic
Cinnamic
Salicylic
m-Hydroxybenzoic
p-Hydroxybenzoic
Anisic
Gentisic
5-Methoxysalicylic

Esters
Cholesteryl oleate
Esters of ceryl alcohol
Esters of benzyl alcohol
Esters of phenols
Esters of gentisic acid

Amines
Castoramine, a sesquiterpene base derived from the Nuphar alkaloids (Valenta and Khaleque, 1959) present in water lilies (Bohlmann et al., 1963)

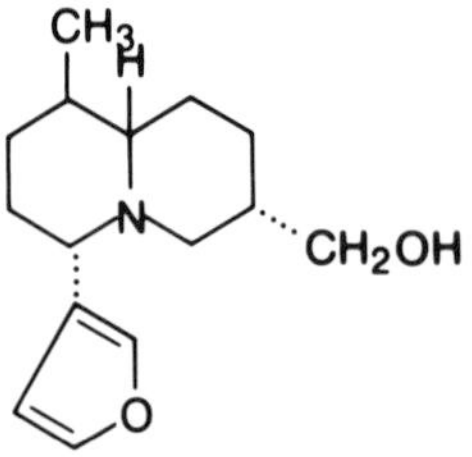

Castoramine

woodchuck, *Marmota monax*, has described sebaceous and apocrine tissues secreting into a sac structure (Smith and Hearn, 1979). Bacteria have been noted in this secretion, dominant species being *Proteus mirabilis* and *Proteus vulgaris*. It is thought that the pasty light-yellow product may be involved in signalling alarm.

The **North American beaver** (*Castor canadensis*) possesses in the anal area two paired secretory structures which are sometimes confused in the literature. Both seem to be involved in chemical signalling. These are the anal sacs and the quite separate 'castor sacs' which lie anterior to them. Svendsen (1978) and Walro and Svendsen (1982) have described both these structures.

Castoreum, the yellowish material contained in the **castor sacs**, has for long been of interest to perfumers because of its strong animal odour. Each sac with its contents may weigh 25–50 g. In a detailed histological study of the paired castor sacs, Walro and Svendsen have shown that these structures are non-glandular and are in no way similar to the preputial glands of other rodents with which they are sometimes compared but are merely glandless pockets formed in the wall of the urogenital sinus. These pockets are not muscular and are not closed by sphincter muscles so that the 'castoreum' which accumulates in them comes into contact with urine whenever the animal urinates. Thus, castoreum is necessarily present in voided urine. Interestingly, early studies of the chemical composition of castoreum have shown that it contains a large number of components of probable dietary (plant) origin (Table 5.7).

In addition to materials derived from urine and desquamated epithelial cells from the sac walls, the castor sacs studied invariably supported a bacterial monoculture. This organism was a Gram-positive facultative anaerobic pleomorphic coccus and it seems very likely that it plays an important part in generating semiochemicals by acting on the castor sac contents. One function might be to break down the water-soluble, involatile urinary conjugates, in which form many substances of dietary origin are excreted, to yield the water-insoluble volatile components in which the castor sac abounds. Experimental evidence on this point is, however, not available. A castoreum–urine mixture is employed by beavers (with anal sac secretion) in scent mound marking (Müller-Schwarze and Heckman, 1980; Svendsen, 1980), but while bladder urine alone elicits no response from beavers when applied in the environment, bladder urine plus castoreum is behaviourally active. Although this environmental marking behaviour is seasonal in the beaver, the production of castoreum is invariant throughout the year. Even so, it is possible that seasonal chemical differences may occur in its composition.

This species also possesses a pair of large muscular **anal sacs** lined with sebaceous secretory cells. These release into the urogenital sinus or cloaca a brown odorous, viscous, oily fluid. This anal sac secretion is actively produced throughout the year by adults of either sex and by juveniles, and is distributed over the body surface by self-grooming to increase the water repellancy of the fur of this aquatic mammal, much as the preen gland functions in water birds. This activity has an additional function, however, for it seems that the secretion thus displayed is a source of semiochemicals.

Observations on the closely related **European beaver (*Castor fiber*)** indicates gross differences in the nature of this secretion between the sexes. A chemical study using chemical ionization mass spectrometry (Grönneberg, 1978/79) has shown that male beaver anal sac secretion includes a major fraction corresponding to the wax esters. Some 60 different esters were noted, the acid moieties having carbon chain lengths of C_5–C_{22} and the alcohol moieties C_{14}–C_{19}. Straight chain and (ω-1)-branched (iso-), saturated and unsaturated moieties were present. These wax esters were not noted in the anal sac secretion of female beaver. No data on other fractions of beaver anal sac secretion were reported. Female anal sac secretion has quite a different appearance, contains much less lipid and wax esters seem to be largely absent (Grönneberg, personal communication).

As in other species, beaver anal sacs support an abundant microflora, here consisting of two microbial species, the facultative anaerobe *Escherichia coli* and the strict anaerobe *Bacteriodes fragilis*. API 20A profiles indicated that a range of biochemical variants of *B. fragilis* were present (Svendsen and Jollick, 1978). Again, the absence of a wide range of bacterial species is noteworthy.

Together with the products of the castor gland, this secretion seems to be involved in scent mound construction. It could also be valuable for a swimming mammal such as the beaver to present chemical signals in the form of lipid substances which would concentrate at the air–water interface.

5.8 CARNIVORE ATTRACTANTS AS DEER REPELLANTS

(Rochelle *et al.*, 1974; Oita *et al.*, 1976; Tiedeman *et al.*, 1976)

Damage to Douglas-fir seedlings due to browsing deer (*Odocoileus* and *Cervus* spp.) is an important forestry problem in parts of North America. In Britain similar problems are caused by red deer and roe deer and in some areas deer cause damage to garden plants. As a result, there is much interest in developing safe, cheap, effective deer repellant materials. In some places, faeces (particularly predator faeces) have been used with effect (Müller-Schwarze, 1972). The possibility of using deer alarm semiochemicals is interesting, but faces many difficulties including the fact that no such compounds have been identified. However, some hope is provided by the observation that certain materials which attract carnivores also repel deer. In particular behavioural studies have shown that putrefied animal tissue and specifically putrefied egg and fish products are quite effective deer repellants, although only at very close range; putrefied animal tissue is also a carnivore attractant, as has been indicated (see Section 5.3). One result is that a putrefied egg fermentation product is now marketed commercially as a deer repellant by an American company. The effects are complex and have not been fully elucidated but it seems possible that repellency is related to

(1) materials derived microbially or enzymatically from the putrefaction of animal tissue;
(2) materials derived by air oxidation of certain lipid compounds.

Here C_6–C_{12} aldehydes, known lipid oxidation products (Frankel, 1980), seem to be important and it is interesting that these substances also occur in various deer gland secretions (see Section 4.2).

Even the simple expedient of using balls of human hair has been used with

some effect as a deer repellant in the USA. This is probably effective for a
similar reason.

An interesting article from the unlikely quarter of the *American Anthropologist* (March, 1980) has again illustrated the remarkable phenomenon
of carnivore attractant/deer repellant association, for it is observed that
human menstrual odours have the effect on carnivores and particularly on
bears of arousing aggressive curiosity whereas herbivores, particularly deer,
display inquisitive avoidance. Preliminary experiments indicated that menstrual odour was indeed very effective at causing deer to avoid a food source,
much more so than other human odours (for example urine odour).

5.9 FAECAL ODOURS; MATERNAL PHEROMONE IN THE RAT

While behavioural studies have shown that faeces are widely employed in
scent marking (e.g. Franklin, 1980; Macdonald, 1980), few detailed semiochemical studies have been reported. Faeces contribute their own odours,
which may be expected to have a large microbial component, but they may
also be vehicles for secretions from the anal sacs, or from anal glands, as in the
case of the rabbit (see Section 4.3). One interesting case of a semiochemical
faecal odour which has been studied is that of the maternal pheromone of the
rat.

Maternal pheromone in the Norway rat

(Leon, 1974, 1978, 1980; Moltz and Leidahl, 1977; Moltz, 1978; Moltz and
Lee, 1981)

A very interesting example of microbial involvement in mammalian semiochemistry is provided by the example of maternal pheromone in the rat.
Similar processes may operate in some other rodents (Breen and Leshner,
1977).

Maternal pheromone, a chemical attractant operating between mother and
preweanling young, 16–27 days post partum, a period when the young are
able to leave the nest but remain dependent on their mothers, derives from
microbial products present in the partially digested food which is in the
mother's caecum and subsequently excreted. The rat produces two types of
faeces, one being hard, dry pellets and the other soft and wet (caecotrophe,
material derived from the caecum, a blind pouch at the proximal end of the
large intestine). These latter the rat regularly reingests and it is these which
contain the maternal pheromone which marks mother and nest. Caecotrophe
taken surgically directly from the caecum is equally attractive to the young,
whether it originates from a male or a female rat. Yet the observed
pheromone effect is limited to the odours produced by lactating females
16–27 days post partum. Only in this case is the pheromone externalized.

Elevated prolactin levels in the female are essential for maternal pheromone production and one view advocated by Leon is that in essence this acts to produce an excess of caecotrophe which is only partially reingested as the result of the high level of eating (hyperphagia) which this pituitary hormone is known to induce. However, there are other occasions when hyperphagia and pheromone emission are not associated with each other and an alternative view is that the main effect of prolactin is on the caecal processes themselves as the result of its known influence on liver function. There is evidence that bile from the lactating female is critically important here (although advocates of the first view would say that this turns merely on its laxative properties) for it was found that male rats which have been castrated and treated both with prolactin and oestrogen still do not emit maternal pheromone, yet when maternal bile is injected directly into the caeca of intact males these males became as powerful emitters of maternal pheromone as are lactating females. The hypothesis offered by Moltz and his associates is as follows. The female liver is much more responsive to prolactin than is the male liver and during lactation therefore produces greatly elevated levels of steroidal bile acids so that as yet unidentified volatile microbial degradation products of these bile salts produced in the caecum constitute the maternal pheromone. The pheromone is emitted only 16–27 days post partum because at this time the bile salt and its products are not reabsorbed by the gut, possibly as the result of gut pH changes.

Much remains uncertain about this fascinating effect but it seems certain that microbial products are centrally involved.

Although faecal odours can be semiochemically important, they have not generally been studied chemically. One exceptional case is that of the odour of **fox faeces** which has been found to elicit a fear response in the Norway rat. This behaviour was mirrored by elevated plasma corticosterone levels. GCMS analysis of fox faecal volatiles revealed that this fear reaction was associated with two isomeric highly odoriferous sulphur-nitrogen heterocycles present in trace amounts and reported to be *cis*- and *trans*-trimethyl-Δ^3-thiazoline (Vernet-Maury, 1980).

Chapter 6
Urine

6.1 THE IMPORTANCE OF URINE

In this Chapter, we consider the communicatory significance and something of what is known of the chemical nature of this important semiochemical substrate. In Chapter 7, we go on to examine more specifically the secretions of the reproductive tract, although this division is a little arbitrary as these materials also of necessity contaminate urine.

At first it might seem strange that so much behavioural evidence should show that urine, the principal medium by which metabolic waste is eliminated from an organism, should function so centrally in the lives of so many species as a source of chemical signals. Indeed, the biological literature is so great that no brief summary would be adequate. But, on deeper consideration, it is less surprising, for urine necessarily conveys to the external world in its detailed composition much information concerning the internal physiological state of the animal concerned and thus provides the necessary basis for the evolution of specialized semiochemical systems.

The medical interest in urine as a sensitive indicator of many internal physiological disturbances has a similar basis and the mammalian semiochemist clearly has much to learn from clinical chemistry in this regard.

The signalling potential of the urines of certain laboratory rodents have been particularly thoroughly studied and a multitude of effects documented (Bronson, 1971, 1974). Rodent urinary semiochemicals may have lasting major effects on the reproductive physiology of conspecifics (primer pheromone effects) and they may have more immediate behavioural consequences. Male and female attractants have been noted in urines of the opposite sex while the presence of female mouse urine will, for example, increase social investigation within established groups of male mice (Davies and Bellamy, 1974). It also contains environmentally stable, involatile components (present in bladder urine) which participate in a non-chemical reproductive communication system by eliciting ultrasonic courtship vocalizations in male mice (Nyby and Zakeski, 1980). A fear substance is also eliminated in the urine of mice subjected to electric shock treatment which deters the approach of other mice. Experiments with genetically uniform inbred strains of mice (Yamazaki et al., 1980) have shown that mice can detect by olfaction, either

165

from the animal itself or its urine, slight genetic differences and that these differences affect mating preferences. The odour-determining genes studied are those of the histocompatibility complex (H-2) which also play an important part in regulating the immune response. In passing, it should be noted that the genetic influence on body odour was observed long ago in the relative difficulty trained dogs experienced in distinguishing the odours of identical twins (Kalmus, 1955). Further, urine components are also involved in the control of aggression, as we shall see later.

Odour cues can both elicit and eliminate aggression between mice. One source of odour signals inducing aggression between male mice is the preputial gland (see Section 7.1). There is, however, a separate androgen-dependent urinary cue which also induces aggression.

Aggression in rodents occurs typically between mature males. Females and juveniles are rarely attacked. These effects are odour dependent as is shown by rendering mice anosmic. If an ovariectomized female mouse is treated with androgen it becomes an object for attack by male mice even though the female may show no evidence of aggressive behaviour itself. On the other hand, aggression between males is reduced if one is painted with female urine. Female mouse urine contains aggression-reducing substances.

If a castrated mouse is painted with urine from a donor animal, the level of aggression it will experience on being placed with an intact dominant male will depend critically on the animal from which the urine was taken, as follows:

	Urine donor
Increasing aggression ↑	Dominant male
	Subordinate male
	Castrate or water
Decreasing aggression ↓	Female

6.2 URINE DISPLAY

Similar effects occur in many orders and are not necessarily confined to rodents. In many species urine and other signals associated with the genitalia are inspected regularly, particularly in the breeding season, commonly with the display of 'flehmen' which occurs when the vomeronasal organ is employed to sense the signal. Urine is commonly employed to mark territory in which context it may provide a variety of information (Peters and Mech, 1975; Henry, 1977, 1980; Rothman and Mech, 1979; Macdonald, 1980; Schilling, 1980) and may also be used as a trail marker. Thus, detailed studies (Seitz, 1969) on urine marking in the lorises, *Nycticebus coucang*, *Loris tardigradus* and *Perodicticus potto*, have shown that urine-mediated chemical signals are very important in all three species as a means of indirect communication. Urine marks of all three species dry to leave sticky residues bearing

Figure 6.1   A slow loris, *Nycticebus coucang*, responding to a urine trail. (Photograph courtesy E. Seitz)

distinct species-specific odours. Certain areas within an animal's territory, such as major feeding sites and sleeping sites, are connected by urine trails which provide olfactory orientation marks for these nocturnal animals. Urine marks may also possess territorial and sexual functions.

Self- and conspecific marking

Quite commonly, many species deposit urine on their own or on a conspecific's body surface, thus providing a mobile chemical display. In some primates, a complex, stereotyped behaviour pattern known as '**urine washing**' is observed (Andrew and Klopman, 1974; Doyle, 1974). This occurs in nocturnal prosimians, particularly the Lorisidae, as well as in some new world monkeys, including species in the genera *Aotus, Saimiri* and *Cebus*. The behaviour pattern is remarkably similar in all species and involves an alternating sequence of raising and urinating on the foot and hand on one side of the body, then on the other. In *Cebus*, urine may be transferred the length of the arms and thence to other parts of the body. Such behaviour patterns are elicited on encountering strange objects in the environment or exploring strange areas. There is also a link with courtship. Thus, the male lesser bush baby, *Galago senegalensis*, will urine wash particularly after grooming a female or sniffing fresh female urine. When the female is in oestrus, male urine washing increases markedly and the male may also urine mark the female. In the bushbaby, *Galago crassicaudatus* (Tandy, 1976) urine marking is manifest principally by the dominant male and involves urine washing as well as a ritualized urine marking in which the penis is lowered to the substrate and urine expelled intermittently as the animal moves slowly forward.

Similarly, dominant male rabbits will also spray urine over conspecifics in the context of aggressive or courtship behaviour (Bell, 1980; Schalken, 1976).

Prior to or during the breeding season, the males of most ungulates impregnate their bodies with their own urine in order to advertise both to other males and also to females their physiological condition.

As Coblentz (1976) has summarized, most sheep and goats (Caprini) urinate directly onto their long body hair while red deer and elk (*Cervus*) drench their belly, rump and tail hair with urine (thrash-urination). Chamois (*Rupicapra rupicapra*) shake urine onto their flanks while urinating, whereas bison, *Bison bison*, and moose, *Alces alces*, urinate on the ground and roll in it. Reindeer, *Rangifer tarandus*, and mule deer, *Odoicoileus hemionus*, urinate on their hind legs, while camels distribute urine over their bodies via the long hairs on their tails, which they soak with urine and then flap forwards. Such a tail-urine flapping and defaecation display is characteristic of the male bactrian camel, *Camelus bactrianus*, in rutting condition (Wemmer and Murtaugh, 1980). In this way the hind legs, rump and posterior of the rear hump are drenched in urine. These are but a few examples of a widespread phenomenon.

In spite of the vast behavioural literature implicating urine as a semiochemical source, very little chemical information has been amassed. Later in this chapter I propose to summarize something of what is known about the chemistry of urine which it is hoped will be of some value to those who attempt to bridge the gap between chemistry and biology in this area. It is useful to subdivide the components of urine into those which derive from the reproductive system, and particularly the accessory sex glands, and those which have other origins. It is not unusual to find that bladder urine lacks the behavioural activity of voided urine because it lacks components deriving, for example, from the reproductive tract.

6.3 SOME PRIMER PHEROMONE EFFECTS

In some species, components present in urine have been found to influence in profound and lasting ways the physiological state of a recipient animal. Investigations of these primer pheromone effects have been conducted using laboratory mice for the most part, although related effects have been noted in some other rodents and, indeed, occasionally invoked in mammals of other orders.

Puberty acceleration and delay

The age at which an animal can first reproduce is influenced by a variety of social and environmental factors. In certain species and under certain conditions urine components have been found either to accelerate or to retard sexual maturation.

In the mouse, *Mus musculus*, a heat-labile, androgen-dependent component of male mouse urine accelerates puberty in immature female mice, the effect being greater when urine from a dominant male is employed and lower when that from a subordinate male in which gonadal function is suppressed is used (Vandenbergh, 1973; Colby and Vandenbergh, 1974; Vandenbergh *et al.*, 1975, 1976; Lombardi and Vandenbergh, 1977). Effective urine exposure can either be post-weaning or before weaning, so that the stimulus can precede the final response by a considerable period of time. The active substance seems to be involatile, to be transferred by direct contact with urine and to be associated with a peptide fraction of molecular mass of about 860. Further work using repeated chromatography and electrophoresis has been reported (Novotny *et al.*, 1980a). It is remarkable that in spite of the rigorous purification procedures which include techniques expected to remove volatiles, the active urine fraction is reported to retain a strong odour suggesting that the active substance could possibly be a volatile compound bound to and slowly released from a peptide carrier. In some circumstances applications of urine of 30 μg/day per mouse for 3 days are sufficient to achieve an effect. Activity of urine is not dependent on the presence of preputial glands. Male mouse urine is relatively rich in protein, levels being higher than in the female, becoming elevated at puberty and decreasing following castration. There is some cross-species activity. For example, adult male rat urine will also accelerate puberty in female mice. The effect of the urinary pheromone is augmented by male–female tactile cues (Bronson and Maruniak, 1975).

The urine of singly caged oestrous female mice (Drickamer, 1982a) and of pregnant or lactating female mice (Drickamer and Hoover, 1979) has the capacity to accelerate puberty in young females and to promote oestrus in adults. However, female mice kept in groups of more than four animals and isolated from contact with a male produce urine containing components which have a contrary effect and act to retard sexual maturation in other females, although here the timing and duration of urine exposure is critical (Drickamer, 1977; 1979). Further, exposure to this urine will trigger the production of similarly active urine by other singly-caged females (Drickamer, 1982a), although singly caged female urine will normally elicit no such effect. The puberty retarding components are produced by grouped females whether they are intact adults, spayed or juvenile, although their production is dependent on the presence of adrenal hormones (Drickamer and McIntosh, 1980), and in further contrast with the male effect, their action is not enhanced by non-chemical social interactions. Both the adult male→female puberty accelerating effect and the grouped female→female puberty delaying effect are manifest in bladder urine. However *bladder* urines even of singly caged females also elicits this effect although as indicated, their *excreted* urines do not. It appears that in this latter case a blocking effect is associated with components derived from the urethra, as experiments with bladder urines mixed with urethra homogenates show (Drickamer and McIntosh, 1980). In both the male and grouped female urine effects, there seems to be a

Table 6.1. The effect of urine components on the sexual maturation of female mice (Reproduced with permission from Vandenbergh et al., 1975)

Sexual maturation assessed from mass of the uterus, excised after treating female mice daily for 8 days from 28 days of age with 30–50 μl urine.

Treatment	No. of mice	Mean body wt. (g)	Mean uterus wt. (mg)
Experiment 1			
Intact adult male mouse urine	18	17.4 ± 0.34	60.5 ± 8.74
Intact adult male rat urine	18	17.2 ± 0.30	52.6 ± 6.54
Adult male human urine	12	17.3 ± 0.41	37.5 ± 10.57
Autoclaved male mouse urine	12	17.2 ± 0.39	22.9 ± 3.25
Water control	18	17.4 ± 0.40	31.8 ± 4.81
Experiment 2			
Intact adult male mouse urine	15	17.7 ± 0.22	73.4 ± 8.24
Castrated male mouse urine	15	17.2 ± 0.31	31.0 ± 4.53
Water control	15	17.5 ± 0.26	45.8 ± 6.76

The uterine growth response to such stimuli is seasonably variable.

circadian rhythm in the excretion of active urine, presumably mirroring circadian hormone variation in the excreting animals. Young females also show a circadian variation in their response to male urine, but not to grouped female urine (Drickamer, 1982c). When male and grouped female urine(s) are mixed, the puberty delaying effect of the latter overrides the puberty accelerating effect of the former (Drickamer, 1982b).

As in so many other cases, the inherent complexity of the phenomenon is striking. The detailed significance of these effects in the chemical ecology of the species is largely a matter of speculation, but they do seem to provide an important means by which levels of reproduction can be constrained by social and consequently by ecological factors (Bronson and Coquelin, 1980). Laboratory studies on trapped specimens of wild Mus musculus show that they respond to urine cues in an analogous way (Drickamer, 1979). From another study of wild house mice living under natural conditions (Massey and Vandenbergh, 1980, 1981), it seems that urine from females living under conditions of relatively high population density possesses activity which could assist in limiting population growth.

There is evidence that exposure to conspecific urine odours from an early age can affect the rate of body growth of female mice in a similar way (Cowley, 1980), although the physical growth and sexual maturation effects do not necessarily go together (Vandenbergh, 1973).

Not unrelated effects involving male and female urine components have been observed with prairie deermice, *Peromyscus maniculatus bairdii* (Teague and Bradley, 1978; Lawson and Whitsett, 1979; Lombardi and Whitsett, 1980). Here treatment with adult male urine accelerates the sexual maturation of young females but retards that of young males, while urine from singly caged adult females retards the sexual maturation of young females. Other non-chemical stimuli are also important determinants of sexual maturation in this species.

The Whitten effect; oestrus induction and acceleration

A related phenomenon observed in mice, the Whitten effect, concerns the induction and acceleration of oestrus cycling in females on exposure to mature male mouse urine. Since grouped female mice in the absence of a male tend to have long irregular oestrus cycles, to become anoestrus or to develop pseudopregnancies (Lee and Boot, 1956), states which are endocrinologically similar to the pregnant state, the introduction of a male thus leads to group synchrony of oestrus (Marsden and Bronson, 1964). The phenomenon was first noted when matings were unexpectedly delayed until the third night following introduction of a male to a group of female mice (Whitten, 1959). This suggested that the introduction of the male triggered oestrus synchrony 3 days later. Both the grouped female suppression of oestrus and the Whitten effect were thought to derive from odour cues. Subsequently (Bronson and Whitten, 1968), the Whitten effect was found to be independent of visual, auditory or tactile cues and to be due to chemical signals associated with mouse urine. Active components are androgen dependent, being present in the urines of intact adult male mice and androgen-treated female mice and absent from castrate urine, and are sufficiently volatile to be transmitted over distances of two metres in an air stream (Whitten *et al.*, 1968) although in such experiments it is difficult to rule out the possibility of a potent involatile substance being transported in aerosol particles. The phenomenon is shown by bladder urine and so is not dependent on accessory sex gland or preputial gland additives. Application of 100 μl mouse urine daily to bedding for 2 days can be sufficient to be effective. A similar Whitten phenomenon is also observed in the prairie deermouse (Bronson and Marsden, 1964).

The Bruce effect; pregnancy block

(Bruce, 1959; Parkes and Bruce, 1961; Hoppe, 1975)

The Bruce effect, another primer pheromone effect first observed with laboratory mice, *Mus musculus*, is manifest as a pregnancy block which occurs when a newly impregnated female is exposed to the urine of a 'strange' male. In this circumstance, there is a failure to implant and the female returns to oestrus. Males of different strains differ in their ability to produce urine active in this way. Production of active urine is androgen dependent and can be demonstrated using urine from androgen-treated females. Activity is present in bladder urine, and may derive from the mouse kidney, an androgen-sensitive organ, as well as in homogenates of preputial gland. Fractionation of male mouse urine suggests that activity is associated with (possibly a substance bound to) a number of a urinary non-dialysable peptide fractions. There is some evidence also that volatile components may be active.

Similar effects have also been recorded in the voles *Microtus agrestis* (Milligan 1976), *M. pennsylvanicus* (Clulow and Langford, 1971) and *M. ochrogaster* (Stehn and Richmond, 1975), although here abortion occurs well into pregnancy and not merely prior to implantation, as in the laboratory mouse.

Some general comments

It has been suggested, although without conclusive chemical evidence, for there often seems to be uncertainty even concerning the volatility of the active substances, that many of these primer effects within a given species might be due to the same substances causing the same initial endocrine responses but leading to different consequences depending on the physiological state of the recipient animal. This identity has been suggested, for example, for the substance responsible for the Whitten and Bruce effects (Marchlewska-Koj, 1977, 1980). However, Bronson and Coquelin (1980) have suggested that the Lee–Boot (mutual suppression of ovulation) and Bruce (pregnancy block) effects are 'laboratory artefacts' in the sense that the special social preconditions necessary for their occurrence are unlikely to occur in the natural environment and that the basic primer effect is concerned with cueing or inducing ovulation. There is evidence that male urinary primer pheromone brings about the release of LH (luteinizing hormone) from the anterior pituitary followed by oestrogen from its target organ, the ovary, and this then leads in due course to ovulation or to pregnancy block, depending on the reproductive state of the female. Elevation of serum LH levels may also be responsible for accelerating puberty. Consequently scent-marking behaviour and female urine pheromone production are also stimulated. In turn, female urinary pheromones stimulates the release of LH followed by testosterone in the male, a major consequence of which is to further stimulate scent-marking behaviour and male urine pheromone production. Since the male rapidly habituates to the urinary cues of a particular female, it is suggested that his production of active urine constituents is likely to decrease with time, while more potent urine (leading here to pregnancy block) is likely to be produced by a strange male. The situation is, however, complex and the importance of other endocrine mechanisms cannot be discounted (Milligan, 1980).

Other species

Although such primer effects have been studied principally with rodents, there are indications that similar phenomena may occur in other orders. It has been known for some time that the introduction of a male will hasten and synchronize oestrus in groups of females of a variety of domestic species, including sheep (Schinckel, 1954), goats (Shelton, 1960) and pigs (Hughes and Cole, 1976) while the presence of a vasectomized bull is said to decrease the interval from parturition to conception in post-partum cows (Petropavlovskii and Rykova, 1958). While direct evidence is largely lacking concerning the semiochemical contribution to these phenomena, there is now evidence that cervical mucus from oestrous cows contains components which exert a synchronizing effect on oestrus in other cows (Izard and Vandenbergh, 1982a) while, in addition, bull urine components can accelerate the attainment of puberty in heifers (Izard and Vandenbergh, 1982b).

One indication of a possible human primer pheromone effect is the report of a degree of menstrual synchrony of women living together in an all-female institution. It appeared that male contact significantly shortened the menstrual cycle length compared with women living sexually more isolated lives (McClintock, 1971). A further study of female undergraduates living on a coeducational university campus showed evidence of synchrony between women living in close association (Graham and McGrew, 1980). The cause of this phenomenon remains unknown, and although it is possible that an airborne semiochemical is responsible (Rogel, 1978), as has been demonstrated with oestrus synchrony in groups of female rats (McClintock, 1978), this remains to be demonstrated.

Urine is also involved in chemical signalling in a variety of primate species, and although definite evidence is lacking, it is possible that primer effects may exist. A study of the squirrel monkey, *Saimiri sciureus* (Candland *et al.*, 1980) has indicated the diversity of information which urine can convey in that species. In a variety of primates, behavioural observations suggest that urine can communicate individual identity, group membership, social dominance, and sexual status, while the age of the scent mark is also communicated.

Although it is not always apparent in the laboratory, in the wild many primates—most prosimians and many simians—are seasonal breeders, and while this seasonality may depend on environmental stimuli, interactions between the sexes can play an important part in achieving the close physiological synchrony which is particularly important in species such as *Lemur catta* which have a short mating season. Such communication of endocrine state from female to male rhesus monkeys living in seminatural conditions has been demonstrated by Vandenbergh and Drickamer (1974). Chemical signals may be involved here. In squirrel monkeys also (Baldwin, 1970), it appears that, at the breeding period, chemical signals arising from the female may trigger major physiological and behavioural changes in the male. Although female odours are of interest to the male throughout the year, their attraction greatly increases when the female is sexually receptive. At this time too, the females display urinary signals by showing a marked increase in stereotyped urine washing behaviour. Here, the female urinates on its hand and the urine is washed over the soles of the feet or rubbed elsewhere on the body. Body parts and branches thus contaminated are of particular interest to the male. Urine washing occurs in dominant males too and could have effects which lead to the inhibition of reproductive function in subordinate males.

6.4 GROSS COMPOSITION OF URINE

Urine consists principally of a filtrate of blood plasma produced in the kidneys by a complex process involving a dialysis separation step in microscopic structures called glomeruli, to yield a cell-free, protein-free filtrate of whole blood, followed by selective reabsorption from this filtrate as it passes through the renal tubules of components valuable to the organism. Further secretion of metabolic waste also occurs in the renal tubules. It follows that many urinary components are also found in blood plasma and also that dietary factors have an important influence on the occurrence of many major urine components. In addition, voided urine may be expected to contain other substances which are the direct product of the kidney itself (as this organ is metabolically active and possesses certain endocrine functions of its own) or are secreted by the accessory sex glands or other structures of the reproductive tract.

As indicated, more is known about the chemistry of human urine than that of any other species of mammal. Although most of this information has been gathered in a medical context, it provides an important starting point for studies in mammalian communication while also revealing something of the range of pertinent information which modern chemical techniques can uncover. It is important, however, to remember that man is but one species of mammal among many and that studies undertaken on other species are likely to reveal major differences in chemistry as well as many expected similarities. Indeed, the relatively few chemical studies conducted so far on urines of other species point in that direction.

In one minutes, the two kidneys of a normal man can filter a litre of blood and produce a little over 100 ml of glomerular filtrate. After passage through the tubules, this may yield 1 ml of urine. An average 24 hour urine sample has a volume of 1200 ml and a pH of 6.2, although both figures vary widely. The pH range of healthy human urine is from 4.7 to 8.0, depending on diet. Apart from water and inorganic salts, the major constituent of human urine is **urea**, 25–30 g being excreted per day.

$$O=C\begin{cases} NH_2 \\ NH_2 \end{cases} \quad \text{Urea}$$

Urea is the end product of protein catabolism and the quantity excreted is related to the protein content of the diet. Fresh urine contains little **ammonia** (0.3–1.0 g/24 h) and the well known ammoniacal smell of stale urine arises from microbial hydrolysis of the urea. Excessively acid urine leads to increased ammonia excretion, mainly as the result of the production of ammonia from the amino acid glutamine in the renal tubule cells. Urinary **inorganic salts** include sulphates and phosphates derived from the sulphur and phosphorus present in foodstuffs. The urine of cattle and horses is often turbid with calcium and magnesium phosphates which come out of solution in the alkaline urine (cow urine, pH 7.8–8.6). This phenomenon is observed on occasions in the urines of humans on vegetarian diets, which is often also alkaline. A herbivorous or vegetarian diet leads to the ingestion of quantities of salts of organic acids which are subsequently oxidized metabolically to carbon dioxide. To maintain electrical neutrality, anions of these salts are excreted in urine as bicarbonates—hence the alkalinity.

Another major nitrogen-containing component of human urine is **creatinine**. This is of interest because its output (1.0–1.8 g/24 h) is independent of diet and is unusually constant for a given individual, so that it is more

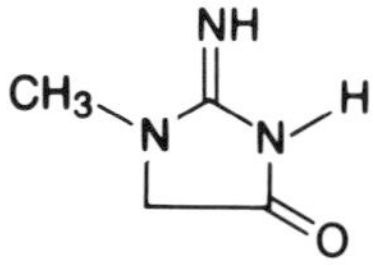

Creatinine

common to refer concentrations of a component in urine to creatinine (mg/g creatinine) than to water (mg/l). There are diurnal variations in urine composition, so excretion rates are best measured on bulked 24 hour urine samples.

Creatinine is a waste product derived non-enzymatically from an energy-rich substance present in muscle, creatine phosphate. The urinary level of creatinine in an individual depends to an extent on body size and muscular development but not on muscular activity. Creatine itself occurs in human urine at levels of 60–150 mg/24 h.

Uric acid is also present in human urine (0.5–0.8 g/24 h) and is partly related to diet, being the end product of purine catabolism. On account of the low water solubility of this substance and its salts, it tends to come out of solution on cooling. In most mammals, apart from man and other higher primates, the end excretion product of purine catabolism is the rather more water-soluble substance, **allantoin**. In man, this occurs in urine at levels of 30 mg/24 h.

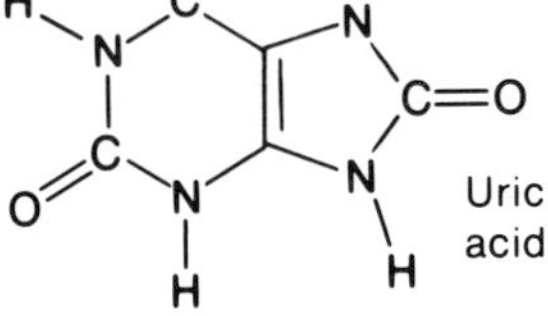

Uric acid

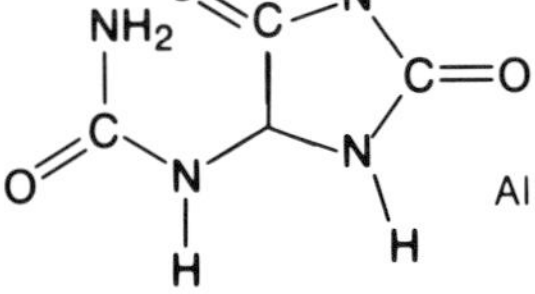

Allantoin

To go beyond these simple facts of gross urine chemistry to consider minor constituents reveals immense complexity with regard to almost every class of compound present for urine chemistry reflects both physiological (endogenous) and dietary (exogenous) factors. Much research has been conducted on the analysis of drug metabolites in urine, but save for the techniques involved, this does not concern us here. Our interest centres particularly on the way in which healthy changes in an animal's physiology are reflected in its urine composition.

6.5 URINE VOLATILES

One approach to the study of urine which is particularly relevant to our interest in olfactory signals has been to examine the profiles of volatile substances which vaporize from the surface of a urine sample. The method most commonly used employs a solid absorbent trapping material to collect these substances, although as usually practised this has the limitation of being unsatisfactory for components of molecular mass greater than 150–200 amu. High efficiency capillary GC columns are normally required and complex profiles of large numbers of volatile components are obtained. Compounds commonly noted are listed in Table 6.2. Profiles are distinguished more by

Table 6.2. Volatile components of normal human urine

Compound (1)	Occurrence (2)
Ketones	
C_3 Acetone	b, c, d
C_4 Butan-2-one	b, c, d
Butan-2,3-dione	b, c
C_5 Pentan-2-one	a, b, c, d
Pentan-3-one	c
3-Methylbutan-2-one	a, b, c
Pentan-2,3-dione	c
Pent-3-en-2-one	a, b, c, d
Cyclopentanone	a
2-Methyltetrahydrofuran-3-one(?)	a
C_6 Hexan-2-one	a, c
Hexan-3-one	a, b
2-Methylpentan-3-one	a
3-Methylpentan-2-one	a, b, c
4-Methylpentan-2-one	a, b, c, d
4-Methylpent-3-en-2-one(?)	a, b
3-Methylcyclopentanone	a, b
Cyclohexanone	a, d
Acetylfuran	b
C_7 Heptan-2-one	a, b, c, d
Heptan-3-one	a, c
Heptan-4-one	a, b, c, d
5-Methylhexan-3-one	a, b
4-Ethoxypentan-2-one	a

Table 6.2—*continued*

Compound (1)	Occurrence (2)
C$_8$ Octan-2-one	a
Octan-3-one	a,b,c
Octan-4-one	c
6-Methylheptan-3-one(?)	b
C$_9$ Nonan-2-one	b, c
C$_{10}$ Carvone	a, b, c
Piperitone	b

Aldehydes

C$_3$ Propanal	b, d
C$_4$ 2-Methylpropanal	c
Butanal	c
C$_5$ 2-Methylbutanal	c
3-Methylbutanal	c
Furfural	b
C$_7$ Benzaldehyde	a, b, c, d

Carboxylic acids

C$_2$ Acetic acid	a

Lactones

C$_5$ γ-Valerolactone	R = CH$_3$	a
C$_6$ γ-Hexalactone	R = C$_2$H$_5$	a
δ-Hexalactone		a
C$_7$ 4-Methyl-5-hydroxyhexanoic acid lactone		a

γ-lactone

Alcohols/Phenols

C$_2$ Ethanol	b, d
C$_3$ Propan-1-ol	a, d
C$_4$ Butan-1-ol	b, d
2-Methylpropan-1-ol	a, d
Butan-2,3-diol	a
C$_7$ p-Cresol	a, b
C$_{10}$ α-Terpineol	a

Alkylfurans

C$_5$ 2-Methylfuran	c
C$_6$ 2,3-Dimethylfuran	a, b, d
2,4-Dimethylfuran	a, b
2,5-Dimethylfuran	c
2-Ethylfuran	c
C$_7$ 2,3,5-Trimethylfuran	a, b
2-Methyl-5-ethylfuran(?)	a, b
C$_9$ 2-Pentylfuran	a, b, c

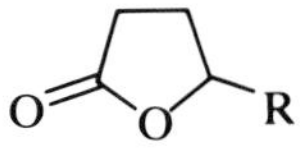

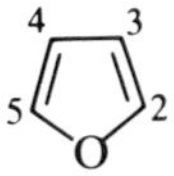

Furan

178

Table 6.2—*continued*

Compound (1)	Occurrence (2)
Pyrroles	
C$_4$ Pyrrole	a, b, c,
C$_5$ 1-Methylpyrrole	b
2-Methylpyrrole	b
C$_6$ Dimethylpyrrole	b
C$_8$ 1-Butylpyrrole(?)	b
Pyrazines	
C$_5$ Methylpyrazine	b
C$_6$ 2,3-Dimethylpyrazine	b
2,5-(or 2,6-)Dimethylpyrazine	b
Vinylpyrazine	b
C$_7$ 2,3,5-Trimethylpyrazine	b
2-Methyl-6-ethylpyrazine	b
2-Methyl-6-vinylpyrazine(?)	b
Sulphur compounds	
C$_2$ Dimethylsulphone	a
Dimethyl disulphide	a, b, c, d
C$_3$ Propylene sulphide	c
C$_4$ Allyl isothiocyanate	a, b
Thiophene	c
Thiolan-2-one(?)	b
Hydrocarbons	
C$_6$ Benzene	c
Hexane	c
C$_7$ Toluene	b, c, d
C$_{10}$ β-Pinene(?)	b
p-Methylpropenylbenzene(?)	b
Limonene	a, d

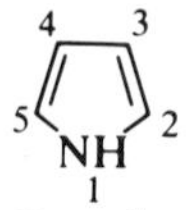

Structures: Pyrrole (5-membered ring, positions 2,3,4,5, NH at position 1); Pyrazine (6-membered ring, N at positions 1 and 4, positions 2,3,5,6).

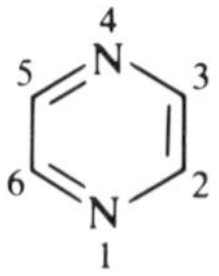

(1) Identified by gas chromatography–mass spectrometry.
(2) a, b and c in urine; d in blood serum. a, distillate of diethyl ether extract of urine; column Dowfax 9N15 (Zlatkis and Liebich, 1971). b, Tenax GC headspace volatiles trap; column Emulphor ON-870 (Zlatkis et al., 1973a, b). c, Chromosorb 101 headspace volatile trap; column, methylsilicone oil SF96 (50) plus 5% Ipegal CO-880 (Matsumoto et al., 1973). d, as b (Zlatkis et al., 1974; Liebich and Wöll, 1977). Conditions as 'c'; many components remain unidentified.

differences in the relative proportions of components than by the total absence or presence of components.

What are the determinants of such profiles? Diet can have a marked effect on urine odour. For example, diet influences the attractiveness of guinea pig ırine to other guinea pigs, as measured by investigation times (Beauchamp, 1976). Odorous compounds, to which a minority of the population are particularly sensitive, are excreted following the consumption of asparagus Lison *et al.*, 1980). These compounds include *S*-methyl thioacrylate and ʃ-methyl 3-(methylthio)thiopropionate (White, 1975). This is perhaps a particularly well known case of the general phenomenon that dietary changes are reflected in urine composition and possibly in urine odour. That being so, it is perhaps surprising that the overall profiles of volatiles obtained from urine samples taken from the same healthy individual over a period of 2 months should remain remarkably constant in spite of probable dietary variations over that period, while profiles from different individuals exhibited major differences, presumably reflecting physiological factors (Zlatkis and Liebich, 1971; Zlatkis *et al.*, 1973a). This individuality in human volatiles' profiles appears not to have been studied in detail, however, and is not similarly apparent in, for example, urinary organic acid profiles.

S-Methyl thioacrylate   S-Methyl 3-(methylthio)thiopropionate

Physiological factors also influence urine odour. This has been most clearly demonstrated in connection with certain pathological metabolic abnormalities in man. It has long been known that characteristic body and/or urine odours attend certain metabolic defects, such as phenylketonuria, maple syrup urine disease, 'cat's urine syndrome', 'fish odour syndrome', etc., and indeed the clinical value to the physician of a knowledge of abnormal scents in the early detection of these and other conditions has been advocated (Liddell, 1976; Mace *et al.*, 1976; Labows, 1979). It is with the principal purpose of putting this type of consideration on a more scientific basis that the tabulation of profiles associated with urine samples has been undertaken. Major changes in profiles of volatiles have been documented, in association, for example, with diabetes mellitus (Zlatkis *et al.*, 1973a; Liebich and Al-Babbili, 1975) and a variety of other conditions (Sastry *et al.*, 1980).

Although the techniques of volatile profile analysis have so far been applied only to studies of the urines of two other mammals, preliminary results already show species differences. The characteristic smell which the countryman and, increasingly the town dweller, associate with the fox is present in fox urine. A study on red fox urine (Jorgenson *et al.*, 1978) showed that this was principally due to two sulphur compounds, isopentenyl methyl sulphide and phenylethyl methyl sulphide. These, together with 6-methylhept-5-en-2-one and *trans*-geranylacetone (the *cis*-isomer is totally absent) appear to be unique to the urine of the red fox. Other compounds were also identified, and although only a limited number of samples were examined, possible correlations of occurrence with sex were indicated.

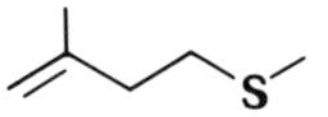

180

Isopent-3-enyl methyl sulphide

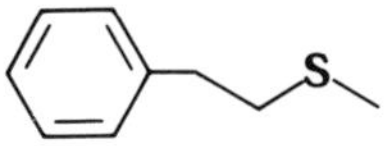

Phenylethyl methyl sulphide

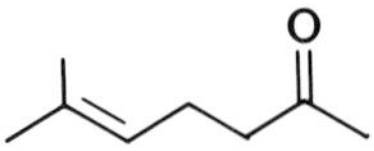

6-Methylhept-5-en-2-one

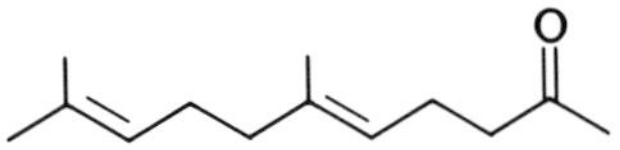

trans-Geranylacetone

As isopentenyl methyl sulphide is closely related chemically to isopentenyl pyrophosphate, an early intermediate in the biosynthetic pathway to the terpenes and steroids, it has been suggested that in the fox these compounds could be biochemically linked so that this sulphide acts as an olfactory indicator of the activity of this biosynthetic pathway in a given fox. Although the phenylethyl methyl sulphide has not been identified in urine previously, amines containing the phenylethyl group are well known.

A subsequent study (Bailey *et al.*, 1980) in England by Ministry of Agriculture scientists has confirmed the presence of these components, with the exceptions of geranylacetone and 6-methylhept-5-en-2-one, which were not detected, in the volatiles associated with red fox urine samples. One difference was that 2-methylquinoline was present in samples from both male and female animals. Samples from adult animals were generally richer in volatiles than those from juveniles, this being particularly evident in male animals. The

Table 6.3. Volatile components of red fox urine (adapted from Jorgenson *et al.*, 1978 *Science*, **199**, 796–798, copyright 1978 by the AAAS) (1)

	Occurrence (2)	
	Male	Female
4-Heptanone	L	L
Isopent-3-enyl methyl sulphide	H	L
6-Methylhept-5-en-2-one	L	H
Benzaldehyde	L	L
Acetophenone	H	H
2-Phenylethyl methyl sulphide	L	L
2-Methylquinoline	H	Zero
trans-Geranylacetone	L	H

(1) Based on samples from wild red fox in January–March period.
(2) H, high level component; L, low level component.

dominant peaks on all gas chromatograms were due to isopent-3-enyl methyl sulphide and acetophenone.

A feature of the British study was that samples were collected for analysis from a limited number of captive animals at intervals over a number of years, so that seasonal variations in volatile profile could be noted. In male animals, two volatile components were found to exhibit seasonal peaks in relative concentration, and this in early to mid-February at just the time when mating activity among foxes is known to be maximal in that part of England. These components were 4-heptanone and a sulphur compound not previously identified in fox urine volatiles, 3-methylbutyl methyl sulphide (see Figure 6.2 and 6.3).

3-Methylbutyl methyl sulphide

It seemed that this compound was produced maximally by the male at the time when females are receptive and that it might have biological importance as an olfactory signalling mechanism. This substance was also observed in female urine volatiles, but its production here was more variable and at considerably lower levels. It is not known whether this compound is biosynthesized via isopentenyl pyrophosphate or isopentenyl methyl sulphide, or

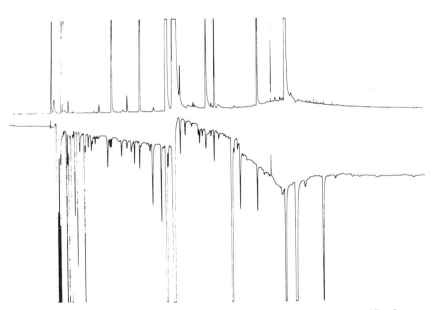

Figure 6.2 Gas chromatogram using simultaneously a sulphur specific detector (upper trace) and a flame ionization detector (lower trace) for the analysis of male red fox urine head space volatiles. (Reproduced by permission of Plenum Publishing Corp. from Bailey *et al.*, 1980)

182

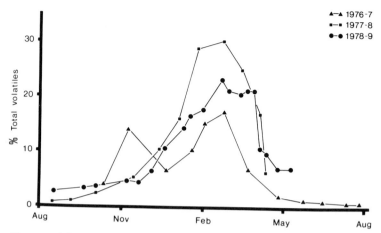

Figure 6.3 Annual variation of relative concentration of 3-methylbutyl methyl sulphide in male red fox urine head space volatiles over a 3-year period, sampling two animals. (Reproduced by permission of Plenum Publishing Corp. from Bailey *et al.*, 1980)

whether its production occurs by some quite other route, such as from an isovaleric acid derivative.

The other mammal to be studied in this way is the laboratory mouse (Liebich *et al.*, 1977; Novotny *et al.*, 1980a). When urine volatiles were examined using capillary gas chromatography, highly complex profiles were again obtained. However, it was reported that profiles from urines of mice of the same sex and strain maintained on the same diet exhibited only minor variations, while major profile differences were noted between the sexes. Major profile differences between the sexes were also noted when attention was restricted to sulphur-containing volatiles by fitting the gas chromatograph with a sulphur-specific detector. The chromatograms thus obtained were, of course, much simpler and it was apparent that two major volatile sulphur compounds which were invariably present in male mouse urine, occurred at the most only in trace quantities in female mouse urine. These sex-specific substances were subsequently identified as 2-isopropyl-4,5-dihydrothiazole and 2-sec-butyl-4,5-dihydrothiazole:

These dihydrothiazoles appear to be characteristic of male mouse urine and have not been identified in other species. Even so, their biological significance remains to be elucidated. They lack activity in puberty accelerating primer pheromone bioassay. These, and a number of other peaks including those due to heptan-2-one, 3,4-dimethyl-3-penten-2-one(?), phenol, a cresol and a toluidine, increase following testosterone administration.

Tom cat urine has a distinctive odour which some have ascribed to sulphur compounds, although it seems that no investigation of the urine volatiles has been published. The odour is present in bladder urine and appears to derive from the kidneys (Short, 1972), as does the sulphur-containing amino acid, felinine (see Section 6.6).

6.6 LESS VOLATILE AND INVOLATILE COMPONENTS

Even in a medical context, studies of urinary volatiles are relatively few. More commonly attention is directed to the occurrence in urine of lower molecular weight components irrespective of their natural volatility. Urine is fractionated and, following the appropriate derivatization procedures to render sufficiently volatile and thermally stable compounds of the classes expected in particular fractions, these fractions are examined by gas chromatography (Jellum et al., 1972, 1973; Lawson et al., 1976). In this way up to 1000 components may be examined in a few millilitres of human urine and at least 40 known metabolic defects detected from their characteristic profile abnormalities (Watts et al., 1975). Abnormal metabolic states which may be detected from an examination of urine profile are usually reflected in gross changes in the relative proportions of certain components rather than by the appearance of totally new substances.

Steroids

One class of compound which has attracted attention is that of the steroids. The psychiatrist, Kloek (1961), seeing urine as a medium by which the internal physiological state of an animal is displayed to the outside world, has written of the sex steroids:

'the gonads . . . secrete their hormones into the blood. These hormones are partly excreted with urine, unchanged or in slightly modified form . . . the organism could hardly achieve a better, more direct way of expressing its sexual status than that which it creates for itself in the hormone level in its blood . . . the same can be said of the level of sex hormones in the urine . . .'

A situation in which steroid hormones behave semiochemically occurs in the relationship between the rabbit flea, *Spilopsyllus cuniculi*, and its host, for rabbit flea reproduction is closely linked with changes in hormone levels in rabbit blood. Unknown volatile substances associated with nestling rabbit urine also have an effect. In this way, primed by high levels of rabbit blood hormones, fleas on a female rabbit which has become pregnant themselves undergo maturation and transfer to the young rabbits at birth, where they mate and lay their eggs. Also rabbit fleas are strongly attracted to rabbit urine and will remain close by for several days, but move rapidly to a host should one appear. Since urine is employed by rabbits as a territorial marker, this provides a semiochemical means by which rabbit fleas are more likely to encounter hosts. (Vaughan and Mead-Briggs, 1970; Rothschild and Ford, 1973).

Writing in 1967, Kloek noted that steroids such as testosterone, androsterone, oestriol and oestrone all have odours. Some steroids, such as 5α-androst-16-en-3α-ol, have strong musky or urinous odours. He refers to a preliminary observation which suggests that a police dog may be able to detect trace quantities of progesterone which a pregnant woman imparts to an object she has handled. Steroids are, then, substances of considerable semiochemical potential although components of very many other compound classes of mammalian secretions and excretions also correlate with internal hormonal states, and signalling potential is in no way confined to this class of substance.

Steroids occur widely in urine, mainly as water-soluble (and thus easily excreted but odourless) conjugates (sulphates and glucuronides). However, the free steroids could presumably be released in quantity in stale urine by the action of microorganisms.

Much of the steroid excreted is, however, in the form of liver and kidney metabolites of the active sex hormones (Table 6.4).

The **17-hydroxycorticosteroids** (17OHCS), cortisol metabolites, are also present in urine at levels which are related to the activity of the pituitary–adrenal cortical system. This system is sensitive to psychological parameters and reflects states of emotional arousal, elements of novelty, uncertainty and unpredictability being particularly potent influences. As a result, urinary 17OHCS levels increase following stress and again provide an external indicator of internal physiological state (Mason, 1968a; Mason et al., 1968a; Dey et al., 1972).

The principal human urinary steroids may be examined by capillary GCMS following initial hydrolysis of the steroid conjugates (Horning et al., 1974; Pfaffenberger and Horning, 1977). Complex steroid profiles are obtained which, while showing individual differences, also correlate with sex. Individual steroid levels range up to a maximum mean for adult men of 5.5 mg/g creatinine of androsterone. Low levels of testosterone (in conjugate form)

Table 6.4. Some steroid hormones and their urinary metabolites

Hormone	Urinary excretion product (conjugated)
Androgens	**Androgen metabolites**
Testosterone	Androsterone
Androstenedione	Epiandrosterone
Dehydroepiandrosterone	Etiocholanolone
	Dehydroepiandrosterone
Oestrogens	**Oestrogen metabolites**
Oestrone	Oestriol
17β-Oestradiol	Oestrone
	17β-Oestradiol
Progestogens	**Progestogen metabolites**
Progesterone	Pregnanediol
	Pregnanetriol

185

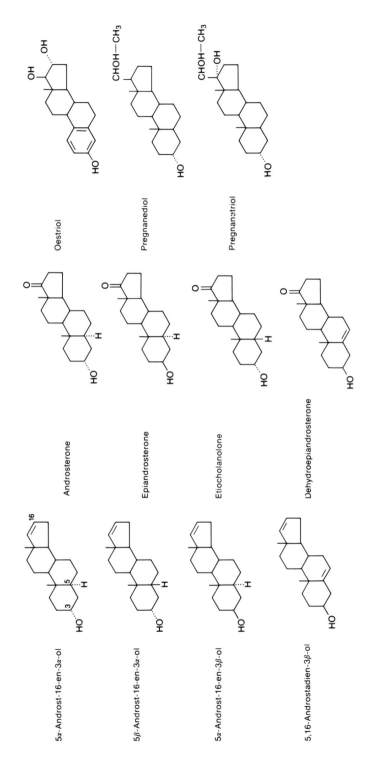

Oestriol

Pregnanediol

Pregnanetriol

Androsterone

Epiandrosterone

Etiocholanolone

Dehydroepiandrosterone

5α-Androst-16-en-3α-ol

5β-Androst-16-en-3α-ol

5α-Androst-16-en-3β-ol

5,16-Androstadien-3β-ol

Figure 6.4   Formulae of urinary steroids referred to in the text

occur in human male urine (Okamoto *et al.*, 1971; Harkness, 1974; Puah *et al.*, 1978), individual levels varying from 15 to 100 μg/24 h. Racial differences in urinary testosterone levels have been noted even though plasma levels may be the same. Elevated plasma and urine testosterone levels accompany sexual activity in the human male.

Thus, the occurrence of steroids in human urine has been studied in considerable detail and the picture which emerges is complex. However, of particular semiochemical interest is the occurrence of **androst-16-en-3-ols** and **androst-16-en-3-ones** in urine. 5α-Androst-16-en-3α-ol, 5β-androst-16-en-3α-ol and 5,16-androstadien-3β-ol (as glucuronides) occur in human urine at levels varying from person to person (Brooksbank, 1962; Gower, 1972). An average value for the former steroid is 1 mg/24 h in men and 0.4 mg/24 h in women. Excretion is very low in children, reaching a maximum following puberty and thereafter falling slowly throughout life. In the free state, as indicated, this steroid has a musky odour and is known to act as a sex semiochemical in the pig. It could also have such a function in man particularly since it has no other known biological function. The 5α-3β compound also has a musky odour while the 5β-epimers are odourless to man. Human urine contains no more than trace quantities of the 5β-3α compound.

Boar urine is of interest as these androstene steroids have been studied as sex pheromones in that animal (Booth, 1980). However, here they are transmitted via saliva (see section 8.2). Boar urine in contrast contains none of the active 5α-androst-16-en-3α-ol identified in saliva, but contains instead the 5α-androst-16-en-3β-ol (as glucuronide) at levels of some samples of 0.25 mg/l (compared with conjugates of androsterone, 1.75 mg/l; of etiocholanolone of 2.76 mg/l and of dehydroepiandrosterone of 10.75 mg/l). This steroid when liberated from its conjugate (e.g. by bacterial action) also has a musky odour, and could have a semiochemical function.

Acids and phenols

The patterns of occurrence of urinary organic acids have been the subject of much study. Both exogenous and endogenous factors contribute to the acid profiles obtained, although dietary variation seems to have less impact on the overall profile than one might have expected (Chalmers *et al.*, 1976).

First, however, it should be noted that many organic acids, amines and alcohols may be excreted in urine either in free or in combined (conjugated) form. This has already been alluded to in the case of the steroids. Conjugation commonly involves combination with an amino acid (particularly glycine or glutamic acid) or excretion as a glucuronide or a sulphate.

Such conjugation provides a means by which the organism may enhance the water solubility of a metabolite and so aid its excretion. Water-soluble conjugates, such as glucuronides, lack volatility and odour. However, it is expected that they will hydrolyse in stale urine, particularly if bacteria are present, and so generate the free component. This could be one reason why behavioural tests sometimes show that aged urine samples possess greater activity than fresh ones.

The mode of conjugation may vary from species to species. This has been documentated following the administration of radiolabelled phenylacetic acid to representatives of 24 different species of mammal (James *et al.*, 1972). Conjugation of the excreted phenylacetic acid in different proportions with a variety of different amino acids and with glucuronic acid occurred according to species.

Toxic compounds are commonly eliminated as conjugates. One example concerns the Koala bear, *Phascolarctos cinereus*, which feeds preferentially on eucalyptus foliage which is rendered toxic to most species due to its high terpene oil content, especially the volatile terpenes, α-pinene, β-pinene, 1,8-cineole, *p*-cymene and cryptone (Eberhard *et al.*, 1975). The Koala has adapted to deal with these substances and the terpenes are voided through urine, faeces and probably through breath and the skin, for Koalas emanate a general eucalyptus-like body odour. Up to 30% of the ingested oils are detected in faeces, although gas chromatography shows that major transformations have occurred. Only 1% of free terpenes are detected in urine. However, urinary glucuronide could account for elimination of more than 50% of the ingested oils. The following lactones have been reported in Koala urine following hydrolysis (Southwell, 1975). They possibly arise from hydroxy acids released on hydrolysis of the glucuronide esters.

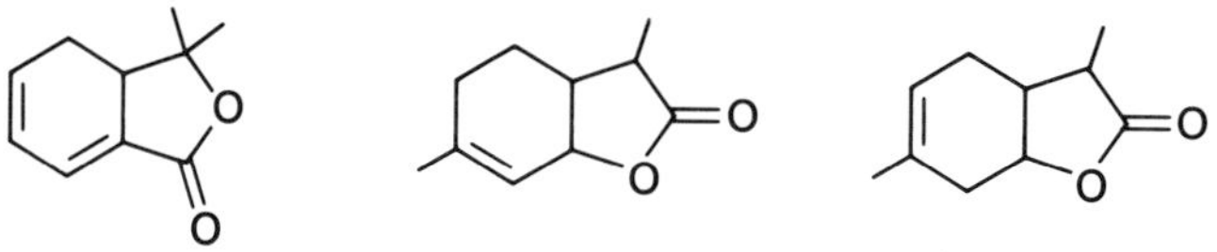

A wide range of organic acids have been detected in normal human urine. These include the following classes (Jellum *et al.*, 1972; 1973; Knights *et al.*, 1975; Thompson *et al.*, 1975; Lawson *et al.*, 1976; Brown *et al.*, 1978).

Monocarboxylic aliphatic acids (saturated)

CH_3COOH	Such as acetic acid and isovaleric acid,
$\begin{matrix} CH_3 \\ CH_3 \end{matrix}>CHCH_2COOH$	which also occur as a glycine conjugates.

Dicarboxylic aliphatic acids (saturated)

straight chain
$HOOC-(CH_2)_n-COOH$

n	
0	Oxalic (27 ± 8 mg/24 h)
	(Charransol *et al.*, 1978)
2	Succinic
3	Glutaric
4	Adipic
6	Suberic
7	Azelaic

branched chain n
HOOCCH₂CH(CH₂)ₙCOOH   0   Methylsuccinic
         | 1   3-Methylglutaric
        CH₃   2   3-Methyladipic
                                          3   3-Methylpimelic
                                          4   3-Methylsuberic

The major acid of this group is 3-methyladipic acid (16–30 mg/24 h). Other acids are excreted in much smaller quantity. Excretion patterns seem insensitive to dietary changes (Pettersen and Stokke, 1973).

The cyclopropane dicarboxylic acid, 3,4-methylenehexanedioic acid is present in human urine, 16 mg/24 h and is probably of exogenous origin (Lindstedt et al., 1974).

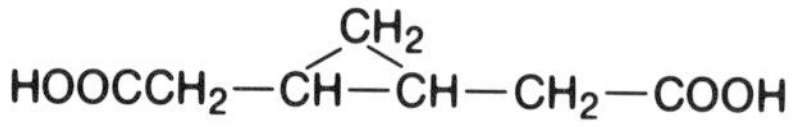

$$\underset{\displaystyle HOOCCH_2-CH-CH-CH_2-COOH}{\overset{\displaystyle CH_2}{}}$$

Unsaturated aliphatic acids

These may sometimes be artefacts. For example, crotonic acid, $CH_3CH=CHCOOH$, may be observed as a work-up product from urine samples containing high levels of 3-hydroxybutyric acid, $CH_3CHOHCH_2COOH$ (Gompertz, 1971):

$$\underset{\displaystyle R}{\overset{\displaystyle }{HOOC-CH_2-C=CH-COOH}}$$

β-methylglutaconic (R=CH₃) and *cis*-aconitic (R=COOH) acids have been reported.

The acetylenic acid, 5-decynedioic acid, with a little 4-decynedioic acid, has been observed (2–20 mg/24 h) and is of exogenous origin (Lindstedt and Steen, 1975):

$$HOOC(CH_2)_3C\equiv C(CH_2)_3COOH$$

Aromatic acids

formula, A—B

—A	—B	Acid
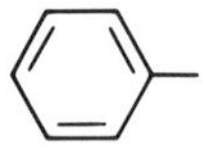	—COOH	Benzoic, occurs as glycine conjugate (hippuric acid, up to 1 g/24 h)
⌬—	—CH₂COOH	Phenylacetic, occurs as glutamic acid conjugate
	—CH₂CH₂COOH	3-Phenylpropionic
	—CHOHCOOH	Mandelic

formula, A—B—*continued*

—A	—B	Acid
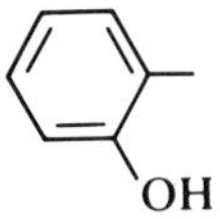 (benzene ring with OH, ortho)	—COOH —CH$_2$COOH	Salicylic, occurs as glycine conjugate (salicyluric acid) *o*-Hydroxyphenylacetic
(benzene ring with HO, meta)	—CH$_2$COOH —CH$_2$CH$_2$COOH —CHOHCH$_2$COOH	*m*-Hydroxyphenylacetic 3-(*m*-Hydroxyphenyl)propionic 3-(*m*-hydroxyphenyl)-3-hydroxy-propionic
(benzene ring with HO, para)	—COOH —CH$_2$COOH —CHOHCOOH	*p*-Hydroxybenzoic *p*-Hydroxyphenylacetic *p*-Hydroxymandelic
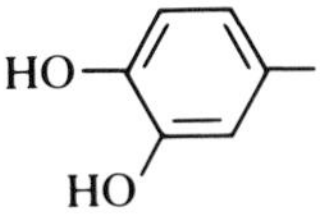 (benzene ring with two HO)	—CHOHCOOH	3,4-Dihydroxymandelic
(benzene ring with HO and CH$_3$O)	—COOH —CH$_2$COOH —CHOHCOOH	Vanillic Homovanillic Vanilmandelic
(indole ring, R—, CH$_3$, NH)	—CH$_2$COOH —CH$_2$COOH	5-Hydroxyindole-3-acetic (R = OH) Indole-3-acetic (R = H)
(furan ring with HOOC)	—H —CH$_2$OH —COOH	2-Furoic, occurs as glycine conjugate 5-Hydroxymethyl-2-furoic 2,5-Furandicarboxylic

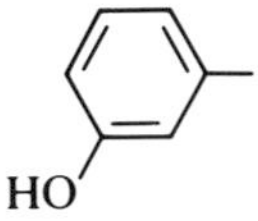

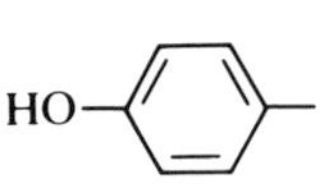

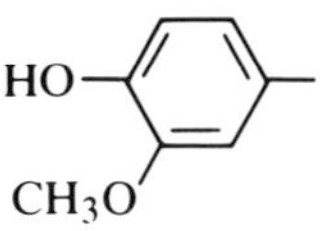

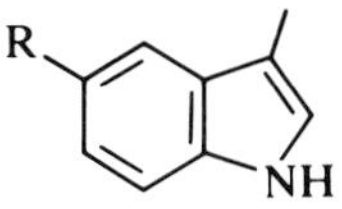

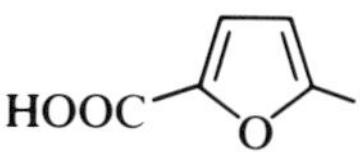

Keto acids

formula X—CO—Y

—X	—Y	Acid
—CH$_3$—	—CH$_2$CH$_2$COOH	Laevulinic
—COOH	—CH$_2$CH$_2$COOH	2-Oxoglutaric
—CH$_2$COOH	—CH$_2$COOH	3-Oxoglutaric
—COOH	—(CH$_2$)$_5$CH$_3$	2-Oxo-octanoic

Aliphatic monohydroxyacids

formula $\underset{Y}{\overset{X}{>}}C\underset{Z}{\overset{OH}{<}}$

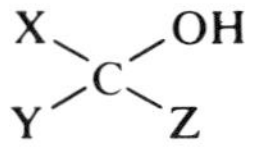

—X	—Y	—Z	Acid
—H	—H	—COOH	Glycollic
—CH$_3$	—H	—COOH	Lactic
—CH$_2$COOH	—H	—COOH	Malic
—CH$_2$CH$_2$COOH	—H	—COOH	2-Hydroxy-glutaric
—CH$_3$	—CH$_3$	—COOH	2-Hydroxyiso-butyric
—CH$_3$	—H	—CH$_2$COOH	3-Hydroxy-butyric
—CH$_3$	—CH$_3$	—CH$_2$COOH	3-Hydroxyiso-valeric
—CH$_3$	—CH$_2$COOH	—CH$_2$COOH	3-Hydroxy-3-methylglutaric
—COOH	—CH$_2$COOH	—CH$_2$COOH	Citric
—H	—H	CH$_3$CH$_2$CHCOOH (with side mark)	2-Ethylhydracrylic (Mamer and Tjoa, 1974)

Aliphatic polyhydroxyacids

formula R(CH$_2$OH)$_n$COOH

R	n	Acid
—CH$_2$OH	1	Glyceric (also 2-methylglyceric acid occurs)
—COOH	2	Tartaric
—CH$_2$OH	2	Tetronic
—CH$_2$OH	3	Pentonic
—CH$_2$OH	4	Hexonic
—CHO	4	Hexuronic

The diastereomeric configurations of the acids present in urine have, for the most part, not been elucidated. Both *erythro*- and *threo*-tetronic acids are excreted. Of the C$_6$ acids, gluconic and glucuronic acids occur.

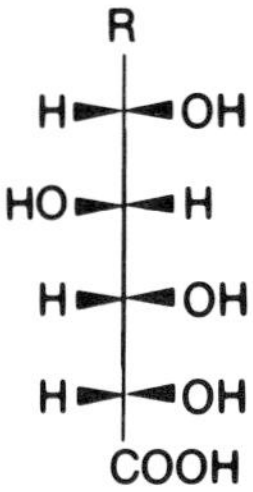

R
H——OH
HO——H
H——OH
H——OH
COOH

Gluconic acid, R=COOH
Glucuronic acid, R=CHO

2-, 3- and 4-deoxytetronic, 2-deoxypentonic and a deoxyhexonic acid are also present. Their formulae are obtained by replacing OH by H on the appropriately numbered carbon atom,. atoms being numbered from the —COOH terminus.

This listing is far from being exhaustive. Many of these compounds are of exogenous origin. For example, many of the aromatic acids are derived from dietary plant substances (deEds, 1968) and the furan derivatives may be removed from human urine by following a restricted diet (Pettersen and Jellum, 1972). Others are, at least partly, of endogenous origin.

In addition to phenolic acids, free **phenols** also occur in substantial quantities in urine, up to 0.5 g being excreted daily. These include

Phenol

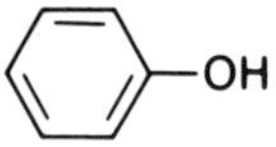

p-Cresol

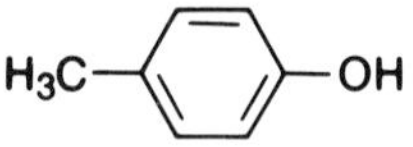

Pyrocatechol

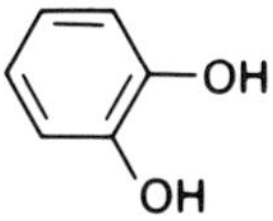

Resorcinol

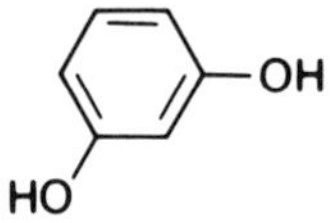

Studies on such compounds in other mammals are limited. Pregnant mare's urine was of commercial interest at one time as a source of steroid hormones and early studies revealed many of the acids and phenols already mentioned as well as numerous C_{13} alcohols and ketones presumed to be derived from herbage which the animals had consumed (Prelog *et al.*, 1948; Lederer, 1949, 1950). The following are among the components noted in pregnant mare's urine.

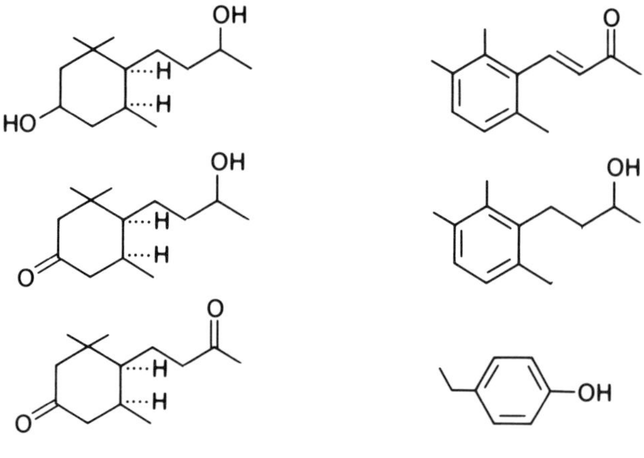

OH

HO

OH

O

O

OH

OH

O

(also in goat urine)

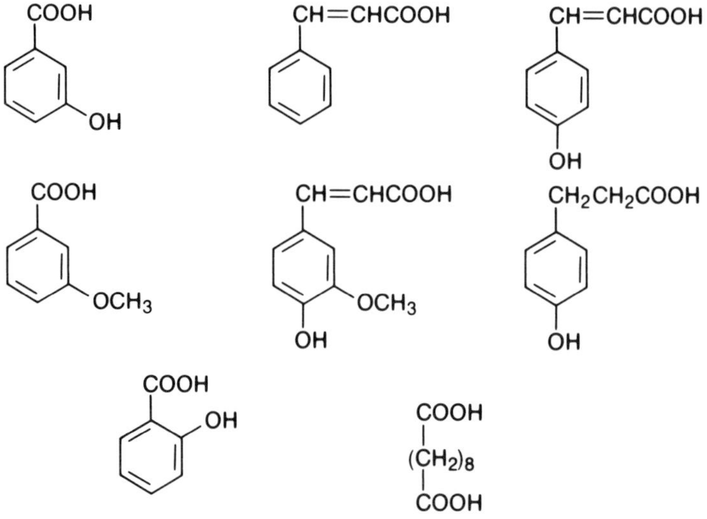

COOH

OH

CH=CHCOOH

CH=CHCOOH

OH

COOH

OCH$_3$

CH=CHCOOH

OCH$_3$

OH

CH$_2$CH$_2$COOH

OH

COOH

OH

COOH

(CH$_2$)$_8$

COOH

A more recent gas chromatographic survey of the oestrogens present in pregnant mare's urine has been published (Zweig *et al.*, 1980).

A preliminary study of coyote urine (Murphy *et al.*, 1978) revealed the presence of C$_2$–C$_{10}$ volatile fatty acids, aromatic acids (benzoic, phenylacetic, 3-phenylpropionic and anthranilic), dimethylpyrazine and acetophenone.

Cow urine has been extracted (Suemitsu *et al.*, 1974) and organic acidic, phenolic and neutral fractions obtained. Complex gas chromatograms of the free acid fractions were dominated by peaks corresponding to benzoic and phenylacetic acids. GC retention time suggested the presence of the following aromatic acids.

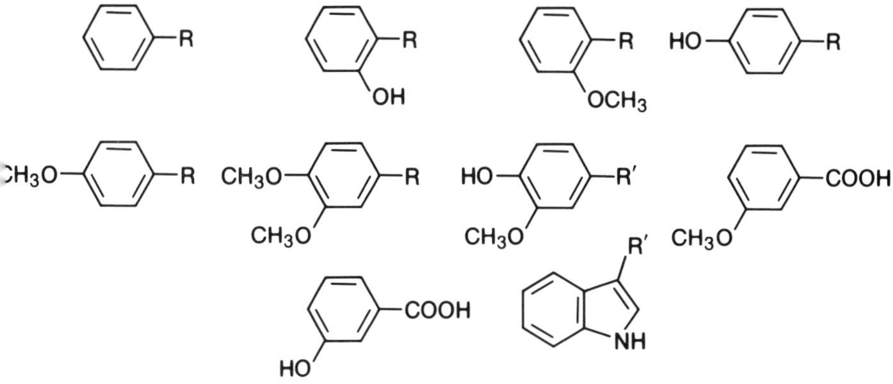

where $R = -CH_2CH_2COOH$ or $-CH_2COOH$ or $-COOH$ and $R^1 = -CH_2CH_2COOH$ or $-CH_2COOH$.

Conjugates of benzoic and phenylacetic acids with glycine (hippuric and phenaceturic acids) also occurred in large quantities. Subsequently small quantities of cyclohexanecarboxylic acid, pentanoic acid and 2-methylbutanoic acid (principally as glycine conjugates) were noted. The principal phenol was p-cresol, although other phenols included guaiacol, phenol and p-ethylphenol. Volatile fatty acids have also been noted in cow urine (Hradecký, 1978).

The odorous, stagnant urine which accumulates in the preputial diverticulum of the male pig is rich in phenols, ammonia and volatile fatty acids (see Section 8.2).

Among related odorants of urinary origin for which a chemical communicatory role has been advanced is the γ-lactone, (Z)-6-dodecen-4-olide (Müller-Schwarze, 1977).

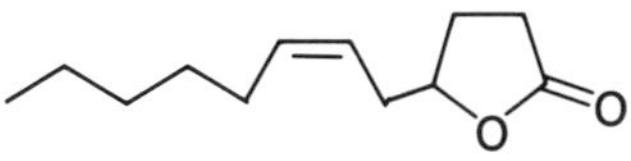

This substance has been observed in the secretion associated with the tarsal scent gland of the black-tailed deer, *Odocoileus hemionus columbianus*, and appears to be involved in social signalling. Until recently, this lactone was thought to be a product of the gland itself, but it is now known to be a urine component which is transferred to the tarsal region by a rub-urination behaviour pattern in which the area is soaked in urine and this compound extracted from the urine into the lipid which coats the specialized tarsal hairs. Such a urine-derived extract will contain other urine lipids also.

As already mentioned, similar transfer and relocation of urinary scents occur about the bodies of other mammals also. For example, the rutting odour of the red deer stag (*Cervus elaphus*) emanates from an area of belly hair which it sprays with urine during the breeding season. A preliminary study (Albone, unpublished results) has revealed the presence of numerous substances in an extract of this hair which are probably urine-derived, including phenol, p-cresol, ethylphenol, benzoic acid, phenylpropionic acid and possibly cyclohexanecarboxylic acid. There is also a number of fluorescent lipid components. A more detailed study is clearly required. In this and other similar cases, it might be simpler to conduct chemical studies on aged urine samples from the animals in question rather than on extracts of hair clippings. A similar mixture of phenols and carboxylic acids, including cyclohexanecarboxylic acid, has recently been reported in the 'strong smelling tar-like mixture' which coats the underside of the tail in male and female red deer (Bakke and Figenschou, 1983).

Two **oximes**, *syn*- and *anti*-phenylacetaldehyde oxime, substances previously unknown in mammals, have been identified in guinea pig urine (Smith *et al.*, 1977).

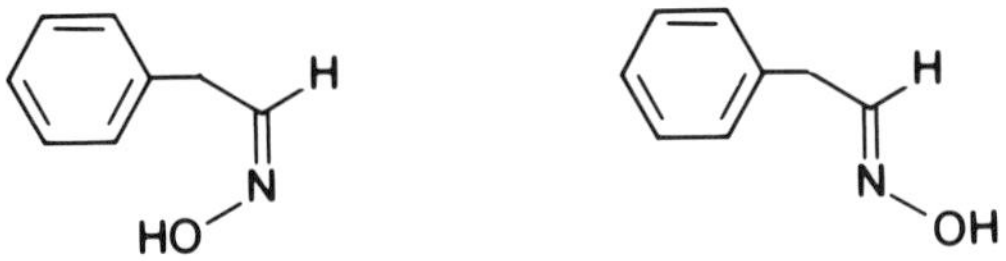

Their excretion appears to be testosterone dependent; they occur in the urine of the intact adult male, become absent following castration, and reappear if castrates are treated with testosterone. Their presence in adult male guinea pig urine also correlates with the aversion which other adult male guinea pigs show to this urine, as compared with their attraction to the urines of females and immature animals. However, behavioural tests on the pure oximes show that these compounds are themselves not responsible for these effects.

Amino acids

Free amino acids are excreted in normal human urine at levels averaging 1.1 g/24 h. Of this, approximately 70% is contributed by just four amino acids, glycine, taurine, histidine and methylhistidine (Stein, 1953). The levels of the last of two of these are increased by a meat diet, and so these amino acids tend to be elevated in the urines of carnivores (Evered, 1956). Two isomers of methylhistidine are present in human urine. Acid hydrolysis shows that amino acids (2 g/24 h), principally glycine, glutamic acid and aspartic acid, are also excreted in conjugated form. For example, 1 gram of hippuric acid (benzoylglycine) may be excreted daily, depending on diet. The level of free glutamic acid in fresh urine is low, but increases on standing, presumably following the decomposition of labile conjugates. The excretion of amino acids in the form of protein in human urine is of minor importance as total urinary protein levels are normally of the order of 70 mg/24 h (McGarry *et al.*, 1955). Higher levels may, however, occur in other species' urines.

A statistical analysis of variations in urinary amino acid excretion patterns in healthy human urine has revealed a strong association of pattern with sex, peaks due to glutamic acid and to arginine having a tendency to be elevated in urine from women, as compared with men, and peaks due to histidine (plus methylhistidine), to taurine and to tyrosine (plus phenyalanine and homocysteine) being elevated in men (Dirren *et al.*, 1975). Another study has revealed an association of urinary free amino acid profile with sex, although here glycine appeared to be the most important discriminator (elevated in women relative to men) (Liappis, 1973). An increase in free amino acid excretion in urine (especially threonine and histidine) has been observed in pregnancy urine (Miller *et al.*, 1954).

Additionally, urine also contains a number of uncommon or unidentified free amino acids. In the aliphatic basic amino acid fraction, for example, in addition to lysine, ornithine and arginine, peaks were observed corresponding to a series of *N*-methylated derivatives of arginine and lysine and some glycosylated derivatives of 5-hydroxylysine (Kakimoto and Akazawa, 1970). These substances appear to be endogenous, probably arising from body protein and collagen. Pyroglutamic acid also occurs in urine.

	Mean excretion (mg/g creatinine) (Kakimoto and Akazawa, 1970)
Arginine, and derivatives	
$HOOC \diagdown \atop NH_2 \diagup CH(CH_2)_3NH-\overset{\overset{NH}{\|}}{C}-NH_2$	5.4
$-\overset{\overset{NH}{\|}}{C}-N(CH_3)_2$	8.3
$-\overset{\overset{NCH_3}{\|}}{C}-NHCH_3$	8.3
Lysine, and derivatives	
$HOOC \diagdown \atop NH_2 \diagup CH(CH_2)_4 \quad -NH_2$	20.3
$-NHCH_3$	1.3
$-N(CH_3)_2$	5.2
$-N(CH_3)_3$	8.8
5-Hydroxylysine, and derivatives	
$HOOC \diagdown \atop NH_2 \diagup CH(CH_2)_2\overset{\overset{CH_2NH_2}{\|}}{C}-OH$	1.5
$-$galactosyl	3.9
$-$galactosyl-glucosyl	9.7

Another unusual amino acid (Yuasa, 1978) present in human urine hydrolysate at low levels (80 mg/100 l urine) is

$$HOOC-CHOH-CHOH-CH_2-CHNH_2-COOH$$

An example of major differences which may occur between urines of different species is provided by a study of the amino acid composition of **cat urine**. Here, by far the most abundant free amino acid is an unusual sulphur-containing substance which has been called felinine (cysteine-S-isopentanol). This is excreted at levels up to 190 mg/100 ml urine, and it seems to be produced in the kidney (Westall, 1953; Tallan *et al.*, 1954; Greaves and Scott, 1960).

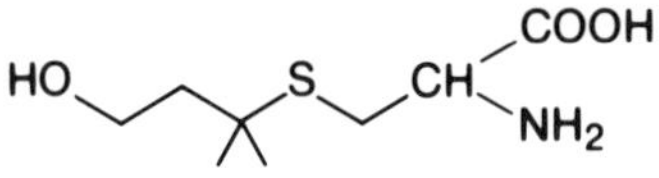

Felinine

Urine also contains compounds containing amino and carboxyl groups which are the biochemical precursors of the porphyrins. 5-Aminolaevulinic acid and the pyrrole, porphobilinogen, are excreted in human urine at levels below 2 mg/24 h. The related compound, hydroxyhaemopyrrolin-2-one, also occurs at much lower levels (0–264 μg/1) (Graham, 1978).

$$NH_2CH_2COCH_2CH_2COOH$$

5-Aminolaevulinic acid   Porphobilinogen

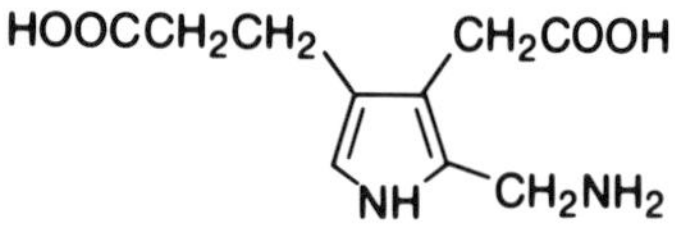

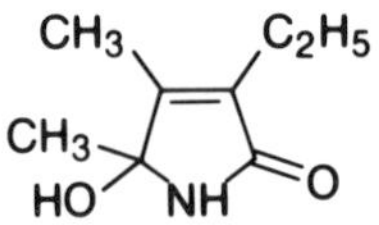

Hydroxyhaemopyrrolin-2-one

Amines

In addition to the amino acids, some 20–100 μg of amines are excreted daily in human urine. Many are biologically important substances of endogenous origin and their concentrations in urine provide evidence concerning the physiological state of the individual in question, extending even to indications of emotional state and mood. Although such substances could well have semiochemical significance, therefore, no studies have been conducted in this area. Even so, it might be useful to summarize something of what has been learnt of them in a medical context.

Frequently, the picture is clouded by the presence of similar, or identical, components of exogenous (dietary) origin. Studies undertaken in the early 1960s revealed the presence in human urine of nearly 50 different amines, in conjugated and/or free states (Perry and Schroeder, 1963). These included

histamine	metanephrine
*N-acetylhistamine	normetanephrine
l-methylhistamine	*synephrine
p-tyramine	octopamine
m-tyramine	serotonin
3-methoxytyramine (3-O-methyldopamine)	tryptamine
*p-hydroxybenzylamine	pyrrolidine
*3-methoxy-4-hydroxybenzylamine	methylamine
putrescine	dimethylamine
cadaverine	ethylamine
2,2'-dithiobis(ethylamine)	ethanolamine
piperidine	2-hydroxypropylamine

Compounds marked * are of dietary origin. Synephrine excretion, for example, follows the eating of oranges (Perry et al., 1965), while substantial quantities of serotonin occur in bananas (Udenfriend et al., 1959) the consumption of which can lead to misleadingly high levels of serotonin metabolites in urine.

Phenylethylamines

Much attention has been given to the presence in urine of biologically active amines, such as the catecholamines and other substances chemically related to 2-phenylethylamine, but such compounds are widely distributed in plants (Smith, 1977).

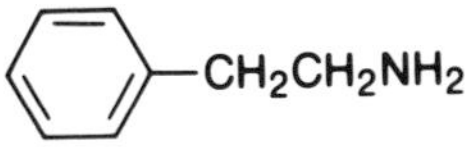

$-CH_2CH_2NH_2$ 2-Phenylethylamine

The situation is complicated, because, until recently, published determinations in urine of some of these compounds, such as those of 2-phenylethylamine itself, have varied widely often leading to greatly overestimated levels. This is probably because of the lack of sufficient compound specificity in the analytical methods employed.

This apart, it remains possible to monitor variations in aspects of the physiology of a mammal from its urine chemistry. Endogenous 2-phenylethylamine and related compounds have attracted research interest because of the involvement of such compounds in central nervous system function. 2-Phenylethylamine occurs in the brain and throughout other body tissues (Durden et al., 1973). Urinary levels of this amine in the free state are influenced by psychiatric conditions, being elevated in mania and schizophrenia and lowered in endogenous depression (Fischer et al., 1972; Slingsby and Boulton, 1976). Tryptamine and p-tyramine also occur in the brain and urinary levels of these free amines may be similarly enhanced in certain

Table 6.5. 2-Phenylethylamines in urine

Levels in human (and rat) urine in μg/24 h (where marked *, in μg/g creatinine)

CH₂CH₂NHR structure with positions X, Y, Z

	Structure				Form		Total
	X	Y	Z	R	free	conjugated	
2-Phenylethylamine	H	H	H	H	47 ± 44[a] 4.9 ± 1.0[b] 6.8 ± 2.9[c]	34–365[a]	
N-Methyl-2-phenylethylamine	H	H	H	CH₃	Trace	20.9 ± 26.0[c]	
p-Tyramine	OH	H	H	H	489 ± 40[b] 312 ± 162*[a] (rat, 6.09 ± 0.66)[f]	0.2–1.0[d] 0–3640*[a]	1000[e] (rat, 200)[e]
m-Tyramine	H	OH	H	H	83 ± 7[b] (rat, 3.69 ± 0.61)[f]		20[e] (rat, 5)[e]
o-Tyramine	H	H	OH	H			<0.5[e]
Dopamine	OH	OH	H	H	207, 292[g] 180 ± 40*[h]		(rat, 1)[e]

[a] Boulton and Milward, 1971.
[b] Slingsby and Boulton, 1976.
[c] Reynolds and Gray, 1976.
[d] Reynolds and Gray 1978.
[e] King et al., 1974.
[f] Boulton and Dyck, 1974.
[g] Christensen, 1973.
[h] Naftchi et al., 1975.

psychiatric states. Urinary free tyramine levels are said to correlate with cerebral electrical activity, as measured by surface electrodes. In its conjugated form, 2-phenylethylamine occurs in human urine principally as the glucuronide (Inwang *et al.*, 1973). *N*-Methyl-2-phenylethylamine has been detected in human urine, principally as conjugate (Reynolds and Gray, 1978) and 2-hydroxy-2-phenylethylamine, a metabolite of 2-phenylethylamine which has been detected in the brain, also occurs in urine.

In many cases, the urine levels of the conjugated amines are considerably more variable than those of the corresponding free amines, although the uncertainty of some of the earlier measurements previously referred to must be borne in mind. In a number of cases, e.g. tyramine (Boulton and Marjerrison, 1972) and 2-phenylethylamine (Reynolds *et al.*, 1978), evidence suggests that the conjugated amines arise, at least partially, from dietary sources while the free amines are endogenous. Although *p*-tyramine may be formed from the amino acid, tyrosine, *m*- and *p*-tyramine also arise endogenously from 2-phenylethylamine or from dopamine (Boulton and Dyck, 1974).

Catecholamines and their metabolites

Compounds such as dopamine, in which the benzene ring contains two OH groups positioned *ortho* to each other, are termed catecholamines, because of their structural resemblance to catechol,

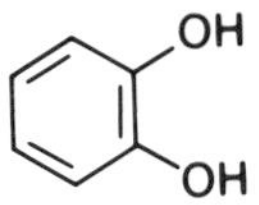

Dopamine occurs in urine in considerable quantities, both in free and conjugated forms. Dopamine is an intermediate in the biosynthesis of noradrenaline from *p*-tyrosine and also functions as a neurotransmitter in certain brain neurons. Dopamine sulphate is formed from dopamine in such tissues as the brain, and is excreted in part in this form (Bronaugh *et al.*, 1972). The major metabolite of dopamine, however, is homovanillic acid, and this, together with iso-homovanillic acid has been detected in the brain, cerebrospinal and various tissues as well as in urine (Dziedzic *et al.*, 1973). The kidney itself appears to make a major contribution to urinary free dopamine levels (Christensen, 1973).

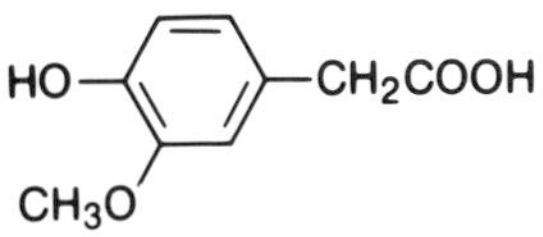

HVA (homovanillic acid)
Urine levels, mg/g creatinine
mean (range) 3.72 (1.70–6.26)
(Dziedzic *et al.*, 1973).

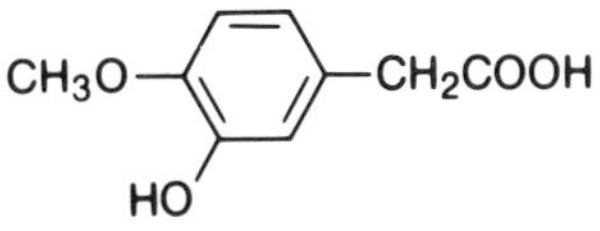

iso-HVA (iso-homovanillic acid)

0.28 (0.022–0.966)

Greater attention has, however, been given to investigations of the urinary concentrations of the catecholamines, **epinephrine** (adrenaline) and **norepinephrine** (noradrenaline), which contain an hydroxylated side chain, even though these substances occur at lower levels than dopamine. This is because of correlations which exist between the amounts of these substances present in urine and physiological state.

Epinephrine, R=CH₃

Norepinephrine, R = H

The presence of the hydroxy group at the 2-position introduces chirality into the molecule, the naturally occurring, biologically active forms of these substances being the (R)- enantiomers. Norepinephrine is a neurotransmitter substance in the sympathetic nervous system which controls glandular activity and involuntary muscle functions, while epinephrine is a hormone secreted by medulla of the adrenal glands. Epinephrine is biosynthesized from norepinephrine which in turn is produced from dopamine.

In man, normal urine levels of norepinephrine and epinephrine are of the order of 30 μg/24 h and 5 μg/24 h respectively, although great variations occur between individuals and these may be related for example to individual differences in renal function. Physical excercise also increases excretion rates. Seasonal and diurnal variations in excretion have been observed in various mammals and factors such as body weight and strain influence excretion rate differences between individuals within a species.

The sympathetic and adrenal medullary systems are closely linked and are highly sensitive to emotional and environmental stress factors, and together constitute a mechanism by which the physiology of an animal is rapidly prepared for action. The activity of this system is reflected to some extent in urine chemistry. Circulating hormone levels can change very rapidly and this adds to the difficulties of chemical studies. In man, the turnover time for these catecholamines, i.e. the time required to completely replace the circulating hormones from endogenous supply, has been calculated as 30 seconds (Cohen et al., 1959). One adrenal-related effect which is revealed in urine chemistry and which has been known for many years is that of 'emotional glycosuria' (Cannon et al., 1911). Here, for example, high urine sugar levels are observed in the urine of a cat following the stress of being held immobile and threatened by a barking dog. More recently, particular attention has centred on the relationships between the levels of catecholamines in urine and various stress situations. Two- to threefold increases in urinary catecholamine levels are commonly encountered as the result of psychological stress in man. Higher levels occur mainly in the urines of severely disturbed psychiatric patients.

In a healthy person, increased urinary catecholamine levels reflect increased catecholamine metabolism associated with such emotions as fear, anxiety, anger and sexual arousal, the intensity of the response being related to the catecholamine excretion rate, although not necessarily in a linear fashion. Changes in excretion have been documented in relation to differential responses to films eliciting various emotional reactions, and to the psychological stress experienced, for example, in anticipation of various demanding, competitive ordeals. The subject is one of considerable complexity (Levi, 1972). It has not been possible to distinguish simply between emotional states which raise urinary epinephrine levels and those which raise norepinephrine levels. Studies involving film viewing suggested that while bland, uninteresting films tended to decrease urinary norepinephrine and epinephrine levels, aggression-evoking films and amusing films both increased epinephrine levels alone while frightening films and sex films raised levels of both catecholamines. All arousing films led also to changes in renal function leading to increased urine volume and decreased urine density (psychogenic diuresis).

In experiments on the effects of prolonged restraint on rhesus monkeys using special restraining chairs, urinary epinephrine levels increased threefold for the first 3 days of restraint, remained elevated for the first week and thereafter slowly returned to normal levels. Urinary norepinephrine levels, in contrast, showed no statistically significant changes throughout (Mason et al., 1973). One hypothesis advanced in the 1950s to explain the differences in elimination of the two catecholamines was that active, aggressive states tend to increase levels of norepinephrine in urine, while tense, anxious states tend to increase epinephrine (Mason, 1968b). Clearly such distinctions are too simple.

Although much attention has focused on urinary epinephrine and norepinephrine, the quantities excreted in urine represent only a small percentage of the metabolic turnover of these compounds. More significant in terms of mass are their metabolites, and these have been studied in relation to the need to detect certain classes of catecholamine-secreting tumours. Radioisotope experiments (Sharman, 1973) show that only 2–7% of epinephrine is excreted in human urine unchanged following intravenous administration, while the major mode of excretion is as metanephrine and vanilmandelic acid (VMA). Some 3-methoxy-4-hydroxyphenylethyleneglycol (MHPG) is also excreted. Norepinephrine similarly yields normetanephrine, VMA and MHPG. A number of minor metabolites are also produced, and sulphate and glucuronide conjugates are excreted as well as the free metabolites.

Norepinephrine differs from epinephrine in being involved in brain function. The brain pool of norepinephrine is isolated from the larger somatic pool of this compound by the blood–brain barrier. However, it seems possible to gain information specifically about brain norepinephrine turnover from a study of urine chemistry. It was shown from studies on dogs in which ^{14}C-norepinephrine was infused into the jugular vein while ^{3}H-norepinephrine was injected into the brain cavity that brain norepinephrine metabolism has a

Table 6.6. Urinary metabolites of epinephrine and norepinephrine

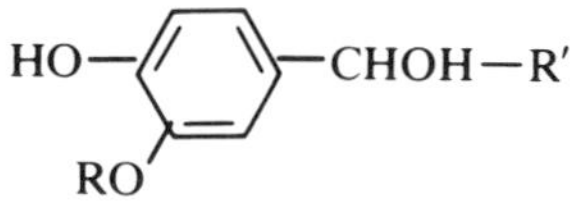

$$HO-\langle\!\!\!\!\!\bigcirc\!\!\!\!\!\rangle-CHOH-R'$$
$$RO$$

Metabolic relationships	Formula R	R'	Typical excretion level (μg/24 h)
Epinephrine[a]	H	CH_2NHCH_3	5
Metanephrine[b]	CH_3	CH_2NHCH_3	240
VMA[c]	CH_3	COOH	3300
MHPG[d]	CH_3	CH_2OH	1500
Normetanephrine[b]	CH_3	CH_2NH_2	300
Norepinephrine[a]	H	CH_2NH_2	30

[a] Mason, 1968b.
[b] Winkel and Slob, 1973.
[c] Stott et al., 1975.
[d] Dekirmenjian and Maas, 1970.

disproportionately large effect on urinary MHPG levels. Thus, some 50–60% of brain norepinephrine metabolized is excreted as MHPG, and while 22–27% of urinary MHPG is of brain norepinephrine origin, less than 1% of urinary VMA is so derived. In the dog, 97% MHPG excreted was conjugated, while 95% VMA was free. MHPG sulphate is formed in the brain in a number of species as a major metabolite both of norepinephrine and of normetanephrine.

Studies on norepinephrine in man suggest that urinary norepinephrine and normetanephrine are largely derived directly from circulating norepinephrine released from sympathetic nerve endings while important sources of norepinephrine-derived VMA and MHPG are the adrenal gland and brain pools of this amine respectively (Maas and Landis, 1971).

In studies on trainee naval air pilots undergoing the demanding and hazardous exercise of practising night landings on an aircraft carrier, it was found that urinary MHPG levels during flight training were elevated 60% relative to levels in urine samples taken over similar periods on non-training days (Rubin et al., 1970). This seems to be associated with the extreme alertness and concentration required. Urinary corticosteroid levels also increased, but did not mirror the elevated MHPG levels in all respects and probably reflected more closely the anxiety stress factor. The interpretation of urinary MHPG levels is, however, complex. Levels are increased in a variety of environmental stress situations and following physical activity. There is also an association with certain psychiatric states although this may be little more than a reflection of increased physical activity also. As with VMA excretion, circadian rhythms of MHPG excretion have been described (Cymerman and Franesconi, 1975).

In addition to psychological stress, direct physiological response to the physical conditions of the environment can also result in changed urinary catecholamine excretion, both in man and in other mammals. A well studied example of this is the physiological reaction to cold exposure (Leduc, 1961). It was found, for example, that when laboratory rats of particular weight and strain which had been kept at 22 °C were transferred to a 3 °C environment, urinary norepinephrine levels increased with the first 24 hours of cold exposure from less than 4 μg/kg body weight/24 h by a factor of four or five. If exposure was maintained, urinary norepinephrine fell slowly to reach a stable but still elevated level at 2 months' exposure. The magnitude of the norepinephrine increase is related to the degree of cold stress because the norepinephrine release from the sympathetic nerve endings plays a central role in the physiological mechanisms by which normal body temperature is maintained on exposure to cold. Norepinephrine acts both to stimulate vaso-constriction, and so increase thermal insulation, and to stimulate biochemical metabolic processes leading to increased chemical heat production (non-shivering thermogenesis). In mammals in which catecholamine release is eliminated, cold-stress survival is greatly reduced.

It seems that epinephrine is of secondary importance to norepinephrine in this situation. Urinary dopamine levels also showed only a small percentage increase and returned to normal within about a month of continued cold exposure.

Norepinephrine metabolite excretion reflects cold stress too. Thus, it was shown that rats transferred from a 28 °C to a 4 °C environment exhibited initial changes in metabolite excretion as shown in Table 6.7.

Intermittent cold exposure induced a stronger stimulation of the adrenal glands in rats and both epinephrine and norepinephrine levels in urine increased and were maintained throughout such a period of cold stress (Bibbiani and Viola-Magni, 1971).

Similar effects of cold stress on catecholamine excretion have been noted in other mammals. An interesting example is provided by the lemmings, *Lemmus trimucronatus* and *Dicrostonyx groenlandicus*, the most northerly distributed small mammals in America. In the cold adaptation of these species, epinephrine plays a more significant part. Cold exposure at 3 °C of animals previously maintained at 18 °C led to urinary elevations of both catecholamines of four to six times their initial levels. In *Lemmus*, the

Table 6.7. Urinary norepinephrine metabolic excretion following cold stress in rats (Reproduced with permission from Shum *et al.*, 1971)

(μg/kg body wt./24 h)	28 °C	4 °C, day 1	4 °C, day 2
Norepinephrine	0.82	2.29	5.74
Normetanephrine	6.77	21.9	31.5
Epinephrine	0.17	0.20	0.59
Metanephrine	5.78	10.8	20.8
MHPG	187.9	530.7	411.4

epinephrine level remained elevated and did not return to normal during a period of continued exposure (Berberich *et al.*, 1977). Also, the basal urinary levels of epinephrine and norepinephrine (per kg body weight) in both species were higher by factors of 20 and 2–3 respectively than levels in temperate mammals (rat, hamster) previously studied. This could be related to the behavioural hyperactivity shown by these arctic species.

Indole compounds

Urine can provide evidence concerning the metabolism of the indole amine, serotonin. This amine is involved in the functioning of the central nervous system and also acts on involuntary (smooth) muscle. It is present in the brain, although some 90% of the body total occurs in cells lining the gastrointestinal tract. Information concerning serotonin turnover is provided by urine levels of its major metabolite, 5-hydroxyindole-3-acetic acid (5HIAA).

In man, urinary 5HIAA levels are normally around 2 mg/24 h, although elevations occur in a variety of circumstances such as in the third trimester of pregnancy. In the rat, stress caused by enforced immobilization leads to a doubling of urinary HIAA levels (Kundrotas and Gregg, 1977). The related compound, 5-methoxyindole-3-acetic acid, has also been detected in human urine (Hoskins *et al.*, 1978), but at much lower levels (7–150 μg/24 h). Its presence could be linked with the metabolism of *O*-methylserotonin deriva-

Table 6.8. Formulae of indole amines and metabolites

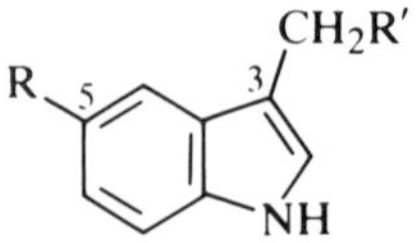

		R	R'
	Serotonin	OH	CH_2NH_2
	5HIAA	OH	COOH
	5-Methoxytryptamine	OCH_3	CH_2NH_2
	Melatonin	OCH_3	$CH_2NHCOCH_3$
	Tryptamine	H	CH_2NH_2
	Indole-3-acetic acid	H	CH_2COOH
	3-Methylindole (skatole)	H	CH_3

tives, such as 5-methoxytryptamine and N-acetyl-5-methoxytryptamine (melatonin) which occur in such organs as the hypothalamus and the pineal, although there is also likely to be a dietary contribution.

Tryptamine itself is excreted in human urine at levels of about 100 μg/24 h free amine, and up to 100 μg/24 h conjugated amine (Boulton and Milward, 1971; Slingsby and Boulton, 1976).

Indole-3-acetic acid also occurs in urine, typical levels in man being 5.8 ± 2.9 mg/24 h free acid, 11.2 ± 3.7 mg/24 h total acid (Austad et al., 1967). Although this substance has plant growth regulator effects, for the mammal it is but a simple metabolite of the amino acid, tryptophan. The urinary acid arises from two sources, direct mammalian metabolism and resorbed products of gut bacterial metabolic products. Indole-3-propionic acid is also excreted. However, the major indole of human urine, even if one of the least interesting, is indoxyl sulphate (indican), which one study showed to be excreted at levels of 77.8 ± 21.7 mg/24 h (Rylance, 1969). This substance arises from indole produced from tryptophan by gut bacteria.

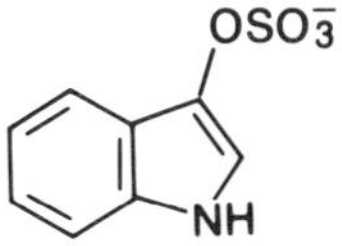

Indoxyl sulphate

In a relatively crude paper chromatographic analysis of the indoles present in the urines of 22 different species of primate (Smith and Lerner, 1971), this was the only indole that could be detected in human urine. However, this study did reveal major gross differences in urinary indole composition be‑ tween species, although the compounds involved were largely unidentified. Hydroxyskatole sulphate was important in some species. This component does occur in human urine although not at levels sufficient to be revealed in that investigation. It has, in fact, been shown to consist of at least three isomers, namely the sulphates of 5-hydroxy-, 6-hydroxy- and 7-hydroxy-3-methylindole (Mahon and Mattok, 1967). Skatole (3-methyl-indole) is formed in the gut also.

Free 5-hydroxy-3-methylindole has been detected in normal human urine at levels of 34.6 ± 26.9 μg/24 h (Mori et al., 1978). Conjugated oxindole also occurs in human urine (Voeltler et al., 1971).

The polyamines and histamine

The polyamines, spermidine and spermine, together with their precursor, putrescine, occur in healthy human urine at levels of 1.2 ± 0.18, 0.04 ± 0.007 and 2.1 ± 0.62 mg/g creatinine in bulked 24 hour samples (Russell, 1977). Although present in serum in their free form, spermidine and putrescine occur in urine principally as their acetyl derivatives. The presence of the

206

related diamine, cadaverine, in a urine sample suggests bacterial contamina-
tion as this substance is not found in mammals to any significant extent.
Putrescine levels may also be elevated by bacteria.

$NH_2(CH_2)_3NH(CH_2)_4NH(CH_2)_3NH_2$ Spermine
$NH_2(CH_2)_3NH(CH_2)_4NH_2$ Spermidine
$NH_2(CH_2)_4NH_2$ Putrescine
$NH_2(CH_2)_5NH_2$ Cadaverine

High polyamine levels are associated with tissues undergoing high cell proliferation and growth and much medical interest has been given to investigating
the possibility of using elevated urinary polyamine levels as indicators of
cancer (Jänne *et al.*, 1977).

Urinary putrescine levels have another context more central to our present
concerns, for they illustrate how gonodal hormones may influence kidney
function and thus modify urine composition indirectly. These observations
are based on experiments with laboratory mice (Henningsson and Rosengren,
1975). Although the magnitudes of the effects were dependent on the strain
of mouse used, it was generally observed that the level of free putrescine
excreted in urine by intact male mice ($\sim$90 μg/24 h) was three-fold greater
than that excreted by females of the same strain, while free histamine levels
were higher in the female, in one strain by a factor of ten ($\sim$20 μg/24 h), than
in the male. Endogenous histamine excretion in the rat is known to be similarly elevated in the female. The effects of hormone administration experiments were in accord with these observations. Urinary free putrescine
increased while free histamine decreased on treating female mice with testosterone. Histamine excretion increased with oestradiol. Progesterone had no
such effects.

In intact female mice, the consequences for free histamine excretion of
treatment with oestradiol or progesterone were similar to those noted for
ovariectomized animals, while administration of testosterone (0.5 mg daily
for 10 days) led to reductions of up to 90% in urinary free histamine levels
and tenfold increases in urinary free putrescine. Urinary 1-methylhistamine,
a catabolic product derived from histamine in the mouse, is also reduced
slightly as the result of testosterone treatment. Putrescine returned to normal
rapidly following the termination of testosterone treatment (contrast the situation illustrated in Figure 6.5).

Enzyme assays (histidine decarboxylase for histamine production and
ornithine decarboxylase for putrescine production) showed that this hormone
dependency was located specifically in the kidney and that the activities of
these enzymes in other tissues were not hormone dependent. Other examples
of sex-hormone-dependent kidney enzyme activity are known. In this case, a
consideration of kidney enzyme activity yields a clearer picture than the
results of urine analysis as urine amine composition reflects the contributions
not only of kidney products but of substances formed in other tissues also.

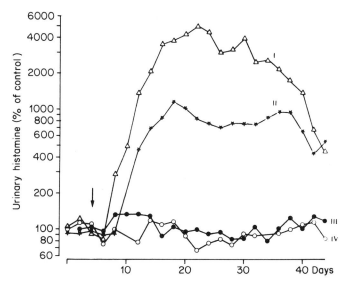

Figure 6.5 Urinary excretion of free histamine in ovariectomized mice as a percentage of the value (1 to 2 $\mu g/24$ h) on day 1 of the experiment. Group I injected daily with 10 μg oestradiol for 10 days. Group II injected with 100 μg oestradiol for 2 days. Group III injected with the oil solvent used to dissolve the various steroids. Group IV injected with 1 mg progesterone daily for 10 days. Injections commenced at the point indicated by the arrow. (Henningsson and Rosengren, 1972, reproduced by permission of The Macmillan Press Ltd.)

Thus is was found that the putrescine-forming capacity of mouse kidney tissue was low in females throughout life, and low also in males until 18–19 days old when ornithine decarboxylase activities increased 100-fold over 3 weeks. In contrast, histamine-forming capacity was high in both sexes at birth, possibly as the result of the action of placental oestrogens, and thereafter decreased throughout life except in the case of the female where a several-fold increase occurred at adolescence (Henningsson and Rosengren, 1975).

$$CH_2CH_2NH_2$$

Histamine, R = H
1-Methylhistamine, R = CH_3

Urine pigments

The amber-yellow pigment of urine, nebulously termed 'urochrome', is a mixture of largely unknown composition. Preliminary experiments indicate

Table 6.9. Urinary pigments associated with porphyrin biochemistry

	A[(1)]	B[(1)]	D[(1)]	Human urine (mg/24 h)
Coproporphyrin I	CH_3	CH_3	CH_2CH_2COOH	0.05–0.23
Coproporphyrin III	CH_3	CH_2CH_2COOH	CH_3	0.01–0.05
Uroporphyrin I	CH_2COOH	CH_2COOH	CH_2CH_2COOH	0.01–0.03
Stercobilinogen				0.6 (0–4.0)

(1) Structures given by substituting in

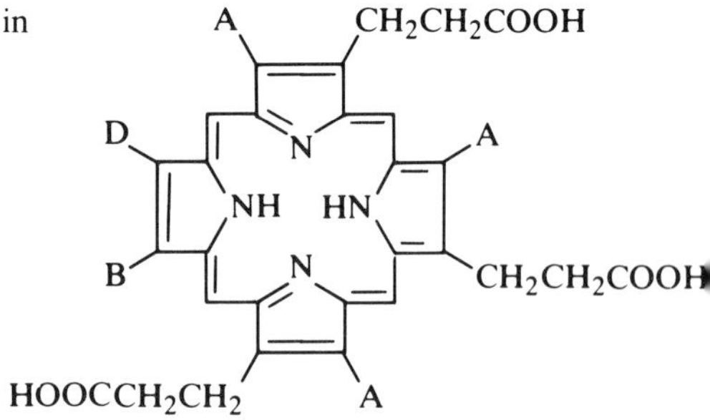

The corresponding porphyrinogens consist of the same four pyrrole rings linked by four CH_2 groups in place of the four CH groups shown here.

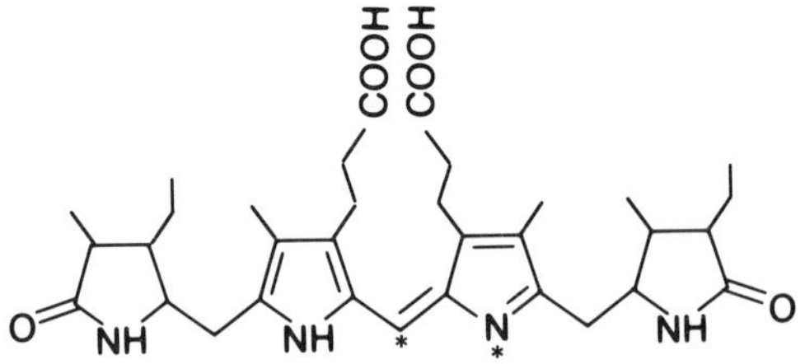

Stercobilin

Stercobilinogen has the structure of stercobilin with two additional hydrogen atom at the positions marked *, together with appropriate changes in the positions of the double bonds. Stercobilinogen is colourless, but is air oxidized to yellow-orange stercobilin.

that amino acids and glucuronic acid are associated with this material (Hanson, 1969). Among the constituents of urine pigment are a number associated with porphyrin biochemistry (Adler, 1975). These are listed in Table 6.9 Porphyrins are conjugated, cyclic tetrapyrroles of the type which include haem. In the biosynthesis of haem, although uroporphyrinogens and coproporphyrinongens of both series (I and III) are formed, only compounds of the III series are utilized, those of series I being excreted. The porphyrinogens are very readily oxidized to the corresponding coloured porphyrins, and are detected in urine in this form. Uro- and coproporphyrins are also excreted in faeces (0.3–1.0 mg/24 h). Urine also contains substances derived from the breakdown of haem. Each day, a human being degrades 1% of the circulating 750 g of haemoglobin in the liver with the formation of biliverdin (green) and subsequently of bilirubin (red-yellow), the major pigment of bile. This bilirubin is excreted as the glucuronide into the duodenum. Subsequently, principally in the colon, bacterial action transforms it into stercobilinogen and related compounds, 40–280 mg of which are excreted daily in faeces. A portion of this material is resorbed back into the blood stream from the gut and appears in urine. On air oxidation, the yellow-orange pigment, stercobilin forms. This is the principal pigment of faeces.

Urine is also fluorescent under ultraviolet light, a phenomenon which has been used elegantly to study patterns of urine marking in caged mice (Desjardins et al., 1973). The fluorescent components are many and abnormalities in composition (e.g. red fluorescence due to high porphyrin content) are indicators of disease. Healthy human urine exhibits a light blue-green fluorescence (Young, 1973). Although a number of urine components are known to fluoresce, as in the case of the pigments, the major contributors to the observed fluorescence have yet to be defined.

Several components play an important part. 1% Urine in buffer (pH 7) fluoresces maximally at wavelength 373 nm on irradiation with 285 nm ultraviolet light and at 404 nm on excitation at 325 nm. The intensities of both bands are highly dependent on pH, being maximal at pH 10 and 14 respectively. In contrast, undiluted urine (pH 6) shows negligible fluorescence on excitation at 285 nm, probably as the result of inner filter and concentration quenching effects. But on increasing the excitation wavelength from 325 to 400 nm, a series of fluorescence spectra are observed having maxima ranging from 426 to 454 nm (with a shoulder at 520 nm), the most intense occurring with a maximum at 438 nm (excitation, 370 nm). These effects appear to be largely insensitive to dietary variation.

Chapter 7
Secretions of the Reproductive Tract

Scents and secretions associated with the reproductive tract are likely to have great semiochemical importance in many species, particularly although not necessarily exclusively in the area of sexual behaviour and reproduction (Cowley, 1978; Keverne, 1978). An understanding of the nature of such urogenital chemical signals necessarily involves a consideration of the various accessory sex organs as likely sources of these semiochemicals. Valuable background information concerning mammalian reproduction is given in texts edited by Hafez (1968, 1980) and Hafez and Evans (1978).

7.1 MALE SECRETIONS

The male accessory sex organs themselves

The accessory sex organs are well described elsewhere (Eckstein and Zuckermann, 1956; Mann, 1964). In the male, their secretions comprise the bulk of the semen, while in the female homologous secretory structures also occur. Considerable differences exist between species with regard both to the patterns of occurrence and development of the various secretory accessory sex structures and to the compositions of their composite products, which in the males are the seminal plasmas of the species concerned. In the male, the following are among the secretory structures which occur.

(1) **The prostate**, present in all mammals, save the monotremes. This is a complex structure involving more than one type of secretory tissue. In rodents, bats and certain catarrhine monkeys a part of the prostate (**the coagulating gland**) provides enzymes which are responsible for the formation of the vaginal plug following coitus. In some species (mustelids, dog, bear) the prostate is the sole accessory sex gland while in others (e.g. the bull) the prostate is small.

(2) **The vesicular glands** (**seminal vesicles**), paired secretory structures greatly developed in some species (pig, hedgehog, rat) and absent in others (carnivores, monotremes, marsupials). In the bull they are large and in the pig they may weigh up to 250 g each and contribute substantially to the bulk of the ejaculate.

210

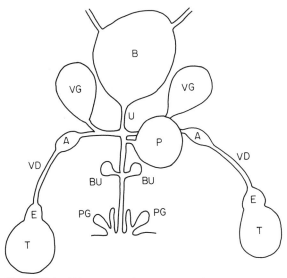

Figure 7.1   Diagrammatic representation of the major male accessory sex organs. A, Ampulla; B, bladder; BU, bulbourethral glands; E, epididymis; P, prostate; PG, preputial glands (rodents); T, testes; U, urethra (containing urethral glands); VD, vas deferens; VG, vesicular glands

(3) **The ampullae**, secretory thickenings of the walls of the lower end of each vas deferens.

(4) **Urethral glands**, including the paired bulbourethral or Cowper's glands present in many species, greatly developed in the insectivores (mole, hedgehog) and well developed in the boar. The cat possesses a prostate and bulbourethral glands. Also glands in the walls of the urethra (Littre's glands).

(5) **Glands associated with the external genitalia** (e.g. rodent preputial glands).

Seminal plasma components are also contributed by such other structures as the epididymis, the vas deferens and the testis.

Seminal chemistry and the semiochemical dimension

Seminal plasma is the aggregate product of these structures. In its chemical composition it exhibits many unusual features most of which have been interpreted in relation to the need to facilitate the survival and transport of the spermatozoa. It remains the case, however, that in the absence of ejaculation, the products of the accessory sex organs are voided insensibly and continuously in urine and, since their activity is under hormonal control, their contribution to urine reflects the endocrine status of the individual animal. It is even suggested that male hares (*Lepus*) involved in target-urine spraying as part of aggressive or courtship behaviour may store its urine in its seminal vesicles prior to emission, thus making it very possible that accessory sex gland secretions are present in that material (Flux, 1970). As a result, the accessory sex glands have considerable semiochemical potential. As long ago as 1934, Allanson noted that in the breeding season almost 10% of the body weight of the male hedgehog is due to its accessory sex glands and pondered why this animal should require such very large secretory structures unless they fulfilled functions additional to those to which they were conventionally assigned. Evidence of insensible voiding of accessory sex gland secretion in urine has been obtained surgically in the dog where the mature male is found to release up to 2 ml/h prostatic secretion into the urethra, levels being elevated at times of sexual excitement. In man, indirect evidence has been obtained by noting that levels of urinary acid phosphatase, an enzyme largely of prostate origin, are up to five times higher in men than in women or children. As would be expected from the continuous, slow release of secretion into the urethra, the initial urine voided at a given time exhibits by far the highest level of this substance (Huggins, 1945).

Careful examination of semen reveals its complex origin. It is possible to see for example in the human ejaculate that the products of the prostate tend to be voided ahead of those of the testis, vas deferens and ampulla while the secretion of the seminal vesicles emerges last. Following ejaculation, the products of the various accessory sex glands mix and a number of enzymatic reactions commence. As the result, the properties of semen change with time. The volume of the ejaculate varies widely from species to species, ranging from up to 500 ml (mean 250 ml) in the pig and 300 ml (mean 70 ml) in the horse to 10 ml (mean 4 ml) in cattle and 2 ml (mean 1 ml) in sheep. The ejaculate is close to neutral pH, but becomes alkaline (pH 9–10) on standing as the result of enzyme action.

The high level of human prostatic acid phosphatase has been mentioned. This enzyme reacts in the mixed ejaculate with vesicular phosphorylcholine (and much less readily if at all with the less abundant glycerylphosphorylcholine) to yield, after a short reaction time, high levels of free choline (zero at time zero; 900 mg/dl after 10 minutes and eventually 2500 mg/dl).

$$HOCH_2CH_2\overset{+}{N}(CH_3)_3 \quad \text{Choline}$$

This choline may then generate trimethylamine, a fish-like odorant to which the human nose is very sensitive (Amoore and Forrester, 1976) (see Table 7.1). In a similar way rat semen yields high choline levels although in many species (e.g. cattle, sheep, pig, horse, goat, rabbit), phosphorylcholine is replaced by the more stable glycerylphosphorylcholine of vesicular and/or epididymal origin.

Seminal plasma is rich in other enzymes also, including fibrinolytic enzymes, glycosidases, such as β-glucuronidase, diamine oxidases and numerous other enzymes. As the result of enzyme action between components arising from different accessory sex organs a sequence of viscosity changes commonly occurs in the ejaculate. In many species (including man) coagulation is followed by liquefaction. In some species (certain rodents, bats, and primates), the ejaculate forms a vaginal plug to prevent loss of spermatozoa from the female tract. In the rat this arises as the result of the action of a small, basic protein, an enzyme produced by the coagulating gland (the anterior lobe of the prostate which occurs in close association with the seminal vesicles) on seminal vesicle components. The products of the rodent coagulating gland are known to have a semiochemical function.

Let us now examine some further chemical peculiarities of male accessory sex organ secretions.

Polyamines

Urinary polyamines have already been discussed. However, semen itself is rich in the polyamine spermine (50–350 mg/dl) (Williams-Ashman and Lockwood, 1970). As long ago as 1677 it was reported that small crystals, now known to be spermine phosphate, were deposited when semen was allowed to stand. The related polyamines, spermidine, putresine and 1,3-diaminopropane are also present, although in much smaller quantities (mole ratio relative to spermine = 1, spermidine 0.08, putrescine 0.01 and diaminopropane 0.03). These minor amines may arise at least in part by the action of seminal diamine oxidase on spermine.

The odour characteristic of semen is also associated with samples of spermine, spermidine, putrescine and a number of related compounds, and in all cases is due to the same trace contaminant, 1-pyrroline, formed by polyamine oxidation. 1-Pyrroline is the cyclic Schiff base of 4-aminobutanal, the oxidation product first formed. It is a very weak base (pK_a 6.7) and is present to a substantial extent at physiological pH in the volatile, non-ionic form. In semen, its oxidative formation is facilitated by the presence of diamine oxidase enzymes.

It is noteworthy that the human nose is unusually sensitive to 1-pyrroline, in comparison with related amines. Also, it appears that some 20% of the population exhibit a well defined specific anosmia to this substance. These two pieces of evidence suggesting olfactory adaptation have led to the proposal that this compound represents a vestigial human sex semiochemical (Amoore *et al.*, 1975).

This semen odour is said also to be associated with male pubic sweat. 1-Pyrroline is found to be associated with other scent sources which produce putrescine and related compounds (e.g. the anal sac of carnivores, the oral cavity and other fermentative scent sources, see Section 5.3).

Seminal spermine is of prostatic origin. Highest tissue levels occur in prostate tissue and although polyamines occur widely in other tissues also their levels are commonly low. There are considerable species differences, however. For example, while human and rat prostate secretion is polyamine rich, no such compounds are present in cattle or horse semen.

Table 7.1. Odour thresholds (ppm in water) of 1-pyrroline and some related compounds (Amoore *et al.*, 1975; Amoore and Forrester, 1976)

3-Pyrroline	 (structure) NH	62.8
Pyrrole	 (structure) NH	49.6
Pyridine	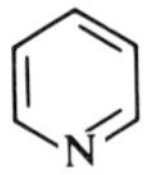 (structure) N	4.3
1-Pyrroline	 (structure) N	0.022
Trimethylamine $N(CH_3)_3$		0.00047

Prostaglandins

(Bergström *et al.*, 1968; Goldberg and Ramwell, 1975)

Although prostaglandins are ubiquitous in tissues at very low levels, they are present at unusually high concentrations and in rich diversity in human semen. Here at least 13 different molecular species are present, classified as PGEs($\sim$50 μg/ml), PGFs($\sim$8 μg/ml), PCAs + PGBs, possibly artefacts, ($\sim$50 μg/ml) and 19-hydroxyprostaglandins ($\sim$200 μg/ml). Physiologically active 19-hydroxy-PGEs (19-OH PGE_1 and 19-OH PGE_2) were found by GC and GCMS to be the principal prostaglandins of human semen and of the semen of three other primates. The production of these compounds is androgen dependent and varies greatly from species to species (Taylor and Kelly, 1974; Jonsson *et al.*, 1975; Kelly *et al.*, 1976).

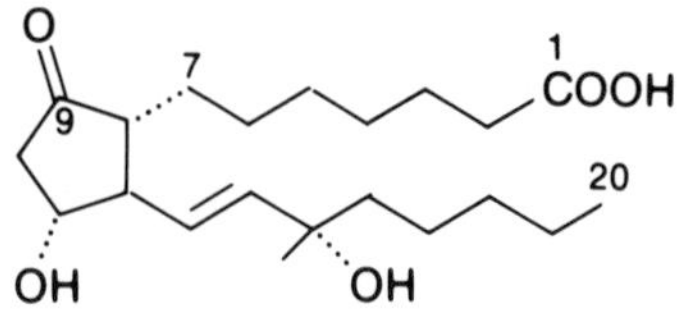

Prostaglandin PGE_1
(PGE_2 has in addition a *cis* double bond at C_5.

Human and ovine studies have shown that seminal prostaglandins arise principally in the seminal vesicles, although these substances are also present in the prostate. In some species (horse, cattle, pig), semen contains only trace quantities of these substances.

The prostaglandins exhibit a wide range of powerful physiological effects.

Different prostaglandins induce different and, in some cases, contrary effects. The biological role of seminal prostaglandins is ill defined. Men with low seminal PGE levels are infertile and it seems likely that prostaglandins exert important effects on the spermatozoa. However, we also know from studies with radiolabelled prostaglandins that if deposited in the vagina, these substances pass through the vaginal wall into the female blood stream. If in fact prostaglandins produced by the male are absorbed by the female at coitus in this way and induce significant physiological changes, then these substances would certainly rank as semiochemicals, and even as pheromones. Although this idea was discussed as long ago as 1963, very little attention has been directed to it by those concerned with mammalian semiochemistry (Horton *et al.*, 1963).

Miscellaneous other compounds present in seminal plasma

Among the numerous other compounds which characterize this complex biological material the following may be listed.

Carbohydrates and polyols   Fructose is the characteristic sugar in semen. In most species this arises in the vesicular glands (rat prostate) under the stimulus of androgen. Typical levels are in the bull 100–600 mg/dl, and in man 150 mg/dl, while in the horse and dog they are less than 1 mg/dl semen. In some species, other sugars are also present. Lactic acid is also commonly present, its level increasing with time following ejaculation as fructose is consumed anaerobically to provide energy for the spermatozoa.

The polyol, sorbitol, occurs in the semen of a number of species, as does *meso*-inositol. In the pig, vesicular inositol is present in semen at unusually high levels (400–600 mg/dl). Its principal function is thought to be to regulate the osmotic pressure of the seminal plasma.

Citric acid   High levels of citric acid, produced in many species (cattle, sheep, pig) in vesicular glands but in man in the prostate, commonly characterize semen (bull, 0–2400 mg/dl; man, 500–1100 mg/dl). The formation of citric acid is stimulated by androgen.

Ergothioneine   This betaine occurs in boar, horse and zebra semen and acts

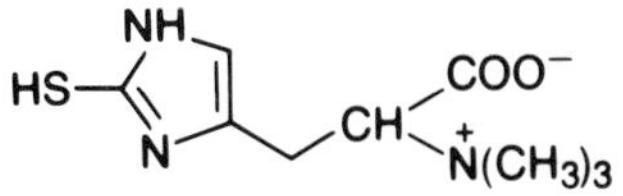

to protect the spermatozoa from oxidation. In other species this reducing function is commonly undertaken by ascorbic acid.

A wide variety of other substances have been noted in semen. These range from the hydrocarbon **heptacosane** (1 gram of which was isolated from 18 litres of human semen), include cholesterol, creatine and creatinine, and extend to the hormone **prolactin**, detected in human semen by radioimmunoassay at levels of ~165 ng/ejaculate (Sheth *et al.*, 1975). Seminal fluid is rich in free amino acids, polypeptides and small proteins, the levels increasing on standing as the result of enzyme action.

The rodent preputial gland, a semiochemically important accessory sex gland

(Clevedon Brown and Williams, 1972)

Simple glands are present in the prepuce of many species. For example, a dense accumulation of sebaceous glands has been noted in the preputium of the hunting dog, *Lycaon pictus,* and it has been suggested these may contribute semiochemicals to the urine with which the male increasingly marks female urine marks as oestrus approaches (Van Heerden, 1981). The microflora of the prepuce (Ling and Ruby, 1978) could also play a part. In some species, such as the musk ox and the pig, preputial sacs are also present (see Section 8.2).

However, the specialized sebaceous preputial glands of rodents are of quite a different kind.

Although most studies have been conducted using laboratory animals, preputial glands of the type discussed here appear to be widespread among the numerous myomorph rodent species (rats/mice). However, the castor gland of the beaver, *Castor fiber,* is not of this kind. For a long time, the function of the rodent preputial gland remained obscure although for 50 years these conspicuous structures have been known to be sensitive to androgen stimulation and have even formed a basis for an androgen assay technique. It is now clear that the preputial gland supplies important semiochemical constituents to rodent urine. As so often elsewhere, behavioural studies have proceeded independently of chemical studies so that we know something of each aspect of the gland without having a clear idea of how these aspects are related together.

Rodent preputial glands are paired, specialized tubuloalveolar sebaceous structures located, depending on species, either between the prepuce and the penis or between the pubic skin and the body wall. Each secretes not into the urethra but via a duct which opens at the free margin of the inner surface of the prepuce close to the tip of the penis. Although glands are best developed in the male, homologous structures, also known as clitoral glands, occur in females.

Molecular aspects: the rat (see Table 3.7)

In the laboratory rat, the preputial glands are up to 2 cm long and account for 10–20 mg per 100 g body weight. Almost half their dry weight is lipid extractable in chloroform–methanol. Histochemical studies have shown, however, that although mouse secretion is almost entirely lipid, that of rat is more complex and contains phospholipid–protein inclusions.

Rat preputial lipid contains sterols predominantly in their esterified forms. In this tissue, cholesterol is formed via the Kandutsch–Russell pathway (typical of sebaceous tissue) which utilizes sterols as esters, and not by the Bloch pathway, typical of epidermal and other tissues. In contrast with the latter, the formation of cholesterol by the Kandutsch–Russell pathway is accelerated by the action of testosterone. The sequence passes through 7-dehydrocholesterol (provitamin D_3) so that relative to other sources this tissue contains unusually high levels of this sterol (2 mg/g) (Ward and Moore, 1953).

Incubation experiments (Patterson, 1960) with radiolabelled acetate also show that the preputial tissue of young male rats tends to accumulate labelled squalene at the expense of labelled sterol. In this it appears to resemble human sebaceous tissue (see Section 3.5) and differs from mouse preputial tissue.

The preputial gland of the laboratory mouse also exhibits macroscopic features which clearly distinguish it from that of the rat. In addition it accounts for a higher fraction of the animal's body mass (120–180 mg per 100 g body weight). However both contain unusual alkyl acetates of possible semiochemical significance (see below).

Molecular aspects: the mouse

In its composition, mouse preputial lipid also differs from that of the rat and each differs from its respective skin surface lipid (see Table 3.7). For example, neither gland contains the type 2 diester waxes which characterize the skin surface in these species. Mouse preputial gland is further distinguished by two compound classes not noted in other mammalian sebaceous sources. These are:

(1) **n-alkyl-(1) acetates** (Spener *et al.*, 1969), R-O-CO-CH$_3$, where R includes a range of C_{14}–C_{18} alkyl groups, but is principally nC_{16} (75%), $nC_{16:1}$ (10%) and nC_{18} (5%). Alkyl acetates have subsequently been identified also as volatile components of male rat preputial gland lipid extract (Stacewicz-Sapuntzakis and Gawienowski 1977). Here, using GCMS, ethyl, propyl, isopropyl, pentyl and decyl acetates have been observed. The technique used was insensitive to the less volatile compounds. Behavioural studies, however, showed that female rats responded to acetates of saturated and unsaturated alcohols up to C_{20}. Such compounds have not been observed in other mammalian sources, although examples are known of acetates of this mass range functioning as insect semiochemicals.

These acetates account for some 5% of the lipid content of the male mouse preputial gland. However, the female mouse preputial gland while being much smaller than that of the male also contains a much lower proportion (<1%) of these substances. Not only is the male gland size decreased by castration and increased by testosterone administration, but so also is its alkyl acetate content.

(2) **alkyl- and alk-1-enyl glyceryl ether diesters** (Sansone and Hamilton, 1969; Snyder and Blank, 1969):

$$CH_2-O-COR^1$$
$$CH-O-COR^2$$
$$CH_2-O-CH_2CH_2R^3$$

$$CH_2-O-COR^1$$
$$CH-O-COR^2$$
$$CH_2-O-CH=CHR^3$$

The pattern of chain lengths and unsaturation and the absence of chain branching of R^3 (in both alkyl- and alk-1-enyl glyceryl ether diesters) are closely similar to the pattern of the alcohol moieties of the preputial wax monoesters, suggesting a common biosynthetic origin. In all cases $R^3CH_2CH_2O-$ was principally nC_{16}, followed by nC_{14} and $nC_{14:1}$ (mainly $\Delta6$). The acyl moieties R^1CO- and R^2CO- included a wide range of branched, unsaturated and odd- and even-carbon chain residues, and differed markedly from the acyl moieties of the gland triglycerides which were principally straight chain even-carbon residues.

Alkyl glyceryl ether diesters are also present in the mouse Harderian gland lipid (see Section 4.5). There is some histochemical evidence that such compounds may be present in rat preputial gland lipid.

Treatment with testosterone stimulates the synthesis of alkyl acetate, wax monoester, and alkyl- and alk-1-enyl glyceryl ether diesters in mouse preputial glands (Sansone-Bazzano et al., 1972). This is reflected in changes in gland lipid composition occurring with maturity in the male and also in male–female differences. Female and immature male glands tend to contain relatively low levels of lipid and this lipid is largely triglyceride followed by sterol ester. Incubation experiments using preputial tissue and radiolabelled glucose reveal *de novo* biosynthesis under the influence of testosterone with incorporation of label into all lipid classes and particularly into components utilizing fatty alcohol moieties (wax monoesters, alkyl acetates, etc.).

Molecular aspects: musk rat

(Van Dorp et al., 1973; Brüggemann et al., 1981; Ritter et al., 1982a,b)

The preputial glands of the musk rat, *Ondatra zibethica*, (125 mg per 100 g body weight) are particularly interesting in containing a very complex mixture of musky macrocyclic ketones closely similar to those encountered in civet, although civetone itself was not observed here (see Section 5.6). The production by the male of large quantities of the yellow musk oil with its very persistent odour is, however, seasonal, being maximal at the mating season. A number of saturated, monounsaturated and diunsaturated fatty acids of chain length in the region $C_{12}-C_{20}$ (principally $C_{14:0}$, $C_{16:0}$, $C_{16:1}$, $C_{18:0}$, $C_{18:1}$, $C_{18:2}$) are present too and Ritter et al. have restated Stevens' (1945) earlier view that these acids could, by cyclization and decarboxylation, prove to be precursors of the musky macrocyclic ketones, for in a year-long survey high levels of ketone were found to be accompanied by low levels of the corresponding fatty acid, and viceversa. Thus, macrocyclic ketone production was maximal in the autumn, when fatty acid (including ester) content was lowest, and lowest in February, when fatty acid content was highest. Similarly female preputial gland secretion was low in macrocycles but correspondingly high in such fatty acids. Among the other lipid classes present (triglycerides, sterol esters) is a fraction containing similar quantities of the odourless long-chain esters of the related macrocyclic alcohols. Some of these alcohol moieties are chiral. For example, the dextrorotary 5-(Z)-cyclopentadecenol has the S configuration. Cyclopentadecanol and cyclopentadecenol were also encountered in the free state by Ritter *et al*.

Following isolation of up to 500 mg quantities of these ketones by a process employing extraction, TLC and preparative GC from up to 100 g quantities of glandular tissue, the ketones shown in Table 7.2 were identified.

Endocrine aspects

We have seen that the rodent preputial gland is sensitive to endocrine status. In this it resembles skin sebaceous tissue (see Section 3.3). However, it also shows certain peculiarities in its pattern of hormonal responses which distinguishes it from these tissues. The gland is extremely sensitive to the action of androgens and enlarges greatly with the coming of maturity, particularly in the male, so that the gland size in the mature, dominant male is close to maximal. Testosterone administered to intact male and female rats results in substantial enlargement of the preputial gland (except where the gland size is already maximal) while castration causes a reduction in the gland mass which can itself be offset by testosterone treatment (Ebling, 1977a).

In the female, ovariectomy causes preputial gland reduction although oestrogen administration generally has little effect apart from tending to antagonize the effect of androgen. Under certain circumstances, progesterone treatment results in some slight enlargement of the gland.

As with skin sebaceous tissue, pituitary factors also influence gland function, although in this case the response to testosterone is not totally dependent on the presence of the pituitary, for the response is not abolished but only halved by hypophysectomy (pituitary removal). In this the preputial gland resembles other rat accessory sex organs, such as the prostate and the seminal vesicles, the testosterone response of the latter being only slightly reduced by hypophysectomy. But which pituitary hormones are involved? Pituitary α-MSH acts synergistically with (and not independently of) testosterone in increasing preputial gland, prostate and seminal vesicle weights. However, these organs differ from each other in their response to pituitary growth hormone. This hormone acts largely synergistically with testosterone

Table 7.2. Macrocyclic ketones in the musk rat preputial gland

	n	m	Percentage (1)
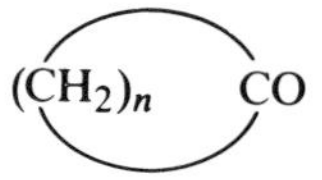 (CH$_2$)$_n$ CO	12	—	0 (2)
	14	—	21
	15	—	0.5
	16	—	41
	18	—	1
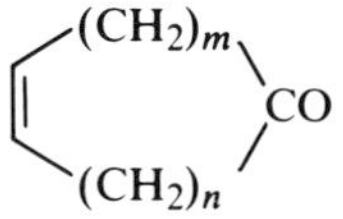 (CH$_2$)$_m$ CO (CH$_2$)$_n$	3	9	16
	3	11	10
	5	9	4
	7	9	0 (2)
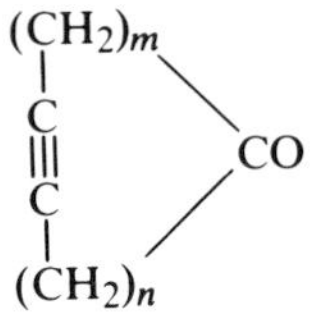 (CH$_2$)$_m$ C≡C (CH$_2$)$_n$ CO	3	9	0.5
	3	11	1 (3)
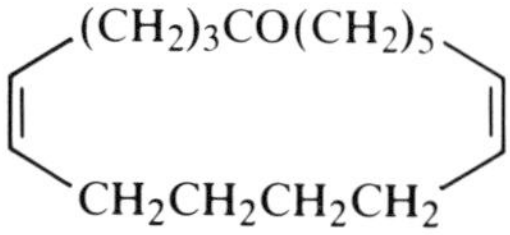 (CH$_2$)$_3$CO(CH$_2$)$_5$ CH$_2$CH$_2$CH$_2$CH$_2$			2
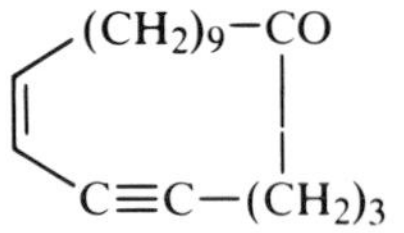 (CH$_2$)$_9$−CO C≡C−(CH$_2$)$_3$			3 (3)

(1) Percentage of total macrocyclic ketones (Van Dorp *et al.*, 1973). A similar pattern was noted by Ritter *et al.*, 1982a, b.
(2) Detected at low levels by Ritter *et al.*, who also noted a further cyclohexadecenone and cyclononadecenone.
(3) Not observed by Ritter *et al.*, 1982a,b.

in increasing the weights of preputial glands in hypophysectomized-castrated rats but is without effect on the prostate and seminal vesicles. As with the preputial gland, both of these organs are sensitive to the stimulating effect of testosterone alone whether in castrated or in hypophysectomized-castrated rats. In the musk rat, as expected, the preputial glands are much larger in the male than in the female and enlarge greatly during the breeding season at which time the odorous musk oil is produced in quantity (Ritter *et al.*, 1982a).

The sensitivity of the preputial gland to endocrine factors is further manifest in that considerable variations in gland size occur in mature males and that these are related to the experience of social stress and to differences in social status. Both subordinate males and stressed, crowded animals of either sex show evidence of depressed reproductive function—in the most stressed or

Table 7.3. Relative rates of 5α-dihydrotestosterone formation from testosterone per unit mass of tissue (Wilson and Gloyna, 1970)

Rate	Species: rat Tissue	Tissue: Prostate Species
Very high	Prostate	Rat
High	Epididymis	Man
Moderate	Seminal vesicle, preputial gland, scrotum, kidney	Baboon, lion, dog
Low	Skin, liver, ovary, adrenal, vas deferens	Mouse, guinea, pig, cat
Very low	Testis, muscle	Bull, rabbit

most subordinate animals reproductive functions are totally suppressed—in parallel with a whole sequence of physiological changes including curtailed spermatogenesis and ovulation, reduction in gonad weight and reduction in preputial gland weight. These effects are associated with the increased pituitary–adrenocortical activity which accompanies stress. Preputial glands are most fully developed in mature, sexually active, dominant males (and also in mature males reared in isolation). In subordinate or stressed males, preputial gland weights are reduced and this effect is closely related to reduced testis weight.

Studies of a different kind have also pointed to the key importance of hormones in the activity of these glands. Following the intravenous administration of testosterone to rats, the highest rates of androgen metabolism per unit mass of any tissue were observed in the accessory sex organs. The principal testosterone metabolite noted was the potent androgen 5α-αihydrotestosterone (formed in the nucleic and the endoplasmic reticulum) with the less active 5α-androstan-3β, 17β-diol (formed in the cytosol) (Wilson and Gloyna, 1970). Comparative studies on prostate tissue show that considerable differences occur with the species studied (see Table 7.3).

Although chemical studies are limited, it remains probable that the preputial gland and other accessory sex organs not only respond to steroid hormones, but may actually secrete odorous steroids of semiochemical importance. Experiments (Brooksbank et al., 1973) with radiolabelled androsta-4,16-dien-3-one injected intravenously into both intact and castrated male rats have shown that radioactivity accumulates rapidly and selectively in certain tissues, most notably in the preputial and the Harderian glands. In this the pattern differs somewhat from that obtained with radioactive testosterone where the highest concentrations occur in the prostate. As the only known function of androsta-4,16-dien-3-one is as an immediate precursor of the musky-odoured steroid, 5α-androst-16-en-3α-ol, a substance of known semiochemical significance in at least one species (see Section 8.2), this evidence suggests that preputial tissue may also yield this steroid. Note, however, that in the musk rat, the major musky-odoured products of the preputial glands are macrocyclic ketones.

Behavioural aspects

In such ways, the preputial glands, and the other accessory sex glands, demonstrate their differing sensitivities to the animal's internal hormonal environment. Such hormonally sensitive organs are well suited to contribute substances of semiochemical significance to urine and could account for many of that substrate's communicatory functions.

Studies already mentioned being undertaken in the Netherlands by Ritter *et al.* on the chemical nature and behavioural significance of musk rat preputial gland secretion have been motivated by the need to develop more effective means of controlling this species. As with the fox, the principal vector for the spread of rabies through Europe, attempts to employ traditional methods of control have proved themselves inadequate. The musk rat was introduced into central Europe at the beginning of the century for its fur, but has spread across the continent and is now causing great concern in the Netherlands as the result of its propensity to burrow from below water into the canal banks and dikes of that low land. Although, as often happens, trappers have their traditional lure mixtures, scientific tests have not been reported. Now Ritter's group's preliminary findings do seem to indicate that musk rat gland homogenate can aid trapping when applied to suitably devised traps.

In nature, the musk rat secretes its musk oil through the penis whence some of the material is picked up by its hind legs and thrown behind the animal. Possible natural functions concern reproduction (sex attractant, priming pheromone function) and territory (nest, burrow) marking, including marking the faecal piles which are important environmental scent points for this species. Marking the female by male has also been observed as part of reproductive behaviour.

Preputial gland lipid exhibits a degree of sexual attractancy not possessed by extracts of other rodent tissues and rats and mice can distinguish between preputial extracts obtained not only from either sex but also from animals of a given sex but of different hormonal states (Bronson and Caroom, 1971; Caroom and Bronson, 1971; Orsulak and Gawienowski, 1972; Gawienowski *et al.*, 1975, 1976; Stacewicz-Sapuntzakis and Gawienowski, 1977).

Investigation time data obtained from controlled behavioural tests show that both rats and mice have a distinct preference for homogenates of preputial tissue obtained from conspecifics of the opposite sex and that this attraction is associated largely with neutral lipid components.

Relative to controls, female rats were attracted to extracts of preputial glands from intact mature male rats, but not to extracts of the following:

> preputial tissue from castrated rats,
> preputial tissue from female rats,
> footpad tissue from intact male rats,
> submaxillary–sublingual gland from intact male rats,
> coagulating gland from intact male rats.

} previously considered to be probable semiochemical sources

Relative to controls, male rats were attracted to extracts of preputial glands from oestrous, sexually receptive females, but **not** to extracts of

preputial tissue from anoestrous female rats,
submaxillary–sublingual gland tissue of female rats, ⎤ random oestrous
footpad tissue of female rats,   ⎦ state
preputial tissue from male rats.

Components associated with this attractancy in either sex could be distilled (80 °C, 0.1 torr) from the preputial lipid and subjected to gas chromatography. Fractionation using preparative gas chromatography showed that activity varied from fraction to fraction. Behavioural studies with alkyl acetates of the type found in mouse preputial lipid demonstrated that substances of this class elicit considerable attraction for female rats, but little interest for male rats (Table 7.4).

Table 7.4. Responses of male and female rats to the odour of saturated alkyl acetates (Reproduced by permission of Plenum Publishing Corp. from Stacewicz-Sapuntzakis and Gawienowski, 1977)

Acetate $R-O-COCH_3$ where R is	Investigation time		Frequency	
	♂	♀	♂	♀
nC_2				
$isoC_3$	+			
nC_3		+++		+++
nC_4				
nC_5		+++		++
nC_6		+++		++
nC_7		+++		++
nC_8		+++		
nC_9		+++		++
nC_{10}		+++		
nC_{11}	− −	+++		+++
nC_{12}		++		+++
nC_{13}		+++		+++
nC_{14}		+		
nC_{15}				
nC_{16}				
nC_{17}	++			
nC_{18}		+++		
nC_{19}				
nC_{20}	−			

+ indicates attraction; − indicates avoidance; no sign indicates indifference.

Dimethyl disulphide and dimethyl sulphite, reported in female rat preputial tissue extracts, said to be attractive only to the male rat (Gawienowski, 1977).

Similar experiments have shown that female mice are attracted by an odour associated with male mouse preputial tissue, but not to male submaxillary or Harderian-lachrymal tissue. This attraction is evident in sexually experienced females whatever their hormonal state. In sexually naive females, attraction was only apparent when oestrogen levels were elevated. Attraction was eradicated with progesterone treatment and at pregnancy. These effects were mirrored in the female's attraction to male mouse urine. Females preferred the urine of an intact male to that of a preputialectomized animal.

Studies on mice also suggest that preputial odours elicit aggression between dominant males (Mugford and Nowell, 1971). It seems that males advertise and maintain their social status at least in part with these scents. Totally subordinate, non-reproductive males, like females and immature animals of either sex, will not elicit aggression from a dominant male, nor will a dominant male from which the preputial glands have been removed even though this animal may itself behave aggressively. Similarly a female treated with testosterone will, after a lag period and without itself manifesting aggressive behavior, begin to elicit aggression from males while showing a simultaneous enlargement (up to fivefold) of its preputial (clitoral) glands. Secretion taken from the preputial gland when applied to the coat of a mouse stimulates aggressive behaviour on the part of another mouse (Jones and Nowell, 1973b). Products of the testosterone stimulated preputial gland thus elicit aggressive behaviour from other males.

However, experiments using urine from preputialectomized female mice treated with testosterone show that this urine retains a further substance which induces aggression in male mice. This can not be of preputial origin and must arise from another source. So, the preputial gland cannot be the only androgen-dependent source of aggression-inducing signals. In contrast to the urine-mediated signal, the preputial cue is in fact released only in agonistic encounters (or experimentally following electric shock treatment)—and may operate independently of the urine medium. This view is supported by experiments using α-MSH, the pituitary hormone which together with testosterone stimulates preputial gland activity (Nowell et al., 1980a,b). Injection of this hormone into one of a pair of male mice causes it to emit an odour cue after a lag period of about 15 minutes which elicits aggressive behaviour in the other mouse. This response and the emission of the substance causing it occur whether or not the mouse injected has had its preputial glands removed surgically. However, preputialectomy shows that the preputial gland does have a longer term effect on the generation of aggression-promoting signals.

Aggression-promoting components in intact male urine have been found to be lipophilic and entirely volatile, being extracted into dichloromethane and also removed completely from a sample of acidified urine maintained at 40 °C by stream of nitrogen and subsequently condensed in dichloromethane at −78 °C (Lee, 1976; Lee *et al.*, 1980).

The rodent coagulating gland, a source of semiochemicals

At least one further male mouse accessory sex gland has been implicated in semiochemical functions. This is the coagulating gland (Jones and Nowell, 1973a), an organ which produces a semen coagulant and leads to the formation of the vaginal plug following copulation.

These glands also produce a urinary 'aversive pheromone' which deters prolonged investigation and inhibits aggression in other males entering a urine-marked area (Jones and Nowell, 1973c). Production is androgen dependent, being eliminated by castration and re-established by testosterone administration. Neither bladder urine nor the product of the gland alone, nor these two materials placed in close proximity, has any such effect. The aversive function arises from contact between the product of the coagulating gland and bladder urine. This would indicate that the aversive signals arise from a chemical reaction between components of these two materials. A comparison of materials from highly aggressive, isolated males showed that these yielded urine which contained the aversive signal while urine from grouped, less aggressive males (the dominant male of the group was excluded) did not. However, studies with bladder urine and coagulating gland tissue taken from these two groups showed that no difference existed in generating the aversive signals between any combination of bladder urine and coagulating gland tissue from the two groups. All did so. So the strength of the aversive signal, which varies with the levels of aggression and so with androgen levels in the donor mice, must be connected merely with whether the mouse in question releases signal-generating material within its coagulating gland.

Studies in which urines (from various testosterone-treated castrates which had undergone different surgical operations) was applied to the fur of castrates paired with aggressive intact male mice revealed that the mice for which the coagulating gland had not been removed produced urines which led to much lower levels of aggression (attacking time, number of bites) on the part of the intact male than was observed when urine from animals lacking the coagulating gland was employed. Thus, the coagulating gland is also a source of an aggression-inhibiting signal.

Both these coagulating-gland-dependent signals—that of deterrence from entering marked territory and that of inhibition of attacking behaviour—can be seen to be important in territoriality. The latter would seem to account at least in part for the fact that intruders in a strange territory seldom initiate agonistic encounters and are at a disadvantage when encountering the resident males.

Female laboratory mouse urine also has the capacity to reduce the aggression shown by a male mouse toward another smeared with this urine. The components are present in female bladder urine and, although their identity is unknown, they appear (Evans *et al.*, 1978) to be resistant to environmental degradation, activity being retained by urine kept at room temperature for a week, to be hydrophilic and not readily extracted into organic solvents. Active components here seem to be relatively involatile, being resistant to distillation at 50 °C under low vacuum (12 torr), but since they are dialysable, they are not macromolecular. The components responsible are present in female urine whether the animal is juvenile or adult, oestrous, dioestrous or ovariectomized, and are distinct from the urine components which elicit sexual behaviour in the male.

7.2 FEMALE SECRETIONS

Sources and context

Female urogenital signals, particularly those arising from the oestrous female, play a most important role in the semiochemistry of many mammals. It is not unexpected, for example, that in certain prosimians the oestrous female excites the male and informs him of her receptive state by means of semiochemicals arising from her vaginal discharge which, with urine, she employs in scent marking. Clearly an understanding of the chemical nature of similar potent reproductive signals employed by very many species could have great scientific interest, and in some cases considerable economic importance. Again, however, although our biological understanding is considerable, our chemical knowledge remains slight.

One area of practical importance concerns oestrus detection in cattle, for every oestrus missed, as increasingly tends to be the case with large herds, delays insemination for 3 weeks with the result that milk yield and beef production are reduced overall to a significant extent. Now, the cow signals its oestrous condition semiochemically and dogs can be trained to detect the oestrous odour (Kiddy *et al.*, 1978). Cervicovaginal mucus from oestrous cows possesses an odour which excites the bull (Paleologou, 1977) but no chemical identification of the active components has been reported. Nor is it clear what the precise origin of this signal is since urine and various reproductive tract secretions naturally mix with each other. One gland of interest is the major vestibular (Bartholin's) gland which in the cow weighs about 2 g and is about the size of a walnut. These glands, which are homologous with the bulbourethral gland in the male, lie on each side of the vestibule of the vagina, some 6 cm from the vulval lips and each connects with the vestibule by means of a short duct. The glands have been found to be very sensitive to oestrogen and it is reported that their histological appearance and secretion rate varies with circulating oestrogen level and thus with the stage of the oestrous cycle (Cribb and Flood, in prep). Just such a gland would be expected to contribute in some way to oestrous signalling, but unfortunately no behavioural or chemical studies on its products have been reported.

As with many other semiochemical sources, female urogenital substrates derive from a multiplicity of sources, for as well as urine, there are the secretions of the reproductive tract and of the accessory sex organs to consider. In addition, these materials may be further modified by the action of microorganisms which inhabit the vagina. Many species also possess well developed skin glands in the anogenital area (e.g. labial glands, circumgenital glands) which further contribute to the chemical ambience of the region.

In broad summary, and bearing in mind the ever-present phenomenon of interspecies diversity, we can say that the female urogenital substrate from which any semiochemical materials must arise may be classified as follows.

(1) **Bladder urine** components.

(2) **Materials arising from the upper reproductive tract**, such as oviductal and endometrial fluid and cervical mucus (Gibbons and Sellwood, 1973; Elstein, 1978; Gibbons, 1978). Cyclic changes are evident in composition of cervical lipids which include hydrocarbons, glycerides, cholesteryl esters, free fatty acids and phospholipids (Singh *et al.*, 1972; Singh and Swartwout, 1972).

(3) **Materials deriving from the vaginal walls**. The vagina is devoid of glands but receives exfoliated cells from its surface and a transudate of blood plasma. This contributes blood-plasma-derived components and the mean 18 μg/ml human vaginal protein is presumed to arise at least in part from that source (Raffi *et al.*, 1977). In man this transudate flows rapidly at times of sexual arousal as the result of increased blood supply to the permeable vaginal tissues. This permeability was referred to when we noted that prostaglandins, acting pheromonally, are absorbed through the vaginal walls and may cause physiological effects in the female (such as oviductal contractions) within minutes. Vaginal tissues are rich in enzymes and contain exceptionally high levels of glycogen. The proliferation of vaginal tissue, the deposition of glycogen and the breakdown of glycogen to glucose enzymically are all stimulated by high circulating oestrogen levels, so that as ovulation approaches, the turnover of vaginal glycogen can be rapid. There are, as always, considerable differences between species. Unlike man and certain primates, many species (e.g. rodents) lack high vaginal glycogen levels.

(4) **The secretions of the vulval glands**, the functions of which remain obscure. In man, these include the products of the paired greater vestibular glands (Bartholin's glands) some 1.25 cm in diameter, homologous with the male bulbourethral glands and Skeen's glands or periurethral glands, homologous with the male prostate. The vulval area also includes apocrine, sebaceous and eccrine skin glands, which may be highly developed in many

species. The rodent's sebaceous clitoral (female preputial) glands were discussed previously (see Section 7.1).

(5) **Odorous substances derived from the presence of male ejaculate** in the vagina. Such substances as propylamine and butylamine are formed enzymically from seminal polyamines when human semen is incubated (Eskelson *et al.*, 1978; 1979). Goldfoot *et al.* (1976b) reported an increase in volatile fatty acids in rhesus monkey vaginal secretions containing stale ejaculate.

(6) **Materials derived from any or all of the above as the result of activity of resident vaginal microflora** (Hurley *et al.*, 1974; Brown, 1978). In man these include lactobacilli and facultitative and strict anaerobes. The human vagina is acidic (pH 4.0–5.0) (Rakoff *et al.*, 1944), the pH varying regularly throughout the cycle and also with age, more acidic conditions being associated with high oestrogen levels. In man and other primates studied this seems to be linked with turnover of glycogen itself stimulated by high oestrogen levels in vaginal tissue, and its breakdown by enzymes and the resident microflora to acidic products, principally lactic acid. However, near neutral vaginal secretion and a quite different vaginal microflora has been reported in rats which do not produce high levels of vaginal glycogen, so interspecies differences are significant (Larsen *et al.*, 1976). In contrast, fresh semen has a pH of about 7.0 (depending on species) while human cervical mucus varies in pH from about 5.0 to 8.5 (Moghissi *et al.*, 1972). Endocervical pH measurements have shown cyclic variations in the range 6.5–7.5 (Macdonald and Lumley, 1970). Acidity has considerable importance in determining odour whenever odorous acids or bases are concerned, as they are here, because as the vagina becomes more acidic, any volatile odorous bases, such as amines, will increasingly form involatile salts and so contribute less to the perceived odour, while any odorous acids, such as volatile fatty acids, will increasingly form such salts as the medium becomes more alkaline. It is useful to remember that the pK_a values of such substances give the pH values at which they are half free (volatile) and half ionized (involatile salt):

pK_a butanoic acid = 4.81,
pK_a butylamine = 10.61.

It should be noted as well that quite commonly measured levels of odorous acids and bases in such materials give values for the total (volatile plus involatile forms) and do not take account of this effect.

Although a substantial amount is known about the chemistry of 'vaginal secretion' and the fluids of the cervix and the oviduct, in the main this knowledge concerns involatile materials, such as sugars, amino acids, enzymes, enzyme inhibitors, immunoglobulins, metal ions, proteins, polysaccharides and glycoproteins, materials which are not themselves odorants, the functions of most of which are clearly concerned with the survival and transport of the spermatozoa rather than with semiochemistry.

More recently volatile components have been examined, although chemical studies have been confined to a very few species, and principally to humans.

Vaginal volatile fatty acids

(Goldfoot *et al.*, 1976b; Keverne, 1976a; Michael *et al.*, 1976a)

Considerable interest has centred on the odorous volatile fatty acids which occur as major constituents of the vaginal secretions of a number of primate species, including man (Michael *et al.*, 1974, 1975), as well as of some non-primate mammals. Examples which have been studied include secretions of the American mink, the golden hamster and the cow. These compounds are commonly encountered naturally as microbial metabolites (see carnivore anal sac secretion, Section 5.3), and those studies which have been undertaken on vaginal secretions have pointed to microbial origins here also. Such an origin in no way invalidates the possible semiochemical value for the mammal, as we have discussed elsewhere. The human vagina is very sensitive to circulating oestrogen levels, with the result, as already noted, that high levels of oestrogen stimulate the turnover of glycogen and other materials thus supplying the microorganisms with increased nutrient (Michael *et al.*, 1976a). In this context it is interesting to note that the counts of bacteria which inhabit the vagina of the rat also vary cyclically, high oestrogen levels leading to high bacterial counts, even though the rat's vagina contains neither glycogen nor acid (Larsen *et al.*, 1977).

Interest first arose in the volatile fatty acids as the result of studies on the contribution of olfaction to sexual signalling in the **rhesus monkey**, *Macaca mulatta*, a microsmatic species like man in which the role of olfaction was thought to be relatively unimportant. However, behavioural tests using intact and ovariectomized hormone-treated females, and males which were on occasion rendered anosmic, indicated that males were attracted to females, indeed performed quite lengthy behavioural sequences to gain access to females, which were either oestrogen-treated or to which vaginal secretion from oestrogen-treated females had been applied, and that the cue to which these males responded was olfactory. Subsequent fractionation of oestrus vaginal secretion samples coupled with behavioural testing showed that this activity was associated with odorous volatile fatty acids present in secretion at quite high levels, identified (GCMS) as a mixture of nC_2, nC_3, nC_4 and $isoC_4$, nC_5, $isoC_5$, nC_6 and $isoC_6$ saturated fatty acids (Curtis *et al.*, 1971; Michael *et al.*, 1971). Acetic acid (C_2) was always present in high concentration although the powerfully odorous isovaleric acid ($isoC_5$) was commonly present in sufficient quantity to dominate the odour. These components were absent from the vaginas of ovariectomized females but were produced rapidly on treatment with oestrogen. They seemed to constitute a good example of a primate sex pheromone, the possibility that they had a similar function for man was considered, and they acquired the name 'copulins'.

However, subsequent investigation showed that the situation was not simple and the interpretation of the initial work was challenged (Goldfoot *et al.*, 1976a,b; Keverne, 1976b; Michael *et al.*, 1976b). Among the areas of debate was the finding that response varied considerably from male to male, depending not only on the test regimen but also on which particular female, irrespective of odour cue, was involved. Clearly, the factors which led to an overt sexual response were complex. Also, the signal was complex. Where a response was noted, the volatile fatty acid mixture was often not as effective as whole secretion. Other components were also involved. The addition of 3-phenylpropionic acid and its *p*-hydroxy derivative (known to be common microbial products although this was not tested here), also identified in vaginal secretion and apparently inactive alone, increased the effectiveness of the volatile fatty acid mixture.

$$\text{C}_6\text{H}_5-\text{CH}_2\text{CH}_2\text{COOH} \qquad \text{HO}-\text{C}_6\text{H}_4-\text{CH}_2\text{CH}_2\text{COOH}$$

It was also reported that the presence of stale ejaculate in the vagina as the result of previous testing increased the levels of volatile fatty acids (which would be understandable as more nutrient would be available for the resident microflora) quite apart from any circulating oestrogen effect. Further it was disconcerting to learn also that volatile fatty acid levels present throughout the entire cycle of intact females not infrequently increased during the luteal phase following ovulation and so would be of little reproductive value as a sex attractant. One possibility which has been considered is that in spite of the irregularities in the levels of volatile fatty acids, their *presentation* to the environment is hormonally dependent. Thus, at oestrus, if the vaginal secretion is more acidic, then the volatile acids will be present to a greater extent in their volatile free acid form rather than as involatile salts. Also, at oestrus the increased through flow of secretion will 'externalize' vaginal secretion and make it more accessible as a semiochemical substrate for males than at other times (Bonsall and Michael, 1980).

Other vaginal components

A further limitation of the work just discussed is the restricted nature of the chemical investigation. Subsequent GCMS studies (Preti and Huggins, 1975; Huggins and Preti, 1976) of human vaginal secretion samples revealed the presence of a wide range of low mass, relatively volatile compounds (Table 7.5). Comparison of samples obtained from 12 women at intervals over a number of ovulatory cycles revealed considerable variation between individuals and also between samples taken at different times from the same individual. Apart from acetic acid, the volatile fatty acids discussed previously were found to occur only in a minority of women ('acid producers'). In addition a number of GC peaks coincident with authentic volatile fatty acid peaks were shown not in fact to be due to these compounds when their mass spectra were obtained. Lactic acid was a major component, 1–5 mg/g secretion, and with urea and acetic acid was present in all secretions, showing sharp cyclic variations with maxima at about the time of ovulation in midcycle although other maxima were observed additionally at other times in a number

Table 7.5. Low mass organic compounds identified by GCMS in human vaginal secretions (Huggins and Preti, 1976; Preti *et al*., 1977)

Carboxylic acids
Saturated
nC_2, nC_3^*, nC_4^*, nC_{14}, nC_{15}, nC_{16}, nC_{18} acids
$isoC_4^*$, $isoC_5^*$, $isoC_6^*$, $isoC_{14}$, $isoC_{15}$ acids
$anteisoC_5$ acid

Unsaturated
$nC_{16:\Delta 9}$, $nC_{18:\Delta 9}$, $nC_{18:2 \, \Delta 9,12}$ acids

Hydroxyacids
$CH_3CHOHCOOH$ (lactic acid)

Aromatic acids

benzene—COOH   benzene(OH)—COOH   HO—benzene—CH_2CH_2COOH

Alcohols	**Hydrocarbons**
Alkan-1-ols	nC_{17}, nC_{18}, nC_{19}
nC_{12}, nC_{14}, nC_{16}, nC_{18}	

Other
CH_2OHCH_2OH (ethylene glycol), $CH_3CHOHCH_2OH$ (propylene glycerol),
$CH_2OHCHOHCH_2OH$ (glycerol),   furan—CH_2OH cholesterol

Aldehydes
nC_9,   benzene—CHO   benzene—CH_2CHO furan—CHO

Phenols   benzene—OH CH_3—benzene—OH   benzene(CH$_3$)—OH

Hydroxyketones
$CH_3CHOHCOCH_3$, CH_3COCH_2OH

Other
pyridine, 2-piperidone,[*] indole,[*] uracil, maleic anhydride, urea, dimethylsulphone, 2-isopentylfuran, squalene

[*] Restricted to 'acid producers'.
Also reported (Keith *et al*., 1975) but without GCMS data: acetaldehyde, acetone, ethanol, 3-methylbutan-1-ol, ethyl formate, ethyl acetate, 1,3-dioxolane, 2,5-dihydrofuran.

of cycles. This was particularly the case for lactic acid, which was the major acid component of all subjects at the time of ovulation.

Linked with this work, a scientific assessment (Doty *et al*., 1975) of vaginal odours by a panel noted variations in intensity and pleasantness ratings of vaginal odour samples taken at different times in the menstrual cycle. Considerable variations in odour pattern, however, occurred between different cycles even from the same donor. On all occasions the vaginal odours were rated unpleasant, but were least unpleasant and least intense at the time of ovulation. The considerable variation in acid and the unpleasantness of the subjective assessments seems to suggest that human vaginal odours do not provide the sexual attraction or reliable information concerning ovulation which they do in other species.

Human vaginal odours have been analysed by gas chromatography (Keith *et al.*, 1975) in experiments in which the effluent from the GC column is continuously sniffed and the resulting chromatogram annotated with subjective descriptors of the emerging component odours (odourograms). A limited number of components have been identified by GCMS. Although the great complexity of the odour profile becomes apparent, similar cyclic changes to those of Doty *et al.* (1975) were noted.

GCMS showed that sexual stimulation (Preti *et al.*, 1979) increased the secretion concentrations of glycerol and stearic acid to a statistically significant degree. No qualitative changes in gas chromatograms were noted. Increases were also noted in mean concentrations of lactic acid, squalene and a number of long-chain fatty acids, but because of the scatter of individual data points, these increases were not statistically significant. A simple supposition would be that concentrations of substances *not* derived from the transudate would decrease and those of transudate origin would increase following sexual stimulation. Thus, acetic acid levels tended to decrease; volatile fatty acids, hydroxybutanone, dimethylsulphone, cresol and 2-piperidone are all likely to be of microbial origin. The increased flow of secretion through the vagina could enhance the odour signal by externalizing the secretion to a greater extent, even if the levels of some components in the secretion are somewhat reduced by dilution.

Chemical studies on the following non-primate vaginal secretions have been reported.

American mink: *Mustela vison*

(Sokolov *et al.*, 1974; Sokolov and Khorlina, 1976).

GCMS revealed the following saturated carboxylic acids: nC_2, nC_3, nC_4, nC_5, nC_7, nC_8, nC_9, $isoC_4$, $isoC_5$, $isoC_6$, $isoC_7$, $isoC_8$, as well as 2,2-dimethylbutanoic and 2,2-dimethylhexanoic acids. Volatile amines observed by GC comparison with authentic materials included methylamine, dimethylamine, trimethylamine, ethylamine, diethylamine, triethylamine, propylamine, dipropylamine, butylamine, and the following compounds.

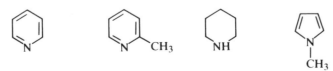

Oestrus samples were neutral, pH 7.2–7.5 while anoestrus samples were reported to be slightly acidic, pH 5.8–6.2.

Domestic dog

(Goodwin *et al.*, 1979)

In an attempt to determine the potent sex attractants produced by the bitch at oestrus, volatiles associated with samples of vaginal secretion taken at different stages of the oestrous cycle were examined by gas chromatography. A number of components were reported in all samples, some of which were only associated with oestrus samples. One of these components, identified by GCMS as methyl *p*-hydroxybenzoate, was reported to act as a powerful sex stimulus and attractant for the male dog when applied to the vulva of an anoestrus bitch. However, subsequent studies have failed to confirm that this substance is a major sex attractant (Kruse and Howard, 1983).

The golden hamster: *Mesocricetus auratus*

(Singer *et al.*, 1976, 1980)

Rather more work has been conducted on the vaginal secretion of the hamster. The golden hamster is quite unusually dependent on chemical signals in its reproduction (Murphy and Schneider, 1970), for a male hamster which has been rendered anosmic will not mate with an oestrous female even though the female advertises its oestrus state in other ways in addition to the production of chemical signals. The chemical signals of interest arise from hamster vaginal secretion which is produced, under oestrogen stimulus, in increasing quantity as oestrus is approached. As well as employing secretion in scent marking the receptive female also produces secretion which the male samples immediately prior to copulation (Johnston, 1975, 1980). This attractancy is considerably reduced when samples are left in the environment for 12 hours.

Chemical cues from the secretion have a number of effects on the male. They reduce levels of aggressive and associated flank marking behaviour and they act as a potent attractant. Volatiles from oestrous vaginal secretion trapped on Tenax yielded a very complicated gas chromatogram. Fractions of the chromatogram were collected on Tenax traps and subjected to a behavioural assay in which male animals dug towards a hidden signal source (vaginal secretion or vaginal secretion components). From this study one component alone stood out as a major sex attractant. This was identified by GCMS as being dimethyl disulphide, present in secretion samples at levels of 5 ng/secretion sample. A peak due to dimethyl trisulphide also possessed behavioural activity but it was concluded that this effect was spurious as it was due to thermal breakdown of some of that compound into dimethyl disulphide (O'Connell *et al.*, 1979). Dimethyl disulphide could account for a substantial part of, but not all, the attractancy of the vaginal secretion. However,

234

volatile acids and alcohols also present in the secretion did not enhance the attractancy of the dimethyl disulphide.

Volatile fatty acids present included saturated nC_2, nC_3, nC_4, $isoC_4$, $isoC_5$, $isoC_6$ and $anteisoC_5$ acids, at levels ranging from 130 μg per female for nC_2 to 0.5 μg per female for $anteisoC_5$ acids. Phenylacetic acid as well as the C_1, C_2, nC_3 and nC_4 saturated alkan-1-ols (up to 1 μg per female) were present.

Although dimethyl disulphide is always present in hamster vaginal secretion at levels sufficient for its short range attraction and thus possibly accounting for the males persistent attraction to the female throughout the cycle under laboratory conditions, concentration does vary cyclically, being maximal at ovulation (O'Connell et al., 1981). It is possible that vaginal dimethyl disulphide is a product of the resident microflora, although this remains to be proved.

However, in addition to its male attractancy, a further property of hamster vaginal secretion which is under semiochemical investigation is its property of inducing mounting and attempted copulation in a male when applied to an anaesthetized animal. This 'mounting pheromone' effect is not associated with the dimethyl disulphide attraction, nor with the volatile acids and alcohols mentioned, but with involatile components in vaginal secretion. Gel permeation chromatography has shown that mounting pheromone activity is associated with protein macromolecules of relative molecular mass in the 15,000–60,000 amu range. Impure active fractions can be separated by ion exchange chromatography and it seems that lower molecular weight components also contribute to mounting pheromone activity.

An interesting experiment (Johnston and Zahorik, 1975) has shown that the male hamster can very rapidly learn to overcome its strong pheromone-like attraction to oestrous vaginal secretion if experience of the secretion is paired with an unpleasant event, in this case an intraperitoneal injection of lithium chloride to produce gastrointestinal illness, severe but brief. This illustrates the importance of context and past experience in the expression of a mammal's response to a chemical signal.

Volatile fatty acids also occur in the vaginal secretions of **cows**, with levels which vary cyclically during the oestrus cycle (Hradecký, 1978). A microbial origin for these acids has been discussed.

Chapter 8
Breath, saliva and the pig (Sus scrofa)

8.1 ORAL ODOUR AND SEMIOCHEMISTRY

With a few notable exceptions, mammalian saliva has been little studied as a semiochemical substrate. Even so, a number of examples are known where saliva is clearly employed in chemical communication, the most thoroughly studied example of this being that of the pig. Some attention has been given to the chemical ecological role of herbivore saliva components acting as plant growth stimulants (Dyer, 1980), although such effects may not be significant (Detling *et al.*, 1981).

Breath cannot be considered in isolation from saliva for breath contains many saliva volatiles, as well as plasma-derived volatiles from within the lungs, and in man the studies of these volatiles may be of medical importance. It also contains substances of microbial origin arising from the degradation of saliva, food particles and other debris in the oral cavity. This is a principal cause of halitosis or bad breath in man and is of concern to dental scientists (Stack, 1979). For example, 5-aminovaleric acid, an involatile component of anal sac secretion (see Section 5.3) of probable microbial origin (Albone *et al.*, 1976), has been observed in saliva (Dreyfus *et al.*, 1968). However, from discussions elsewhere in this text, it is clear that substances arising from such commensal microorgansims can have semiochemical significance and should not be disregarded.

Relatively little is known concerning such microbial breath volatiles. It was assumed that amines (indole, skatole, putrescine, cadaverine) (Hyatt and Hayes, 1975) as well as organosulphur compounds were largely responsible for oral malodour, and indeed all are present in such cases, but at the pH of the mouth it is found by headspace analysis that only the sulphur compounds (principally methanethiol and hydrogen sulphide with some dimethyl sulphide) are important odorants.

The medical dimension is exemplified by the fact that in women the levels of these microbial sulphur-containing breath volatiles vary cyclically with circulating hormone level, much as do the vaginal levels of lactic acid, and this effect can be used to detect ovulation although peaks do occur at other times in the menstrual cycle (Preti *et al.*, 1978; Tonzetich *et al.*, 1978). At ovulation the level of these compounds together is elevated two- to fourfold.

Some bacteriological studies have been undertaken and it is found that strongest odour production is associated with a mixed microflora. A useful survey of work in this area has been provided by Tonzetich (1977).

GCMS studies (Krotoszynski *et al.*, 1977) have shown that healthy human breath contains a most complex mixture of trace organic compounds, some 102 components in the concentration range 0.06–9.5 ng/l having been identified. However, three major trace organics are present at higher concentrations. These are propanone (120 ng/l), isoprene (33 ng/l) and acetonitrile (24 ng/l, mean values).

In a number of species (e.g. rodents) salivary glands are now known to be androgen sensitive and as in the case of the pig to offer semiochemical possibilities. Human saliva for example contains testosterone at levels which mirror circulating free testosterone levels (Smith *et al.*, 1979), so that levels in the saliva of men, as well as showing a circadian rhythm, are significantly higher than those of women and of children (saliva testosterone levels, men, 295 ± 36 pg/ml; women, 195 ± 25 pg/ml; children largely undetectable; Landman *et al.*, 1976).

A GCMS study of the free acids present in healthy human saliva revealed a complex mixture of compounds including lactic, succinic, phenylacetic, 2-phenyllactic, *p*-hydroxyphenylacetic, *p*-hydroxyphenylpropionic, myristic, palmitic, oleic and stearic acids as well as cholesterol (Ward *et al.*, 1976). Alkanols (C_{12}–C_{16}) are present as are phenols (phenol and *p*-cresol), aryl alcohols (benzyl alcohol and 2-phenylethanol), indoles (skatole and indole) and other compounds (Kostelc *et al.*, 1980).

In a number of species there is evidence of the possible semiochemical use of saliva. In the woolly monkeys (Lagothrix: Cebidae), there are accounts (Ullrich, 1954; Epple and Lorenz, 1967) of the male at times of sexual excitement producing chewing movements to accumulate such large quantities of saliva that the saliva drools out of the corners of his mouth, applying this to a surface, and then overmarking with his sternal gland. It has been suggested that the specific occurrence of long dense hairs on the ventral trunk in these animals may be associated with the development of body scents from sternal gland secretion/saliva mixtures (see, axillary organ, Section 4.7).

Similarly hedgehogs, for example *Erinaceus europaeus,* produce a frothy saliva which they apply to their spines (self-anointing behaviour) in the breeding season. This saliva produces 'a sharp, rank smell resembling horse or hedgehog's urine' and is detectable by the human nose up to 10 metres away. Although the functions of self-anointing are uncertain, they are presumably, at least in part, semiochemical (Brockie, 1976).

A rather different example involves gerbil Harderian gland secretion which when mixed with saliva yields a chemical signal (see Section 5.7).

In the bactrian camel (Wemmer and Murtaugh, 1980) saliva may also be a source of chemical signals. The ordinarily clear liquid is worked into a froth by excited animals of either sex and the copious foam distributed over the face and neck by head shaking. This is most particularly the case for the male in rut. However, the semiochemical use of saliva which has been most thoroughly studied is that of the pig.

8.2 CHEMICAL COMMUNICATION IN THE PIG

(Patterson, 1967, 1968a,b,c; Gower, 1972; Stinson and Patterson, 1972; Reed *et al*., 1974; Claus, 1976, 1979; Claus and Alsing, 1976b; Signoret, 1976; Claus and Gimenez, 1977; Booth, 1980; Perry *et al*., 1980)

Saliva provides an important vehicle for chemical communication in the pig and this will provide a useful starting point for the consideration of chemical communication as a whole in this species. It is well known that when it is sexually aroused, the boar will champ its jaws and drools a viscous, glycoprotein-rich frothy saliva which derives from its submaxillary salivary glands and which is quite unlike the more watery parotid/sublingual saliva involved in feeding. This viscous saliva contains two odorous C_{19}-$\Delta16$ steroids at exceptionally high levels. These are the musk-like alcohol. 3α-hydroxy-5α-androst-16-ene and in lower concentration the related ketone which has a urine-like odour, 5α-androst-16-ene-3-one. Only trace quantities of these compounds occur in female saliva and in the male their concentration is age dependent.

Figure 8.1 A boar drooling semiochemically active saliva. (Photograph courtesy W. D. Booth)

These substances play an important part in the reproductive semiochemistry of this species. Boars will pursue females whatever their reproductive state and although female odours may play a part in attracting the male, they are less important than visual cues. However, semiochemical cues of male origin arising from submaxillary saliva are of great reproductive significance. Initial interaction between a boar and a sow involves much head-to-head contact, the male placing its snout close to the female's head, champing its jaws to exude the viscous saliva and puffing its odour in the female's face. Oestrous females are distinguished by responding to this signal with a characteristic immobile mating stance. Interestingly, it is also observed that the same jaw champing and frothy submaxillary salivation occurs when boars meet in aggressive encounters so that it seems probable that the levels of the same chemical signals which make the boar sexually acceptable to the sow also play a part in determining the dominance relationships between males.

Experiments conducted on boars from which the submaxillary salivary glands have been removed surgically at an early age reveal not only that the female's response to these animals is much reduced in comparison to their response to intact boars, but that the surgically treated male's behaviour is itself modified, being unusually passive and suggesting a remarkable loss of male libido. Oestrous females are reported to appear to anticipate a chemical signal deriving from the submaxillary saliva and lacking this they are less ready to proceed to accept copulation but revert to other non-sexual behaviour.

It is of some interest that either or both of these odorous C_{19}-$\Delta 16$ steroids present in the boar's submaxillary saliva will function as such a chemical signal. Both steroid ketone and steroid alcohol will elicit a similar standing response when applied in aerosol form to the head of an oestrous female and indeed this has provided the basis for the first application of a mammalian semiochemical, an aerosol preparation 'Boar Mate' now being marketed for use by commercial pig breeders as an aid to detecting the sow's relatively short oestrous period.

The situation is, however, a little more complicated than it seems. The oestrous sow will 'stand' in response not only to olfactory signals, but also, for example, to the sound of the boar's grunting. In the natural situation the sow is exposed to a simultaneous combination of cues of many kinds, olfactory, visual, tactile and auditory, all of which play some part in stimulating the standing response, although it is clear that among these, olfactory signals are very important. Also, it is found experimentally that the oestrous sow will stand in response to the odour of boar urine or boar preputial fluid, substances in which these particular C_{19}-$\Delta 16$ steroids are either absent, or present at very low levels. Further, the oestrous female will respond to varying degrees to the odours of some other closely related steroids.

To the human nose a number of closely related 3α- and 3β-hydroxy C_{19}-$\Delta 16$ 5α-steroids possess a similar musk odour, while the 5β-steroids are odourless.

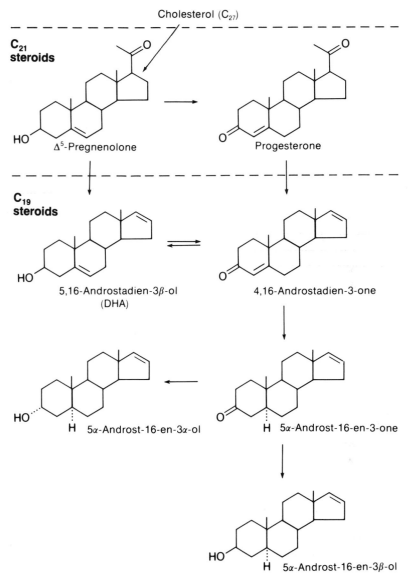

Figure 8.2   Biosynthetic route to 16-unsaturated C_{19} steroids in boar testis

Both the boar submaxillary C_{19}-$\Delta16$ steroids occur in boar testis at very high levels, considerably exceeding those of testosterone. Indeed it was from boar testis that this steroid alcohol was first isolated in 1944. Such high levels do not occur in the testes of many other species. Subsequent biochemical studies have shown that the C_{19}-$\Delta16$ steroids are biosynthesized by a route which is quite different from that leading to testosterone although both routes seem to be stimulated by similar hormonal events and corresponding diurnal fluctuations in plasma levels have been described. This is all the more remarkable, because until the discovery of their semiochemical function, no biological or biochemical significance could be attached to these substances which possess little or no androgenic activity.

Both the 3α-hydroxy- and the corresponding 3β-hydroxy-C_{19}-Δ16 steroids are formed in the pig testis. However, the 3β-hydroxy compound is largely eliminated from the body in urine as the odourless glucuronide conjugate, although as a component of preputial fluid or following deposition in the environment, the free steroid may be liberated by microbial action. This substance itself elicits the standing response in the oestrous female, albeit rather weakly.

In contrast, the 3α-hydroxy steroid is not eliminated in boar urine but is transported via the blood (as the sulphate conjugate) to other tissues and principally to the submaxillary salivary gland. The steroid ketone is similarly transported but as free steroid and by virtue of its more lipophilic nature concentrates in the fatty tissues. Following sexual arousal, blood plasma levels of this steroid and of testosterone increase synchronously and this leads to in the long-term elevation of the steroid ketone in fatty tissues which can be regarded as a reservoir for this boar semiochemical. Indeed our understanding of the chemistry of this system arose as a consequence of the GCMS study of the unpleasant taint which is apparent when boar meat is cooked and for which this compound was found to be responsible (Patterson, 1969).

Although the rich glandular fields of the skin of the pig and their possible role in chemical signalling have not been studied in any detail, numerous apocrine sweat glands which occur in the skin of the boar's back and sides contain these same C_{19}-Δ16 steroids, principally the musk smelling 3α-hydroxy-steroid, rather than the ketone of boar fat. Other secretory structures of pig's skin (Montagna and Yun, 1964) also include specialized serous glands having some resemblance to eccrine glands embedded in the skin of the snout (and by which it is lubricated), as well as clustered around a linear row large pores on the carpus (the carpal organ).

The boar submaxillary gland appears to be much more than an ordinary salivary gland. It seems to be an androgen target organ and its activity is influenced by circulating testicular androgens. As expected, its tissue contains relatively large concentrations of testosterone and its active metabolite, 5α-dihydrotestosterone, as well as very high levels of the C_{19}-Δ16 steroids. Studies for example on the histochemical localization of hydroxy-steroid dehydrogenase enzyme activity and on *in vitro* steroid transformations in boar submaxillary gland tissue show that this tissue possesses considerable steroid metabolizing activity of its own so that it is possible that some of its C_{19}-Δ16 steroid content could arise from synthesis in that organ.

Important as these steroid odours are, boar odour is generally dominated by other components. The principal of these arises from the repulsive dark brown ammoniacal fluid present in the boar's **preputial diverticulum**. Apart from ammonia, the major odorant appears to be *p*-cresol, accounting for

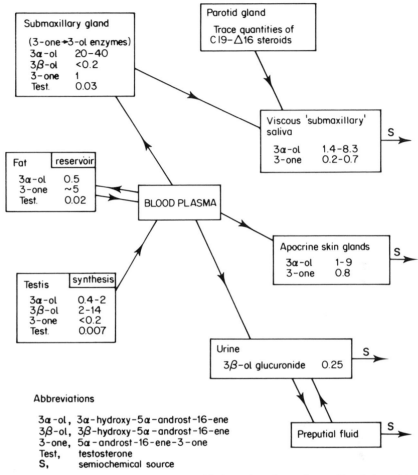

Figure 8.3   16-Unsaturated C_{19} steroids in various boar tissues (levels in $\mu g/g$)

>90% total phenol content (Patterson, 1967). Other phenols are also present at lower concentrations (phenol, *o-* and *m*-methoxyphenol and *p*-ethylphenol) as well as a range of low molecular weight acid components, which, however, contribute relatively little to the odour of the fluid owing to its alkaline pH(8.5–9.5) (Table 8.1). C_{19}-$\Delta16$ steroids may also occur in preputial fluid at low levels.

   The preputial diverticulum in the boar consists of a non-glandular invagination of the skin, a pouch of up to 150 ml volume located through a circular opening in the dorsal surface of the prepuce (Ellenberger and Baum, 1943; Sisson, 1955). Its contents consist of stagnant urine with some seminal fluid and the secretions of the male accessory sex organs (the boar possesses very large seminal vesicles and bulbourethral glands) and epithelial cells from the

Table 8.1. Low molecular weight acids in boar preputial fluid (Patterson, 1968b)

Saturated acids	Unsaturated acids	Aromatic acids
*Acetic	Pent-2-enoic	*Benzoic
Propionic	Pent-4-enoic	*Phenylacetic
Butyric	Hept-2-enoic	*2-Phenylpropionic
Isobutyric	Oct-2-enoic	
Valeric		
*Isovaleric		
Caproic		
2-Methylvaleric		
3-Methylvaleric		
Nonanoic		
Undecanoic		

*Principal components.

diverticulum walls. As in the case of the carnivore anal sac, many of the odorants present are expected degradation products of inputs to the diverticulum. It seems likely that this secretion could be a source of male chemical signals of reproductive importance.

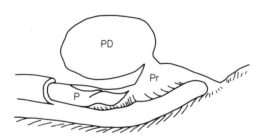

Figure 8.4   Preputial diverticulum (PD) of the boar in relation to the prepuce (Pr) and the penis (P). (Adapted from Ellenberger and Baum, 1943, by permission of Springer Verlag)

Chapter 9
Mammalian Chemoreception

Stephen G. Shirley

9.1 SENSORY SYSTEMS

So far we have concentrated on the sources of chemically coded information. Now we must turn our attention to the receivers. The first section of this chapter is a 'thumbnail sketch' of the sensory systems. Following this, olfaction is considered in more detail as it is probably the most important sense from the communication point of view.

To understand the operation of the sensory systems it is necessary to appreciate how nervous tissue transmits information. For those unfamiliar with the principles, an outline is given at the end of this section.

The olfactory system

Figure 9.1 shows the arrangement of the main parts of the olfactory system in the rat. The sensitive area is the olfactory epithelium, the tissue of which is typically pigmented yellow or brown. The olfactory epithelium extends from the perforated bone known as the cribriform plate over part of the surface of the turbinates. It also extends over part of the nasal septum, the partition between the right and left halves of the nasal cavity.

The purpose of this geometry is presumably to increase the surface area of the olfactory region. The turbinates are not situated directly in the path of the main air flow through the nasal cavity. The air reaching the sensory surface consists of turbulent eddies which break away from the main stream.

The olfactory bulb (the neural structure in which the information received at the epithelium is initially processed) lies within the skull on the other side of the cribriform plate, and is part of the limbic system of the brain.

A section through the olfactory epithelium is shown diagrammatically in Figure 9.2. It can be seen that most of the volume of the structure is accounted for by the supporting (sustentacular) cells, but that the ciliation of the nerve cells increases the sensory surface area. The number of cilia varies from species to species (Table 9.1), but even in an individual animal the number of cilia per cell is variable. For instance, on average the ox has 17 cilia

243

Figure 9.1 Section of a rat's head showing the location of the olfactory epithelium (OE) supported by the turbinate bones, immediately adjacent to the olfactory bulb (OB), from which it is separated by the perforated cribriform plate. The olfactory bulb is part of the brain (B). The scale is in millimetres

per cell, but cells can be found bearing any number from zero to 25. On a larger scale (Figure 9.3), the cilia are seen to possess two regions, a short thick proximal section and a long thin distal section. It has been found recently (in the frog) that different types of olfactory cilia are distributed, in clumps, across the epithelium (Mair et al., 1980). They have been named streamers (70–200 μm long and immotile), strokers (40–150 μm, with a slow beat) and wigglers (10–40 μm with rapid motion). The functional implications of this phenomenon are not known.

In cross-section the cilia show the usual 9(2)+2 pattern of subfibres near the base. In the outer regions some of the fibres may be absent.

Freeze etching studies reveal the presence of particles embedded in the olfactory cilia membranes (Kerjaschki and Hörandner, 1976; Menco et al., 1976; Menco, 1980a,b). These particles are approximately 10 nm in diameter and there are 30,000–150,000 of them per cilium. If these particles are exposed in vivo they could account for much of the surface area of the cilia. It is tempting to believe that these are the olfactory receptors, but this point is not proven. (A 10 nm particle, if a protein, would have a molecular mass of about 300,000 amu, which is not an unreasonable figure.)

In most peripheral nervous tissue, the axons of the nerve cells are myeli-nated; each axon is surrounded by Schwann cells which serve to insulate the

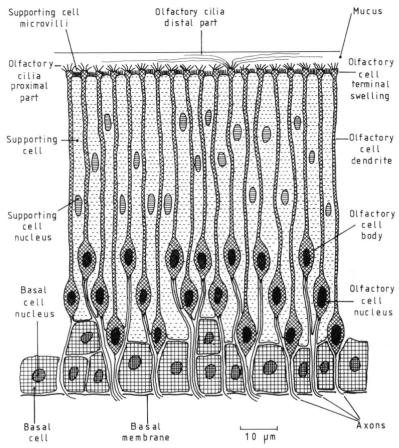

Supporting cell
microvilli

Olfactory cilia
distal part

Mucus

Olfactory
cilia
proximal
part

Olfactory
cell
terminal
swelling

Supporting
cell

Olfactory
cell
dendrite

Supporting
cell
nucleus

Olfactory
cell
body

Basal
cell
nucleus

Olfactory
cell
nucleus

Basal
cell

Basal
membrane

10 μm

Axons

Figure 9.2 Schematic section through the olfactory epithelium. For clarity,
most of the distal segments of the olfactory cilia have been omitted

Table 9.1. Some characteristics of the olfactory epithelia of a number of species

Species	Area of epithelium (cm^2)	Nerve endings per cm^2 epithelium ($\times 10^6$)	Cilia per nerve ending	Particles per cilium ($\times 10^4$)	Particles per cm^2 epithelium ($\times 10^{12}$)
Frog	0.5	4.5	6	16	4.3
Ox	1–4	7	17	3.6	4.3
Rat	1–4	8	16	4	5.2
Dog (Beagle)	75	6	16	6	5.7
Man	3	3			
Rabbit	7	9			
Cat	14	10			

From several sources, principally Menco, 1980a, b.

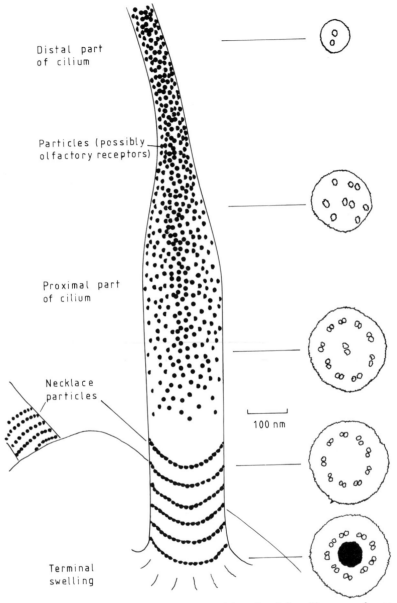

Distal part
of cilium

Particles (possibly
olfactory receptors)

Proximal part
of cilium

Necklace
particles

100 nm

Terminal
swelling

Figure 9.3   An olfactory cilium. The total length of the cilium may be as
much as several tens of micrometres. The particles are contained within the
membrane and are not external to it

axon and increase the speed of conduction of impulses. In the olfactory system, however, this is not the case. The bare axons run in bundles to the brain
through fine holes in the cribriform plate. The first synapse occurs within the
olfactory bulb within the brain itself. This has led some people to conclude
that the olfactory epithelium is an outgrowth of the brain rather than a true

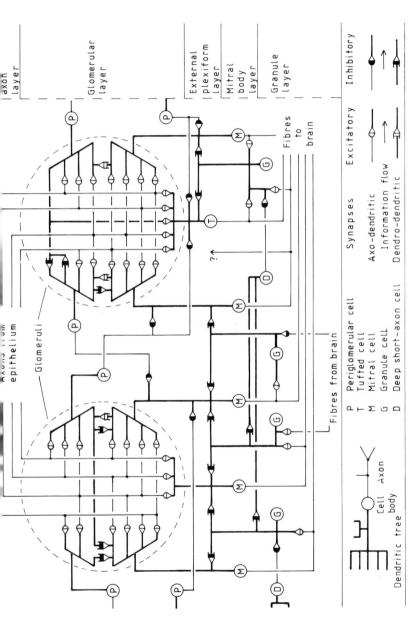

Figure 9.4 Schematic diagram illustrating the kinds of connection which occur within the olfactory bulb. It is not suggested that a group of cells connected precisely as shown would necessarily be functional. Note that despite the apparent complexity, the mitral cells, the main output from the bulb, connect directly via synapses with axons of the primary receptor cells

peripheral sense organ. However, against this view, the cells of the epithelium die and are replaced (the turnover time is about a month) (Moulton, 1974) and this behaviour is unknown for true brain cells. There have been no confirmed reports of centrifugal nerve fibres in the olfactory epithelium. This means that the direction of the information flow is from the epithelium to the bulb only, and there is no nervous control of the peripheral cells.

The primary olfactory cells possess several distinctive chemical features. Thus, they exhibit rather high levels of the dipeptide carnosine and of the enzyme carnosine synthetase. They also possess a unique protein called olfactory marker protein (Margolis, 1972; Keller and Margolis, 1975; Monti Graziadei *et al.*, 1977; Zomzely-Neurath and Keller, 1977). The functions of these substances remain unknown.

The olfactory epithelium within each nostril communicates with its own half of the olfactory bulb. The structure of the bulb is known in considerable detail, but only an outline will be given here.

Within the olfactory bulb, several layers of tissue can be distinguished (Figure 9.4). The axons of the primary receptor cells constitute the outer layer. Beneath this lie the glomeruli. A glomerulus is a roughly spherical region in which some 25,000 primary cells make contact with about 25 mitral cells. When it is remembered that each primary cell may contact more than one secondary, it is apparent that each secondary cell may receive input from 10 or 20 thousand primary cells. This is a very high degree of convergence, even by the standards of the nervous system in general.

The axons of the mitral cells leave the olfactory bulb by way of the lateral olfactory tract.

Other kinds of cells are found in the olfactory bulb. Tufted cells seem similar to mitral cells in that they make contact with the primaries within the glomeruli. The tufted cells are smaller than the mitrals and their cell bodies lie nearer the glomeruli. At one time it was thought that the axons of the tufted cells led to the opposite side of the olfactory bulb, but now they are believed to run to the higher centres of the brain along the same tract as the axons of the mitral cells (Nicoll, 1970). It is possible that the tufted cells serve virtually the same function as the mitral cells and are just slightly displaced.

Periglomerular cells carry information 'horizontally' at the level of the glomeruli and the deep short axon cells do this in the granule layer.

The granule cells have no axons; but they make numerous dendro-dendritic synapses. They form a feedback pathway for the mitral cells (Shepherd, 1963).

There are also fibres leading from the brain which can transmit information to the olfactory bulb.

The trigeminal system

Trigeminal sensations are often confused with olfactory ones (Doty, 1975). The trigeminal nerve has endings distributed over much of the body surface including, in particular, the surface of the eye and the mucous membranes of the mouth and the lining of the nose.

The free nerve endings are chemosensitive and this is noticeable particularly in the regions mentioned where they are most exposed. Trigeminal stimulation is responsible for the stinging sensation caused by salt or acid in a cut on the skin. With extreme stimulation the whole skin is chemosensitive; concentrated formic acid vapour can be felt as a warm tingling sensation on the back of the hand. As mentioned the trigeminal nerve endings are particularly exposed and hence sensitive in the nose. Stimulation of these nerve endings usually results in a sensation of tingling or, in extreme cases, pain. Much of the sensation experienced when smelling such things as hydrochloric acid or ammonia is in fact trigeminal. There have been patients who have lost their sense of smell (usually because of a blow to the head severing the olfactory nerves as they pass through the cribriform plate) but who retain the trigeminal sense.

To be a good trigeminal stimulant, a substance must be chemically aggressive or present in high concentration. It is unlikely therefore that the trigeminal sense would be involved in a semiochemical system.

Taste

In everyday life the word taste is used to describe multiple sensations. The temperature, texture and smell of food all contribute to what is loosely called taste. In sensory physiology the word has a more restricted meaning. It refers to the sensation elicited by water-soluble chemical stimuli acting on the taste buds.

The sensory organs or taste buds are located mainly on the tongue. Each contains a group of about 50 sensory cells. These are surrounded by supporting cells and basal cells. The sensory cells themselves have a fairly short lifetime, about 10 days, new sensory cells being derived from the basal cells. At the top of the bud is a pore of about 20 μm diameter which allows water-soluble substances access to the microvilli of the sensory cells.

Within the bud are synapses which pass information to afferent nerves. Typically each bud is served by about 50 nerves, but there is branching and each primary cell makes contact with several afferent fibres. An afferent nerve is not dedicated to a single taste bud; further branching gives each fibre a receptive area which may amount to several square millimetres of the tongue's surface and include many taste buds.

The taste buds themselves are grouped on papillae. Three different kinds of taste papillae can be distinguished morphologically. A human will typically possess 7–12 circumvallate papillae. These lie at the base of the tongue. Each bears an average of about 200 taste buds, although the range is wide. The foliate papillae lie along the edges of the tongue and the fungiform papillae are scattered over its upper surface, although they are most concentrated near the edges and the tip of the tongue. Nine thousand is a figure commonly

250

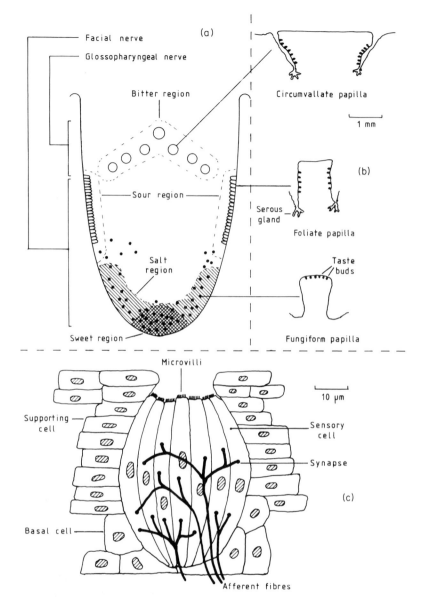

Figure 9.5 Taste. (a) Taste-sensitive regions of the tongue. (b) Cross-sections of various types of papilla. (c) Schematic representation of a taste bud

quoted for the total number of taste buds in man, but the number is very variable from individual to individual and declines with age.

The glossopharyngeal nerve (ninth cranial nerve) gathers information from the buds at the base of the tongue, while the facial nerve (seventh cranial nerve) serves the tip. Taste buds are also found in the palate, the pharynx and the larynx. In humans, such buds are found mainly in children. The afferent nerve fibres from these buds derive from the vagus nerve (tenth cranial nerve).

Four quite different basic taste sensations can be distinguished, namely sweet, sour, salt and bitter, different regions of the tongue showing different sensitivites toward stimuli from the four groups. The tip of the tongue is most sensitive to sweet sensations, the edges to sour and the base to bitter. The salt-sensitive region is at the tip of the tongue, but it extends further than does the sweet region.

When examined closely, the sense of taste loses its apparent simplicity (see also Schiffman and Erikson, 1980). Stimuli which fall into the sour category are acids and it is probable that the hydrogen ion is the main factor responsible. But there is no clear demarcation between the other three basic tastes. For instance, such diverse substances as saccharin, d-leucine and beryllium chloride are all sweet, while nicotine, quinine, l-leucine and magnesium chloride are all bitter and sodium chloride, sodium fluoride and magnesium chloride are all salty. A substance must dissociate in order to taste salty, but dissociation itself is not sufficient to cause saltiness.

Electrophysiological recordings from receptor cells, or from the afferent fibres, show that no cell responds to stimuli of one kind only. An individual cell may be sensitive to sweet or to salty, but it will respond to stimuli of all four classes. The overall sensation depends, therefore, not on the activity of any particular sensory cell or fibre, but on the pattern of activity in all the fibres.

Do the four basic tastes correspond to four different features of the receptor cells, such as four different receptor proteins, or, since the brain is interpreting a pattern, are the four basic tastes expressions of patterns of responses of numbers of receptor cells bearing no simple relationship to the behaviour of individual receptors?

As mentioned above, the sour taste seems to correspond to the hydrogen ion. Also there is a reagent, gymnemic acid which blocks the sensation of sweetness. If this substance is applied to the tongue, normally sweet substances become tasteless. This reagent probably acts by combining with what may be termed a sweet receptor protein.

So, out of the four basic tastes, one has an identifiable stimulus and one seems to have a specific receptor. On balance it seems likely that the four basic tastes do correspond to distinct receptors. It is worth mentioning that the classification of tastes into four basic categories was based on subjective judgement, backed by the spatial patterns of sensitivity which occur on the tongue. Some people confuse sour and bitter and some seem to detect different kinds of sweetness. It may be that a basic taste corresponds not to just one kind of receptor protein but to several related kinds.

The taste systems of mammalian species other than man have also been studied (Kare and Beauchamp, 1977; Boudreau and White, 1978).

The vomeronasal organ (see Section 2.3)

Many animals possess yet another chemical sensory system, the sensitive surface of which occurs in the vomeronasal organ, also known as Jacobson's organ. The structural details of this vary considerably from species to species. For instance, in the guinea pig, it is a pouch located beneath the nasal cavity and communicating with it via a narrow duct. Its opening is about 20 μm in diameter. In some other species the opening is into the mouth and in some the organ is a tube linking mouth and nasal cavity. The vomeronasal organ seems to be most highly developed in reptiles (especially in snakes) and in the lower mammals. In man it is vestigial. It is almost universal in children, but it often disappears in the adult and it rarely if ever has any nervous connections. In other species, it is functional and nerve fibres run, quite separately from those serving the olfactory epithelium, to a structure known as the auxiliary olfactory bulb.

The cells lining Jacobon's organ seem similar to olfactory receptor cells except that they possess microvilli instead of cilia (Bannister, 1968; Graziadei and Tucker, 1968; Kauer and Moulton, 1970; Bannister and Cushieri, 1972). As in olfaction and taste the cells are in a state of continuous turnover (Barber and Raisman, 1978a,b).

A Russian publication (Bronshtein, 1976) contains observations which may allow some speculation about the vomeronasal organ. In the primitive fish (Chondrichthyes and Dipnoi) all olfactory cells possess microvilli. In the other fishes cells with cilia and cells with microvilli are found in mixed populations in the olfactory region. In the land vertebrates the microvilli bearing cells are localized in the vomeronasal organ.

As already mentioned the vomeronasal organ is relatively most highly developed in reptiles. Indeed, in the snake, Jacobson's organ is more highly innervated than is the olfactory epithelium. We may be witnessing the evolutionary development of the chemoreceptive system, with the microvilli cells arising first and the system being extended by the inclusion of ciliated cells. In the land vertebrates, there is a separation into two distinct systems, with olfaction becoming progressively the more important of the two.

A snake will flick its tongue out of its mouth and then place the tip near the opening to the vomeronasal organ. This is believed to bring 'non-volatile odours' to the vomeronasal organ in a kind of sampling process. Since humans do not possess this organ, there is no common term for stimuli perceived by it, and 'non-volatile odour' is probably as descriptive as any other term. A snake prevented from performing this manoeuvre, loses its ability to track prey.

Mammals also transfer material to the neighbourhood of Jacobson's organ by a variety of manoeuvres (e.g. Bailey, 1978). Hamsters have a mechanism for pumping samples into the organ (Meredith and O'Connell, 1979; Meredith et al., 1980), and similar features are suspected in other species.

The vomeronasal organ will respond to gaseous odours, but the consensus of opinion is that its primary function is the detection of material in solution.

Should the vomeronasal system be regarded as an additional sensory modality, or is it a specialized receptor? There is still debate concerning the biological function of Jacobson's organ. It seems to play a role, for example, in the pheromonal control of oestrus in mice (Reynolds and Keverne, 1979), in the determination by males of the sexual status of female cats (Verberne, 1976; Verberne and De Boer, 1976), in the investigation of the female lemur by the male (Bailey, 1978), in feeding in the bat (Cooper and Bhatnagar, 1976), and, in conjunction with olfaction, in the attraction and sexual arousal of male hamsters by females (Winars and Powers, 1977; Powers *et al.*, 1979) and the recognition of young by female hamsters (Marques, 1979) and rats (Fleming *et al.*, 1979).

The organ is known to perform many functions in the snake and, considering the diversity of uses found in mammals it is probably safe to assume that Jacobson's organ tells the animal about almost any chemical in its environment. It should be regarded as a general sensory organ rather than as a specialized receiver for particular semiochemicals.

9.2 PROPERTIES OF NEURONS

This section has been included for those readers lacking a background in basic physiology and may be omitted by those with a knowledge of the subject. The structure of an olfactory primary cell has already been given. It is very similar to an ordinary neuron (Figure 9.6). The neuron consists of a soma or cell body, a long axon and many dendrites. The different parts of the cell perform different functions.

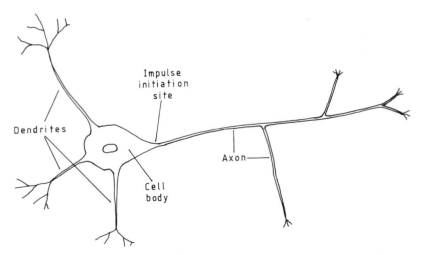

Figure 9.6   Generalized diagram of a nerve cell. The dendrites and the cell body serve as the 'input' region. The 'output' is in the form of impulses of electrical activity which travel along the axon and eventually cause the release of transmitter substance at a synapse

Information is transmitted across the synapses (junctions between cells) in chemical form, one cell releasing a transmitter substance which diffuses across the gap between cells and binds to receptors. This causes an electrical change in the membrane of the receiving cell. This change is graded (or continuous), i.e. its extent depends upon the amount of transmitter received. In the dendrites and cell body these changes are averaged (integrated). At the base of the axon a signal is generated which is of a different kind. It is an on/off (all or nothing) response depending on whether the integrated signal exceeds a threshold value. This signal is propagated unchanged along the axon, its arrival at the output synapses causing the release of transmitter. So, as information passes through the cell it undergoes changes in form; from chemical to continuous electrical to discontinuous electrical to chemical again.

As far as is known the primary olfactory cells behave like ordinary neurons as far as the generation and propagation of nerve impulses are concerned.

The cell membrane and the resting potential

If there are unequal concentrations of an ion on the two sides of a membrane which is permeable to that ion, then an electrical potential difference is set up. The potential difference is given by the equation

$$E = \frac{RT}{zF} \ln \frac{\text{(external ionic concentration)}}{\text{(internal ionic concentration)}}$$

where E is the potential of the interior with respect to the exterior, R is the gas constant, z is the ionic charge, T is the absolute temperature, and F is the Faraday constant.

The system behaves as an electrochemical cell and so can be characterized by two variables, the EMF (which is determined by the ionic concentrations) and the internal resistance which depends on the permeability of the membrane. This is near zero for a perfectly permeable membrane and infinite for a non-permeable one.

The situation in a neuron is rather more complex. Inside the cell there is protein which carries a net negative charge. The membrane is impermeable to protein so this does not of itself create a potential difference across the membrane, but it does affect the distribution of the other ions. Also the cell is equipped with ion pumps which remove sodium ions from the inside of the cell. The membrane is somewhat permeable to sodium, potassium and chloride ions, and each of these three species has different concentrations inside and outside the cell due to the action of the protein and of the sodium pumps. (The other ions also have an effect but, for our purposes, we can ignore it or consider these ions together with the chloride ions.)

The membrane can be regarded as consisting of three electrochemical cells in parallel (one for each ion), each with its characteristic EMF (set by the ionic concentrations) and an internal resistance (determined by the membrane permeability toward the particular ion). This is represented schematically in

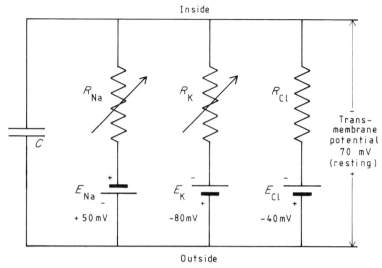

Inside

Outside

Figure 9.7   The equivalent circuit of the membrane of a nerve cell

Figure 9.7. The potential observed is determined by the EMFs and resistances of the three constituent electrical cells and tends towards the value of the EMF of the electrical cell whose internal resistance is lowest, that is, towards the value of the EMF due to the ion for which the membrane is most permeable.

In the resting neuron, this ion is potassium, and the resting potential of the interior of the cell with respect to the exterior is about -70 mV. Deviations from this potential are termed depolarizations if the potential of the cell becomes less negative, and hyperpolarizations if it becomes more negative. It is important to realize that the resting cell is not an equilibrium system but a steady state, and that the resting potential is achieved at the cost of the metabolic energy used to drive the sodium pump.

It is also most important to realize that the potential of the cell can be changed by varying the membrane permeability to one or more ions. Since the membrane is poorly permeable to all three ions, small increases in permeability can cause changes in the potential without setting up ion fluxes sufficiently large to change the internal or external ionic concentrations. In fact a neuron can continue to function for considerable periods of time with the sodium pump disabled, as only tiny ionic fluxes occur during the operation of the cell.

The generator potential

The details of the mechanism which links the binding of the receptor molecules to the initial depolarization of the cell are not known for the olfactory primaries. In neurons there are intermediate enzyme systems which are modulated by the receptors. These systems are adenylate cyclases and the resulting changes in the level of cyclic AMP controls ionic channels in the membrane. The permeability of the membrane is therefore changed and the resulting change of potential is termed the generator potential. The corresponding electrical current is the generator current. In general many synapses will be involved in stimulating a cell.

The initiation and propagation of the nerve impulse

The axon of the cell possesses specialized sodium and potassium channels. In the resting cell, these channels (or gates) are closed and no ions flow through them. The permeabilities of these channels are, however, controlled by the voltage across the membrane. In response to a depolarization the sodium channels open rapidly and then close rapidly; the potassium channels open and then close more slowly, see Figure 9.8.

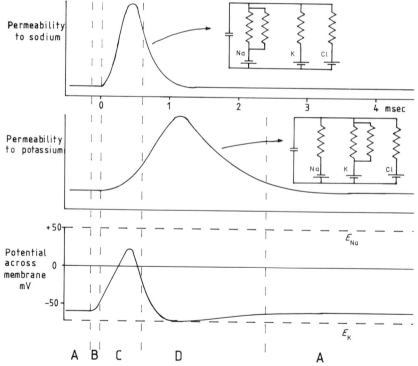

Figure 9.8   The events which occur during the passage of a nerve impulse. A   The membrane potential is at its resting level. B   A slight depolarization is caused by events at a neighbouring part of the membrane. C   This depolarization triggers the opening and closing of sodium channels, so that the initial depolarization is amplified. D   The potassium channels open and close more slowly, bringing the potential past, and finally back to, its resting value

As the sodium channels open, the membrane permeability to sodium increases and the potential within the neuron moves toward that of the sodium cell, that is, the initial depolarization is amplified. As the sodium channels close and the potassium ones open, the potential swings back toward that of the potassium cell. The net result is that any initial depolarization (provided it is large enough to initiate these events) is amplified into a pulse of fixed amplitude and fixed duration. Depolarization at one point in the axon of course affects the neighbouring points. The disturbance is amplified at these points and the result is a wave of depolarization moving down the axon.

It takes a little time for the ionic channels to recover before they can be triggered again. This ensures that the impulse travels away from the point of initiation and that the cell does not oscillate indefinitely.

Release of transmitter

The arrival of the impulse at the end of the axon causes release of a transmitter substance such as acetylcholine or noradrenaline. Prior to release the transmitter is stored in vesicles. Each impulse causes the release of a (relatively) constant number of vesicles and hence an approximately constant amount of transmitter. The transmitter diffuses across the synapse to the next cell where it causes depolarization (or hyperpolarization) of its membrane.

The main events in the operation of a neuron

These may be summarized as follows.

(1) Initial (graded) depolarization caused by transmitter action at the dendrites or at the cell body; or, in the case of an olfactory primary receptor a depolarization caused by the action of odorants.

(2) Generation of an 'all or nothing' spike response at the base of the axon. (This may occur in the cell body in olfactory cells).

(3) Transmission of the action potential along the axon.

(4) Release of transmitter substance.

Since the action potentials are of fixed amplitude and duration, and each causes the release of a fixed amount of transmitter, information is carried either by the frequency of the action potentials or by the intervals between them.

Cells of the central nervous system

The above description, which considers the dendrites as inputs and the axon as an output, is typical of a cell used to transmit information over fairly long distances. Within the central nervous system there is found a variation on the theme. In some cells the axon still serves as an output but the dendrites serve as both inputs and outputs. They can both be stimulated by and stimulate other cells. Cell junctions occur which are termed dendro-dendritic synapses. The olfactory bulb contains many examples of such synapses. Indeed the granule cells of the olfactory bulb possess no axons at all and communicate solely via dendro-dendritic synapses.

Also in the olfactory bulb it is possible to find reciprocal dendro-dendritic synapses, where information apparently flows both ways between the dendrites of two cells.

9.3 STUDIES ON THE OLFACTORY SYSTEM

Why does a particular odour smell the way it does? This really is the central question in olfactory research. A number of techniques have been used to investigate this problem.

Psychophysical methods

The ideas and work described in this section are largely those of Amoore and are conveniently set out in his book *The Molecular Basis of Odour* (1970). The fundamental concept of the stereochemical theory is that there are a number of receptor sites for odour. Stimulation of one kind of site only will give the sensation of a **primary odour**. The primary odour sensation will therefore be elicited by molecules of a certain shape, those which bind the chosen site, and by no other. Much of the work described here was done using rigid or nearly rigid molecules so the concept of shape is clear and there is no need to consider conformational changes which may occur as a molecule passes from the gas to the liquid phases and binds to its site.

Amoore first searched the literature and found that certain terms were used to describe odour quality much more frequently than others. He reasoned that these terms might correspond to the primary odours and he selected a typical member of each group of compounds to act as a standard. A panel of volunteers then compared other compounds to the standard, giving a degree of similarity of smell. The same compounds were then compared to the standard for similarity of molecular shape. Several techniques were used for judging similarity of shape. But with any measure of shape there was a correlation between similarity of shape and similarity of smell. The correlation was by no means perfect but about 30% of differences in smell could be attributed to differences in molecular shape.

Amoore also used the phenomenon of **specific anosmia** (Amoore, 1967, 1977). Certain individuals either lack the ability to smell certain compounds or their sensitivity towards these compounds is very much reduced. The phenomenon has been compared to colour blindness, where some people lack a specific colour receptor. Amoore reasoned that specific anosmia could be explained by supposing that the individuals concerned lacked one or more receptors and so were insensitive to certain primary odour(s).

Amoore set out to establish that the phenomenon was real and collected data on specific anosmics. In the case of one compound, isovaleric acid (isoC$_5$ fatty acid:sweaty odour) the evidence is clear cut.

About 2% of the population have difficulty in smelling this compound. The mean threshold of detection is about 100 times higher for the anosmic group than for the normals. The phenomenon is not just shown by isovaleric acid alone but also by other volatile carboxylic acids; and for the straight chain acids the effect is most marked in the range C$_4$–C$_7$, falling off smoothly outside these limits. Also there is a strong correlation between similarity of shape and similarity of odour for these acids. The phenomenon of specific anosmia is firmly established for isovaleric acid, but does this mean that this acid is a true primary odour? A very few individuals are extremely insensitive to this compound. Instead of detecting it at 100 times the normal threshold they require it to be presented at 10,000 times. We can postulate that this indicates the existence of two receptors A and B. A is 100 times more sensitive than B which is 100 times more sensitive than the rest of the receptors. Normal people possess A and perhaps B. The main group of anosmics possess B only, while the extreme anosmics possess neither. Perhaps A and B are alternative versions of the same protein (a mutation which has partially spread through the population). The situation is not quite as simple as there being a single receptor which is present or not but there is no theoretical objection to this possibility. In the case of isovaleric acid most individuals seem to be stable normals or stable anosmics, i.e. few people seem to change from one group to the other when the tests are repeated. This would be expected if the effect were due to a missing receptor. The really conclusive evidence would be to establish that anosmia is inheritable (like colour blindness). Unfortunately, these experiments have not been undertaken with isovaleric acid, although with another reported specific anosmia (musky, ω-pentadecalactone), there is evidence that the trait is inherited as a recessive gene (Whissell-Buechy and Amoore, 1973).

It has been argued that since isovaleric acid possesses such a repulsive odour, some individuals may be reluctant or unable to admit smelling it for psychological reasons. But the bulk of the evidence suggests that for isovaleric acid, specific anosmia is a real physical phenomenon and probably reflects an absence of receptor protein.

Specific anosmias to other compounds have been reported, for example, to 1-pyrroline (spermous odour), trimethylamine (fishy odour), 1,8-cineole (camphor odour), isobutyraldehyde (malty odour), 5α-androst-16-en-3-one (urinous odour) and ω-pentadecalactone (musky odour), but in no case is the evidence as firm as for isovaleric acid. The effects tend to be smaller, and the data more variable. Individuals may be anosmic on one test, but normal on another, and the psychological arguments carry more weight. It is unfortunate that the only substance to qualify beyond doubt as a primary odour should also be ionizable. It is not known whether the receptor for isovaleric acid responds to the free acid or to the valerate ion, which will be the dominant species at the pH of the mucosa.

At this point it may be relevant to mention some of the problems associated with psychophysical measurements. There is the well known and much ignored phenomenon of haematogenic smell. Many substances, if injected into the blood stream, can be detected by smell (and/or taste). The effect probably results from the substance desorbing from the blood into the nasal cavity during inhalation, or into the lungs during exhalation, and then being smelt in the usual way.

It must also be remembered that the human body is not surrounded by still air but by rising convection currents produced by the body heat. As the human breathes in, much of the inspired air is drawn from these currents. So a large part of the inspired air has been flowing over the skin prior to inhalation.

In these ways the olfactory mucosa is not normally exposed to clean air but to a general background of odour produced by the body. Such low level odours are normally ignored by the brain. But if the substance under test normally exists in the blood or on the skin, it is almost impossible to measure with precision its absolute threshold, since the detectable amount will vary with the background level. Isovaleric acid is not found in any great quantity in blood, but there are appreciable amounts of C_4 and C_6 fatty acids, and these can fluctuate rapidly, being increased by stress. For isovaleric acid, the effects are probably not large enough to create spurious evidence for specific anosmia, although with some substances, similar effects may be misleading.

Androst-16-en-3-one has been shown to have some of the properties of a pheromone in the pig (see Section 8.2), and a similar function has been discussed in man (see Section 4.7). This substance is said to have a urinous odour, and, as indicated above, specific anosmia has been described. However, experimental results are rather varied. Griffiths and Patterson (1970) showed that 8% of women and 44% of men are unable to smell this substance, while Beets and Theimer (1970) demonstrated that about 50% of men and women smell it normally while the remainder cannot detect it at all, register a different sensation, or are inconsistent in their response. However, Amoore et al. (1977) found that about 50% of people cannot smell it at all, there being no difference according to sex, while Koelega and Köster (1974) (also Köster and Koelega, 1976) observed that women are about twice as sensitive to the compound as are men. In a recent publication, Hendricks and Punter (1980) claim that 50% of subjects classified as anosmic to the musk exaltolide, ω-pentadecalactone, were able to smell the substance on being retested.

The results can be seen to be erratic. Should these results be interpreted as indicating a specific receptor protein which some people possess and some do not (a protein which in some cases seems to come and go) or is it more reasonable that the sex steroids normally present in blood are finding their way (leaking) onto the olfactory epithelium and holding the receptors in a more or less permanently adapted state, the sensitivity of the system changing with the fluctuating blood levels of the hormones and with their ease of access to the epithelium?

There are other problems associated with psychophysical measurements. Human olfactometers (machines which can generate known odour concentrations) are rare and difficult to use with high molecular weight odorants. Much testing involves the subject simply sniffing flasks of solution. This technique is relatively cheap and easy to employ but there are drawbacks. A human being will typically inhale about 0.5–1 litre of air in a series of sniffs lasting less than a second. If the nose is placed over a 250 ml flask most of the inhaled air does not come from the flask at all. Neither does the odour leave the flask by diffusion as is often stated. (Perhaps 1 ml of odorized vapour will diffuse from the flask in one second.) More is removed by turbulent currents in the neck of the flask displacing perhaps 10–50 ml depending on the pattern of sniffs and the distance from nose to flask. So for different individuals there is a dilution of the vapour by a factor of between 10 and 100 before the sample reaches the mucosa and individual differences of a factor of ten may be totally meaningless. It is with this sort of uncertainty that an individual whose threshold is perhaps 60 times less than normal will be classified as anosmic. The shortcomings in the experimental techniques can be overcome to some extent by using very large numbers of subjects.

Electrical responses of olfactory cells

Much of the recording of electrical activity of olfactory cells has been done on amphibians rather than on mammals. However, the evidence suggests that there is no great difference in the kinds of behaviour observable.

Most species possess olfactory cells which are too small to allow the insertion of an electrode to record electrical activity directly. As a result two kinds of indirect measurement have been made.

The EOG (the electro-olfactogram)

The generator potential gives rise to an external signal. The generator current flows, as shown in Figure 9.9, and causes a small voltage change at the surface of the olfactory epithelium. The amplitude of the response is of course dependent on many factors other than the true generator potential; in particular it varies with the thickness and electrical resistivity of the mucus. But for any given animal, it gives an indication of the underlying generator potential. This surface response is termed the electro-olfactogram or EOG. The response as measured is an average over the many cells which are situated near the electrode. For this reason the EOG can be a useful tool in the investigation of general receptor properties.

A typical EOG waveform in response to an odour stimulation is shown in Figure 9.10. It consists of three phases; an initial peak, a plateau and a decay.

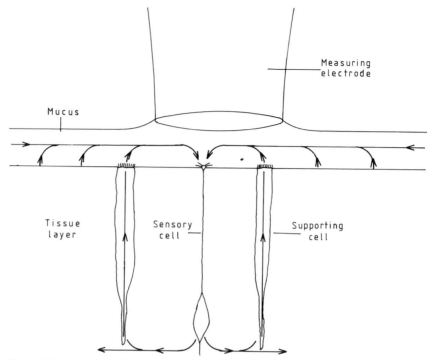

Figure 9.9   The flow of current produced by a stimulated sensory cell. The flow of the generator current through the electrically resistive mucus causes a small voltage difference to be produced between the measuring electrode and a reference electrode, imagined to be in contact with the tissue at a large distance from the measuring electrode

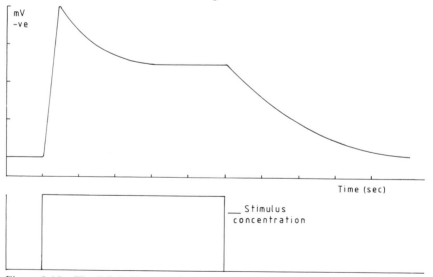

Figure 9.10   The EOG. The waveform measured by the electrode system shown in the previous figure in response to a square-wave pulse of odorant

Extracellular recording

It is possible to position a fine electrode so that its tip lies near an axon. Each time an action potential passes down the axon, the associated currents create a disturbance in the surrounding medium. Since all action potentials travelling along an axon are of the same size and duration, it is only necessary to record the time at which each occurs, and this is possible with the extracellular electrode.

The main problem which may be solved by recording electrical activity is to determine the response characteristics of the primary cells. For instance each cell might respond to a limited range of odour types, to one single odour or to almost any odour; it might also respond to an odour at one concentration and not at another.

Changes in response with concentration

It is well established that the EOG to an odour increases in amplitude as the concentration of the odour is increased. The variation is not linear. In many biological systems the response is proportional to a power (usually <1) of stimulus intensity over a wide range:

$$R \propto (S - S_0)^n$$

where R is the response, S is the stimulus intensity, S_0 is the threshold stimulus intensity, and n is a constant, usually <1.

The amplitudes of EOG responses do fit tolerably well on a logarithmic plot. Unfortunately most data for EOGs cover a rather narrow concentration range. Over that range they fit equally well on a double reciprocal plot. This would make sense if we assume that the initial depolarization is simply proportional to the number of receptors which are bound and that the receptors display simple kinetics. The intercept of the double reciprocal plot would then give the affinity constant for the odour receptor interaction. We are assuming that the EOG simply indicates the degree of binding. This has been done (Tucker, 1962/1963; Poynder, 1974; Senf *et al.*, 1980). The apparent affinity constants seem to correlate with the length of the hydrocarbon chain for the fatty alcohols. Such a correlation lends credence to the idea that the average response of the cells can be described by a simple hyperbolic law.

The peripheral cells show a fairly simple response to stimuli of different concentrations (Getchell and Shepherd, 1978a). As the odour concentration is increased the firing rate of the cell increases. This is in response to short duration odour pulses. With longer or quickly repeated pulses of odour the cells adapt; the firing rate decreases for a given concentration of odour (Getchell and Shepherd, 1978b). Different cells differ in their threshold concentration and in dynamic range.

In the olfactory bulb several different kinds of behaviour are observed (Kauer, 1974). When indirectly stimulated by a brief pulse of odour, cells

264

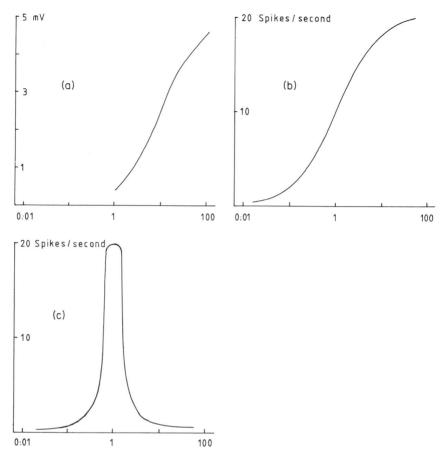

Figure 9.11   The effect of odour concentration at various parts of the olfactory system. The concentration axes are arbitrary, logarithmic and not synchronized. The curves are qualitatively illustrative only. (a) The amplitude of the EOG, which is proportional to the generator potential of the peripheral cells. (b) The spike activity of a peripheral cell. (c) The spike activity of a cell in the olfactory bulb

respond in one of the following ways:

(1) type 1, cease their spontaneous activity.

(2) type 2, cease their spontaneous activity initially and fire as the concentration of the odour falls (an afterburst).

(3) type 3, fire when stimulated.

(4) type 4, fire on the rising edge of the stimulus.

(5) type 5, fire on the rising and falling edges of the stimulus.

(6) type 6, do not respond.

Also cells can be found which change from type 3 to type 5 as the stimulus intensity is increased.

Most of these behaviours can be explained if a particular cell responds to a fairly narrow range of concentrations, see Figure 9.12. Note that the concentration is never assumed to be zero, even with the best precautions, there will always be some background odour perhaps emanating from the sample itself.

Type 4 behaviour perhaps suggests that some cells are sensitive to the rate of change of concentration.

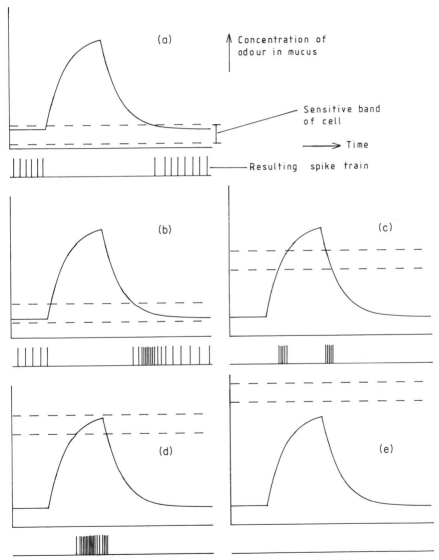

Figure 9.12   Types of activity which might occur when a cell which is sensitive to a band of concentrations of odorant is stimulated with a pulse of odour. The pulse is imagined to be initially square, but to become distorted by diffusion in the mucus. (a) Suppression of spontaneous activity. (b) Suppression with an afterburst. (c) Excitation on the rising and falling edges. (d) Excitation. (e) No response

Responses to odour quality

The shape of the EOG response is different toward different odours, but this may be an artefact. High molecular weight or very absorptive odours tend to give distorted pulses in olfactometers. The apparent differences in EOG shape may arise from this cause.

How do individual cells react to different odours? This problem has been investigated by Gestland *et al.* (1963). The results are anarchic. Each cell responds to many odours which seem unrelated either by structure or by smell. No two cells seem to have the same response pattern. This kind of result has been confirmed and it has led to the conclusion that each receptor cell carries a variety of types of receptor molecule. (This interpretation of the data will be questioned later.)

There have also been studies carried out using recordings made in the olfactory bulb. Døving (1966) found that neighbouring members of a homologous series of chemicals tended to stimulate the same cells. The relationship became weaker for non-neighbouring members of the same series.

MacLeod (1971) found bulbar units which responded to some odours and not to others; in this sense the results are 'well defined response' or 'no response'. However, there was no simple pattern relating the structure or smell of the stimuli to the response. The results were not random as statistical analysis showed that some groups of odours tended to stimulate the same bulbar units more often than by pure chance, but no clear, intelligible pattern emerged.

It is necessary to point out that these experiments which attempt to link the responses of single cells to the structure or smell of odours which trigger them were carried out before the systematic investigation of odour concentration effects. Each odour was presented at a single concentration. It has been mentioned above that a cell shows different responses to the same odour at different concentrations. That a cell does not respond to an odour at a particular concentration does not mean that it does not respond at all; perhaps simply the concentration is wrong. Perhaps much of the apparent anarchy revealed in the qualitative experiments is an artefact caused by neglecting the quantitative factors.

Spatial patterns

With special apparatus it is possible to stimulate small areas of the olfactory epithelium (Kauer and Moulton, 1974). In the salamander, different regions of the olfactory epithelium show differing sensitivities towards any particular odour. Different odours have different sensitive regions. At the time of writing, evidence is beginning to emerge which indicates that these phenomena may be displayed in mannals also. It should be remembered that a particular cell in the olfactory bulb can receive information from widely spaced locations in the olfactory epithelium.

Chemical modification studies of olfaction

Many chemoreception systems have been probed by the use of chemical modification techniques. Gymnemic acid is used in the study of taste, for instance. Three main kinds of chemical modification seem applicable to the olfactory system.

Affinity labelling

An affinity label is a reactive substance which is also designed to mimic the activity of a natural odorant. For instance ethyl bromoacetate is chemically reactive and is similar in shape and size to ethyl propionate.

Ethyl propionate is a fruity smelling liquid; the bromoacetate (which is a tear gas) is very irritating to the eyes and mucus membranes. However, at low concentrations, it is said to be fruity smelling. As the two molecules are alike in shape and size, we might expect the bromoacetate to bind at the same site(s) as the propionate ester and, by reacting, disable them selectively. The overall result would be that the sensitivity of the olfactory system towards ethyl propionate would be reduced while that towards non-fruity smelling substances would be unchanged. There is now evidence that small effects of this type do occur (Persaud *et al.*, 1980, 1981).

There are two fundamental objections to this approach. First, affinity labels simply do not smell like their supposed analogues. Secondly, if the odour is a primary then the technique should work well; but if very few odorants are primary and most odours bind to many sites then the effects will be small and the supposed affinity label will reduce the sensitivity of the system towards many odorants, related or not. With small effects or non-effects we cannot be sure if the site model of the system is incorrect or if the affinity label is simply binding to a great many sites if most odours simply interact with more than one site.

Photoaffinity labelling

This technique overcomes the first of the above objections. A stable chemical species is created which has an odour. It also contains a group which can generate a free radical under the influence of ultraviolet light. Such substances behave as ordinary odours eliciting normal EOGs. If they are activated while at the binding site they should bind to it and cause loss of function. Again there is some evidence that this kind of behaviour can be observed (Menevse *et al.*, 1977), but again the small differential effects observed suggest multisite binding.

Group-specific reagents

Reagents exist which can modify certain amino acid residues in proteins. If the receptors are protein and there are many kinds of receptor site, we would

268

expect them to differ in their amino acid composition. Modifying one particular amino acid, therefore, should disable the receptors semi-selectively. There is evidence that this is so (Shirley *et al*., 1981), for the sensitivity of the olfactory tissue towards different odorants changes differentially on treament with such reagents. In one instance, the sensitivity was particularly reduced towards odorants possessing roughly similar molecular shapes (although exhibiting different smells), suggesting that one particular site was being disabled, this site binding molecules of a particular shape, the characteristic smells arising from the particular combination of sites each odorant could bind. But again, the differential effects produced by group-specific reagents are relatively small.

A modification of the group-specific reagent technique utilizes a chemical protection effect. Here an odorant is applied simultaneously with the reagent. By competing for binding sites with the reagent, the odorant should protect these sites from being disabled chemically. Moreover, the odorant should protect its own binding sites specifically. Some degree of differential sensitivity changes have been observed in such protection experiments, but to evoke these changes the protecting odorant must be present at very high concentration (Getchell and Gesteland, 1972; Menevse *et al*., 1978). This argues that the interaction between an odorant and its site is rather weak.

Some of the comments in the section on structure–activity relationships are also pertinent to chemical modification studies.

A model of the olfactory system

How then does the olfactory system work? Outlined here is a model, a design for an olfactory system. It is certainly not proven. It is not even widely accepted, but it does have many of the features found in the real system.

How can an odour be recognized? One strategy might be to have many different receptor types, each one capable of binding just one odorant molecule. To account for olfaction, this would require a very large number of different kinds of receptor.

There is, however, a second strategy, and, almost paradoxically, it gains its qualitative information (what kind of smell) by performing quantitative measurements. The 'quantity' measured is the 'concentration of odour in the olfactory mucus'. This means the total concentration of **all** molecules, not the concentration of any particular type of molecule.

Any technique for measuring such a concentration is necessarily biased, being more sensitive towards some odorants than towards others.

Biased concentration measures are frequently encountered in chemistry. If we have, for example, a solution of many solutes, we can obtain concentration estimates by:

(1) evaporating and weighing, a measurement biased toward high molecular mass species;

(2) osmotic pressure measurements, a method biased toward low molecular mass species;

(3) pH determinations, a measurement biased highly, but not entirely, in favour of hydrogen ions;

(4) conductivity methods, an approach biased in favour of ionized species;

and so on for other possible approaches (e.g. from refractive index, density, light absorption, etc.).

All these properties, and many others, are concentration dependent. Some are easily related to particular molecular features of the substances present, and some are not. The point is that we can characterize a mixed solution by making a whole range of different concentration-dependent measurements and arrive at a set of numbers which describe the particular solution in question.

It does not even matter which particular set of measurements we choose to make, provided that there are enough of them and we are consistent in our techniques of measurement. Once a solution has been encountered and its 'concentrations' measured, we can 'recognize' another sample of the same solution by repeating the measurements and comparing the new values with the old. Let us choose our olfactory receptors such that each is capable of binding many different odours and, most importantly, let the 'output' (fraction of the receptors which are bound at any one time) depend on the concentration of the odour.

Any particular kind of receptor will then function as a concentration measuring probe. The measurements are biased since any particular type of receptor will display different affinity constants towards different chemical species.

So each kind of receptor will return a different estimate of the total concentration. In Figure 9.13, substance X binds to three receptors in a 4:3:2 ratio. Substance Y binds to the same three receptors but in a 1:5:1 ratio, these two are therefore distinguishable.

Changing the concentration of X results in the point which represents the binding moving towards or away from the origin but the 4:3:2 ratio which characterizes X does not change.

A substance showing affinity constants which are nearly in a 4:3:2 ratio will lie in the 'cone' and should be regarded as being similar to X. A mixture of X and Y will be represented by a point which lies between X and Y on the binding diagram.

It is clear that with three receptors we can distinguish very many more than just three different odours using the concentration measuring approach.

If such a system is to work we must design our receptors to bind to many different odours and we must ensure rather weak binding so that no receptor becomes saturated. In other words we **require** receptors which are not very selective and which bind odours rather weakly.

Normally, binding in biological systems involves hydrogen bonding, ionic

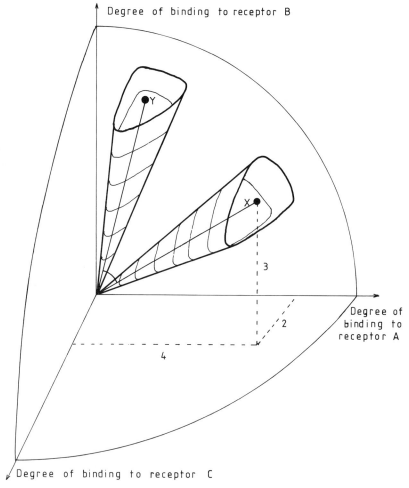

Figure 9.13 The binding of an odorant to three different receptors can be represented by a point in three-dimensional space

interactions or other such point-to-point interactions. This tends to make the binding strong and strongly selective. There is, however, a kind of binding which has precisely the properties we require—hydrophobic binding.

Organic molecules in aqueous solution cause an ordering of water molecules around themselves. Each organic molecule becomes in effect the nucleus of an 'iceberg'. The ordered water is in a low entropy state. The result is an entropic force tending to associate organics in the presence of water. The 'binding energies' involved would be of the order of 1–5 kJ/mol for typical odorants, equivalent to one or a few hydrogen bonds/molecule. This is suffi-cient energy to cause a significant binding of odorants to lipid or hydrophobic regions of a protein. The hydrophobic bond has the required lack of selectivity, while the interaction between any particular odorant and any particular receptor would also depend on features peculiar to the pair, such as goodness of fit of general shape and any non-hydrophobic binding features.

So we choose as our receptors proteins which are constrained so that a hydrophobic region is in contact with the aqueous mucus. (The binding should not be considered purely hydrophobic, just that the hydrophobic force will account for more than half of the binding energy in the majority of cases.) The hydrophobic force alone should result in dissociation constants of about 1 mM for a medium molecular weight odorant.

Since the binding is weak, we require a great many receptor molecules of any one type in order to increase the sensitivity. How are these to be arranged on the receptor cells?

It is at this point that this model deviates most widely from the commonly accepted view. Here each cell is postulated to possess only one type of receptor molecule. This arrangement has several advantages.

(1) It maximizes the sensitivity of the cell.

(2) It allows the cell to pass on the information it receives at a reasonable rate. (The transmission of information is governed by the laws of information theory. Basically a cell which transmits information about two or more different kinds of receptor must be capable of signalling faster than a cell which reports on one kind of receptor only.)

(3) It simplifies subsequent processing.

(4) It allows for easier cell replacement.

The last point has other ramifications. How can the sensory cells be replaced without impairing the performance of the system? If small groups (maybe tens) of like cells are clustered together on the epithelium then a developing cell can 'look' at its neighbours to determine what kind of cell it is to become and can follow the neighbouring axons to the olfactory bulb to make its 'correct' connections. Such processes are common in developing embryos where cell differentiation and the establishment of connections have been studied. On the other hand, if a cell carries many kinds of receptor it can no longer be regarded as being of distinct type and the process of replacement is made very much more complex.

The receptor cells themselves also differ in sensitivity. One may become active with only one receptor molecule bound, another may require 10 thousand. In essence each cell is 'tuned' to be sensitive to a particular range of odour concentration. This allows the cell to transmit precise information about a limited concentration range rather than imprecise information about the whole range.

But this feature of the system defeats experiments which attempt to relate odour quality to cell output without paying proper regard to concentration. A cell which responds to say amyl acetate at a particular concentration may be a 'sensitive cell' whose receptors are rather insensitive to the stimulus, it may be an 'insensitive cell' whose receptors bind the acetate rather strongly, or it may be intermediate on both counts. To regard it as an 'amyl acetate cell' is nonsense.

In the olfactory bulb the information from many primary cells is averaged and filtered; but it remains primarily quantitative information. (We have seen that bulb cells respond to particular concentrations of odour.) There is also probably some differentiation to obtain the rate of change of concentration.

All that remains is to recognize the odour. Referring to Figure 9.13, the line from the origin to the point X can be regarded mathematically as a vector in a three-dimensional space. The line has length and direction. As the concentration of the odour changes the length of the vector changes but the direction remains invariant (we have chosen the receptors so that saturation is not normally a problem).

So we can answer the question 'is odour X the same as odour Y?' by measuring the angle between the two corresponding vectors. In this case the angle is quite large—the two odours are dissimilar. The process of calculating the angle is (mathematically) a simple one. Essentially the angle can be found by calculating the scalar product of the two vectors.

This is the process by which the model system recognizes odours. The first time an odour is encountered its vector is remembered along with any associations the animal gains from its other senses. Subsequently the vector which is perceived at any time is multiplied by this remembered vector. If the angle is small associations are recalled. It is a learning system.

It is not to be imagined that the brain explicitly performs a vector multiplication nor that the output of any primary cell or bulb cell is strictly proportional to concentration. No, the information is coded in a non-linear form and distributed between many cells and it is on this non-linear representation of the data that the brain performs its operation which is analogous to the vector multiplication.

This model is therefore characterized by the following features.

(1) The receptors are protein.

(2) The primary interaction is hydrophobic.

(3) Each receptor can bind many different odours and each odour can bind to several receptors.

(4) The binding is rather weak.

(5) Each primary cell has receptors of only one type.

(6) The cells differ also in sensitivity.

(7) The information flowing through any part of the peripheral system is (non-linearly coded) quantitative information relating to the concentration of odour as perceived by a particular kind of receptor.

(8) Qualitative information can only be obtained by examining all the quantitative information streams.

(9) Recognition occurs by comparing a stimulus with a previously encountered stimulus. It does not depend on any particular receptor property.

It is often stated that the olfactory system is rather insensitive to stimulus concentration. How does this affect the idea of a concentration measuring olfactory system? The answer is not at all. An odour is identified by the ratio in which it binds to several sites. It is the ratio which is important.

At the same time as the direction of the concentration vector is found the length may also be determined. The length carries information about the absolute concentration of the odour. The absolute concentration is of course dependent on the distance of source to nose, wind speed etc. It is not a very informative parameter and the animal largely disregards it for that reason. But the receptors are all bathed in the same medium so the binding ratios can be measured with high precision and it is these ratios that are informative.

Structure–activity relationships

One property of the mammalian olfactory system is its apparent lack of strong structure–activity relationships among odorants. There are examples of structurally related molecules which smell similar, and of structurally related molecules which smell different. There are also examples of structurally unrelated molecules which smell similar and of structurally unrelated molecules which smell different. There are enough examples of each type that none of the four categories can be considered to be 'freaks'.

The kind of structure–activity relationships that are found are vague generalities. Molecules of similar shape tend to smell similarly, and in most cases adding one CH_2 unit to a hydrocarbon chain will not drastically change the smell. In simple small molecules, there seem to exist smells associated with simple functional groups, such as carboxyl, keto and thiol. But these generalities contrast sharply with the kinds of precise structure–activity relationships found in some other chemoreceptive systems, such as the insect pheromone receptors and the acetylcholine receptor.

A plausible explanation for this difference can be advanced. If there is a single kind of receptor molecule, we can describe a precise structure–activity relationship for the ligands (e.g. odorants) which it will bind. Such ligand molecules must possess those features which promote binding to this receptor (and also lack those features which would prevent this binding).

If the binding is of the point-to-point type (hydrogen bonding, ionic bonding or any kind of binding in which a particular group or atom of the ligand interacts with a particular group or atom of the receptor) then the first half of the condition means that the ligand must possess the correct groups in the correct locations and the second half summarizes the absence of any steric hinderance. This gives rise to the precise 'classical' type of structure–activity relationship.

But if we have (as outlined in the last section) a system with many different binding sites, there is a subtle change. Potentially each site may bind many different odorants and each odorant may bind several different sites. (This is not a consequence of this particular model, but a property of any model which has far fewer sites than there are odours and which can discriminate between more odours than there are sites.)

An odour is characterized by the particular combination of sites to which it binds (and in the proposed model, by the relative degree of binding).

The general structure–activity relationship now becomes clear. A molecule must possess those features which promote its binding to some particular sites (and lack those features which would prevent that binding) **and** lack those features which would promote its binding to the other sites (or possess features which would prevent such binding).

This is still a structure–activity relationship. To fix ideas, if there are 12 kinds of receptor and the sensation of say mint is caused by an odour binding to sites 1, 2 and 3, then a 'minty' molecule must bind to sites 1, 2 and 3 **and not** to sites 4–12. This puts constaints on the molecular architecture but a great many of the constraints arise from the condition that the molecule **does not bind** to sites 4–12.

In most systems it is sufficient to explain binding only. But in olfaction we know that almost any chemical sufficiently volatile to reach the olfactory epithelium will elicit a sensation, i.e. it will bind to something. In olfaction binding is the norm and we need to explain both why a molecule binds to some receptors and why it does not bind to others.

Thus the structure–activity relationship has assumed a 'negative' quality. There is no reason to expect the simple and concise 'classical' relationships which stem from a consideration of whether or not a molecule binds to one site.

In the case of the model proposed in the previous section it is also necessary to explain the relative strengths of binding to form the structure–activity relationship.

The absence of clear systematic structure–activity relationships does not prove the multiple-site multiple-binding kind of model but it does arise naturally out of such a model.

Will clear structure–activity relationships ever be found? If the individual receptor types can be separated and their binding properties towards a particular odour measured then it should be possible to predict odour quality from such data. But such experiments lie far in the future.

There are, however, a few special cases where we might expect to see the classical kind of structure–activity relationships. It is probably safe to assume that the different receptor types would differ in size; and that the large sites would tend to bind large molecules and the small sites small molecules. A very large or very small odorant molecule would have intrinsically therefore a very restricted choice of sites. There are simply not enough examples of very small odorous molecules to establish a firm structure–activity relationship, but amongst the very large such relationships might be found.

The danger is that if classical structure–activity relationships are found among very large molecules then they may be held up as examples of something which is 'typical' in olfaction when they are, in fact, a special case.

If the mammalian olfactory system is of the multiple-site multiple-binding type there are implications for the chemical modification studies. So far these have been performed using the EOG to indicate if effects are produced. The EOG is an average measure. Suppose there are 12 receptors and a 'perfect' affinity label (one that selectively disables just one receptor); after the tissue has been treated the EOG will be produced by the other 11 receptors. If most odours do in fact bind to more than one site this EOG will not be very different from the original. What is needed is a reagent which disables 11 sites and leaves just one functional. This is probably beyond the ingenuity of the chemist.

The model proposed in the last section does, however, have just one receptor type on each cell. The perfect affinity label should therefore disable certain cells and leave others unscathed. It is very curious that experiments using affinity labels and single cell recording have not been done.

9.4 OLFACTION IN PRACTICE; SENSORY STRATEGY AND COMMUNICATION

The range of the sense of smell

There seems to be a popular belief that olfaction is primarily a long range sense. In fact smell may be used in two different ways depending on the distance from the source to the nose. There certainly is a 'long range mode' of use of the sense. No one who has stood downwind from a factory or has seen a dog sniffing the breeze can doubt that animals can detect odour sources at long range.

The dog is an animal normally thought to have a very well developed sense of smell. It has been shown that dogs can distinguish between different kinds of buried objects provided they are shown the site of burial or can see the disturbed earth. In the absence of such clues as to the approximate location of burial their ability to detect such objects by smell is rather poor (Ashton and Eayrs, 1970). The coyote (another mammal with a well developed sense of smell) can locate its prey by smell only at a distance of 2 metres in a light (10 km/h) wind and at 5 metres in a 40 km/h wind. Vision is usually more important than smell to the coyote in the location of its prey (Bekoff and Wells, 1980).

It is almost as if there were two uses for the sense of smell. At long range it can inform the animal about general conditions prevailing upwind, but only at short range can it give precise information about the position and composition of the scent source.

This is hardly surprising. If an odour is released into the air it rapidly becomes diluted and mixed with other odours, and much of the information contained in the original odour source is soon lost. As the distance from source to nose increases the reliability of the received information falls off dramatically. An animal may be able to tell that there is an odour source 'somewhere upwind' but for precise and reliable information it needs to approach it closely. In this context closely means centimetres or millimetres. A human trying to identify an odour will bring the source close to the nose.

An odour that is at all times spread throughout the environment can convey no information to an animal. In order to be informative, it must be localized in space or time or both. This localization is usually achieved (deliberately or accidentally) by depositing relatively non-volatile substances on the ground or by carrying a scent source.

Most useful information therefore occurs at or near ground level and, to utilize the information, an animal must have its nose close to the ground. The upright human posture places the nose away from the ground which may in part account for the relatively poor human sense of smell.

Sensory strategy

An animal needs to react to events occurring in its environment in order to survive. However, it can only detect those events by means of its sensory organs.

It is important to distinguish between the event in the external world and the stimulus which the animal receives (the odour which reaches it or the pattern of light which falls on the retina). Potentially there is a vast amount of information available to an animal, the problem is in filtering this information to extract that which is relevant.

There are two basic strategies for dealing with this problem.

(1) Assume that there are rather few kinds of significant events and that each will produce a unique stimulus, detect just those stimuli and react accordingly. In other words the animal would tackle the filtering problem by adopting highly 'tuned' sensory organs. There would be little need for further processing of the information as any stimulus which falls outside of the range of the sensory organs is automatically ignored.

(2) Assume that almost any event may be important, gather as much information as possible, and perform the filtering at a higher level in the nervous system. This approach calls for 'general' sensory organs and extensive further processing of information.

These two strategies represent extremes in a continuous spectrum of possibilities, as any real sensory organ is in some degree tuned and some degree of information processing is always necessary. The difference is in the degree of complexity and in the flexibility.

The first category is approximated to by the insects, the second by the mammals. The male mosquito can detect sound but only those frequencies produced by the female's wing beat. This system could with some justification be regarded as a 'wing beat detector' rather than an 'ear'. The human too can detect sound, but no one would describe the human ear as a 'symphony detector' or a 'speech receptor'; it is a general sense organ which collects information about any sound in the environment. Compared to the mosquito's, the human sense of hearing is 'untuned'. The mosquito filters the useful information from the stimuli at the peripheral sensory level. The human performs the filtration at a higher level.

In insects, there are specific pheromone receptors which react to just one or a few closely related chemicals. These are 'tuned' receptor systems. But no such 'tuned' receptor has yet been found in a mammalian system. The insect needs to know about specific chemicals in its environment, while the mammal has an interest in a very wide range of chemical stimuli in its environment. Each seems equipped according to its needs. For this reason, it is probably fallacious to try to draw close parallels between insect and mammalian chemoreception.

The sensory strategy also has implications for communication. The female insect has 'specialized equipment' for generating a pheromone and the male has 'specialized equipment' for detecting it. Evolutionary changes must affect both transmitter and receiver and such linked changes are rather unlikely. Further, the communication potential of such a system is rather limited. The presence or absence of a particular chemical signals the presence or absence of a female at a particular stage of development. It cannot signal very much more, but this is sufficient for the needs of the insect.

With the mammalian strategy, the 'freezing' of communication systems need not occur. Since the olfactory system is 'general' and can detect almost any odour anyway the 'scent transmitters' can be modified almost without constraint. Each individual animal can have its own characteristic smell which can be learned by those other individuals to whom this matters. Subtly differing physiological states can be distinguished.

Paradoxically it is by discarding the pheromonal type of receptor that detailed communication becomes possible.

Appendix
Key Chemical Methods

Although it would not be appropriate to go into great detail here concerning the range of chemical techniques relevant to mammalian semiochemical studies, some further remarks relating to the scope and significance of the most central chemical methods might be useful both to chemists and to biologists interested in the subject.

A.1 THE ANALYTICAL APPROACH

The chemist may approach the sample which is to be analysed in one of two ways.

Approach 1: attempted identification of what is present

He may seek to identify, and probably to quantify, as many of the components of the sample as possible whose properties fall within the particular 'window' defined by the analytical method selected. To do this he will first need to separate the components, for separation is a general precondition for identification. Because semiochemical substrates are, in general:

(1) highly complex in composition,
(2) available for study only in relatively small quantity,
(3) likely to contain compounds of biological importance as minor or trace components,

the analytical methods appropriate to their analyses require

(1) high resolving power, or ability to separate and distinguish between large numbers of similar compounds,
(2) high sensitivity.

These considerations effectively determine the types of analytical approach appropriate to such a study.

Approach 2: attempted detection of specific compounds

Alternatively, he may seek to detect, and probably to quantify, in the semiochemical substrate a preselected known compound or group of compounds.

278

In such a procedure it is not always necessary to separate the compound or compounds under study in pure form provided one has available a specific assay for that compound with which the other substances present are unlikely to interfere either by reacting themselves or by inhibiting the test reaction. In most of such cases, it is advisable to undertake a preliminary fractionation of the crude sample if only to increase the overall sensitivity of the assay. An example of this approach is the use of radioimmunoassay to measure, for example, the levels of a particular sex steroid in a biological sample. Although the method can be extremely sensitive, and can be used routinely, each assay provides only one piece of narrowly defined information on what can be an extremely complex biological mixture. However, many assays for specific substances (for example, those based on GC peak area measurements) do incorporate the necessity of achieving the separation of the substance under study.

A.2 CHROMATOGRAPHY

Gas chromatography (GC) (McNair and Bonelli, 1969)

Of the large number of chromatographic methods which are available for the separation of complex natural mixtures, gas chromatography is proving itself to be of central importance in semiochemical research as in so many other fields and this ascendency is certain to continue. This is because the technique possesses both high resolution and high sensitivity and is appropriate for the separation of components which volatilize without decomposition up to about 300 °C. Many involatile or thermally labile low molecular weight compounds can also be examined by GC following chemical conversion to suitable, related derivatives (derivatization, for example, acids → methyl esters; alcohols → acetates/trimethylsilyl ethers/methyl ethers; amines → amides) (Blau and King, 1978; Drozd, 1975). Volatile fatty acids, for example, if not subjected to gas chromatography directly, may conveniently be converted into their benzyl esters first (Liardon and Kühn, 1978).
The principle of the method is simple. The sample under investigation is rapidly evaporated in an inert gas stream (nitrogen, helium and sometimes hydrogen) (mobile phase) and carried as a vapour through a tube (column) packed with a solid or coated with an involatile liquid (stationary phase). The column is mounted in a precision oven which may be maintained at a constant temperature (isothermal operation) or operated in a precisely controlled temperature programmed mode. Components pass through the column at different rates depending on their differential interaction with the stationary phase and are eluted from the column at different time intervals following the injection (retention times). On elution, each component is detected electronically and a detector response/time trace (gas chromatogram) is recorded.

Gas chromatography has now developed into a complex, highly sophisticated technique. Its success depends on the following factors:

(1) the high sensitivity of the detectors now available;
(2) the high resolution (efficiency) of the columns now available;
(3) the versatility of the method which through the appropriate selection of operating conditions makes it possible to analyse a wide range of different compound classes from gases to steroids, amino acids and sugars;
(4) the compatability of the gas chromatograph with other analytical instruments and particularly with the mass spectrometer.

The technique is most commonly used *analytically*, that is to obtain a chromatogram from which numbers of peaks (the numbers of components resolved), relative peak sizes (related via calibration with the relative amount of the various components present) and peak retention times (characteristic of specific compounds under the chromatography conditions employed, although not uniquely so as a number of different compounds can share a particular retention time, particularly on low resolution columns) may be obtained. The method may also be used preparatively so that substances corresponding to peaks or groups of peaks may be trapped (Baker *et al.*, 1976; Goodrich *et al.*, 1981b) as they emerge from the gas chromatograph and either subjected to further gas chromatography using a different stationary phase or examined in other ways. Using the appropriate apparatus, preparative gas chromatography may be scaled up in order that gram quantities of components may be collected (Zlatkis and Pretorius, 1971). Equally, because of the high sensitivities which can now be achieved with modern analytical techniques, it is possible to obtain valuable chemical information from the submicrogram quantities of pure components eluting from high resolution capillary column gas chromatography and to that end preparative techniques have been devised appropriate for such tiny quantities (Williams and Vinson, 1980). Gas chromatography may also be used on line with other techniques yielding analytical data on components as they emerge from the gas chromatograph.

The column

Column efficiency (number of theoretical plates), N, is measured by the sharpness of the chromatogram peaks obtained under isothermal conditions, where

$$N = 16 \frac{x^2}{y^2} \quad \text{(see Figure A.1)}$$

The efficiency of a column depends on the operating conditions used and also on the nature of the compound studied. It depends, for example, on the gas flow rate, the choice of carrier gas and temperature of operation. It is related

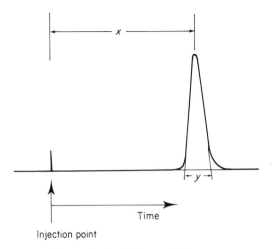

Figure A.1   Column efficiency

to the details of column construction. When packed columns are used, the smaller and more uniformly sized the packing particles, the greater the column efficiency, although very fine mesh particles lead to unacceptably high resistance to gas flow (optimally 100–200 mesh). The choice and the percentage loading of the stationary phase has a crucial effect on the separation obtainable from a particular mixture. Very generally, polar stationary phases separate polar components while non-polar phases separate non-polar components. High stationary phase loadings increase sample capacity but also increase analysis time and may decrease efficiency.

Column efficiency is related to column dimensions. The longer the column, the greater the efficiency obtainable but the greater the resistance to carrier gas flow, while the smaller the column diameter, the greater the column efficiency but the smaller the sample capacity. The larger diameter columns are packed with granular solids. They are commonly (gas liquid chromatography) inert (diatomaceous earth) supports bearing coatings of involatile oils (such as a silicone oil, a polyester or a polyether). Here components are separated as the result of differences in their partition coefficients between the carrier gas and this oil. In some cases (gas solid chromatography), the solid, usually a porous polymer such as Porapak, is selected to act as the separating agent itself as the result of differences in the adsorption characteristics of the various mixture components.

The subsequent introduction of very narrow bore, very high efficiency capillary columns has greatly strengthened the role of gas chromatography in the analysis of complex organic mixtures (Schomburg *et al.*, 1976; Freeman, 1981; Grob, 1979; Jennings, 1980). As Kolor (1979) has graphically demonstrated in relation to food flavour chemistry, a single GC peak obtained using a packed column can be comprised of numerous components which require repeated trapping and rechromatography on a variety of different columns if a pure component is to be obtained. This problem is substantially reduced using high efficiency capillary columns. Capillary columns are of two types, wall coated open tubular (WCOT) and supported coated open tubular (SCOT) columns. In either case the stationary phase is located on the column wall with the result that the column itself presents negligible resistance to carrier gas flow so that very long columns of very high efficiency can be produced.

In WCOT columns, the walls of the column are coated with a thin film of stationary phase. The practical details of the construction of the columns are quite involved and vary with the polarity of the stationary phase selected (Horning *et al.*, 1974; Grob *et al.*, 1978, 1979; Dandeneau and Zerenner, 1979; Verzele, 1979; Verzele and Sandra, 1979). The larger bore WCOT columns have the advantage over the narrower bore WCOT columns of higher sample capacity. The wide bore capillary SCOT columns are similar to WCOT columns but bear on their walls a very thin layer of silica particles with the result that a higher loading of stationary phase can be achieved with advantages for sample capacity but disadvantages for column efficiency (Thomas, 1980; Chauhan and Darbre, 1981). Except where it is important to have larger sample size (for example in coupling GC with FT-IR, Shafer *et al.*, 1980) WCOT columns are clearly the columns of choice. In all cases metal (usually stainless steel) columns have been largely superseded by glass columns. Metal columns tend to adsorb and also to catalyse the decomposition of certain types of compound so that glass columns are generally to be preferred. Fortunately the problem of fragility which has hitherto surrounded the use of glass capillary columns is now being offset by the introduction of flexible vitreous silica glass columns.

An important consideration in selecting a suitable stationary phase concerns thermal stability. In some important applications, when GC is coupled with other techniques (as in GCMS) even slight thermal instability at a particular temperature while not being particularly significant for the GC itself can lead to sufficient 'column bleeding' (the elution of column-derived components), to cause considerable problems for the associated method. An advantage of the use of capillary columns over packed columns is not only that column bleed is less of a problem overall but also that since the emergence of any trace component from the column is concentrated into a very short time interval, its momentary level in the effluent carrier gas is optimized relative to any background bleed.

The detector

A multiplicity of detectors is available. The most commonly used is the flame ionization detector (FID) in which the column effluent is directed into a hydrogen flame and the ions formed in the flame when an eluted component

Table A.1. General characteristics of different types of GC column

Column type	Internal diameter (mm)	Length (m)	Order of magnitude capacity/component	Efficiency (no. of theoretical plates)	Gas flow rate (ml/min)
Capillary, WCOT	0.1–0.8	20–100	Low nanogram	50,000–200,000	0.5–1
Capillary, SCOT	0.5–0.8	20–100	Nanogram	20,000–80,000	2–3
Packed, analytical	2–4	1–4	Low microgram	2000–6000	20–80
Packed, preparative	10–20	3	Milligram	<2000	100–200

The introduction onto capillary columns of very small samples has been accomplished by injecting larger samples while using inlet split to vent the majority of the sample to the air. However, a less wasteful splitless injection of a microlitre of very dilute solution is usually feasible.

burns are detected. This detector has achieved universality both because it is very sensitive (detecting at the subnanogram level) to virtually all organic compounds and because it has a very wide linear range. ($\sim 10^7$). It is unresponsive to oxygen, nitrogen, water, ammonia, hydrogen sulphide and carbon disulphide. Its response per mole differs somewhat for different organic compounds so that calibration is required if quantitative compositions are to be calculated from FID chromatogram peak areas.

For those compounds which possess functional groups (such as halogen or nitro groups) which have a high electron affinity the electron capture detector (ECD) has the advantage of offering a very high sensitivity indeed, being capable of detection at the subpicogram level. The detector, however, has a very narrow linear range (50–500). The carrier gas is generally nitrogen, in which effluent slow electrons are formed by the action of β-rays from a radioactive ^{63}Ni or ^{3}H source. These electrons are collected at an anode and a reduction in the anode current is noted when any substance containing an electron-absorbing group is eluted. The technique has found wide use in the analysis of organochlorine compounds and may be applied to the analysis of a wide variety of compound classes by making derivatives containing ECD-sensitive substituents, for example following perfluoroacylation.

A range of very useful detectors specific for compounds containing sulphur, nitrogen and phosphorus compounds is also available. The resulting simplification of the chromatogram can sometimes be very valuable. Sulphur and phosphorus can be detected selectively using a flame photometric detector in which the light of a flame in which the effluent gas is burnt is monitored at a characteristic wavelength (Brody and Chaney, 1966; Gangwal and Wagoner, 1979). Nitrogen and phosphorus can also be monitored selectively using a modified FID incorporating a caesium rubidium salt pellet in the burner jet (Hartigan *et al.*, 1974; Kolb and Bischoff, 1974; Patterson and Howe, 1978). Parliment and Spencer (1981) have discussed the usefulness in the gas chromatographic analysis of highly complex flavour components of coupling capillary columns simultaneously with flame ionization, specific sulphur and specific nitrogen detectors via a three-way splitting device, thus obtaining simultaneously three corresponding chromatograms of the same mixture, for as they rightly point out, sulphur and nitrogen compounds contribute very importantly to the aromas of many materials. The complexity of the chromatograms of natural materials obtained using high efficiency columns which reveal some hundrends of components is beginning increasingly to require computer techniques to extract biologically significant information from random variations in relative abundance of peaks (McConnell *et al.*, 1979; Zlatkis *et al.*, 1979a; Kowalski, 1980). This is likely to prove an increasingly important area of study as mammalian semiochemical studies develop. An animal, which is itself monitored continuously, has been used with considerable success to provide a continuous physiological monitor of GC effluent in studies on rabbit semiochemicals (see Section 4.3).

Ancillary gas chromatographic techniques

The challenge of determining the structures of new compounds isolated often in trace quantities from natural sources generally requires more than simple GC or even combined GCMS analysis. To this end an array of microchemical procedures has been developed so that, for example, insect pheromone chemists now have available procedures which will enable the structure of tens of micrograms of a new compound of molecular weight less than 350 to be elucidated. Very many of these procedures are undertaken in close association with GC. The outcome of microchemical reactions accomplished in the laboratory may be monitored by GC and GCMS. An example of this is microozonolysis which has particular utility in locating the position of unsaturation in the carbon skeleton (Beroza and Bierl, 1967). Alternatively chemical reactions may be accomplished by design within the GC procedure itself (Beroza and Inscoe, 1969; Beroza, 1962, 1970, 1975; Stanley and Kennett, 1973; Ma and Ladas, 1976; Inscoe *et al.*, 1978; Morgan *et al.*, 1979). For example, 'in syringe' reductions of aldehydes and ketones may be effected using borohydride or the alkylation of acids and other compounds in the heated injector port by coinjection with an appropriate quaternary ammonium salt (Kossa *et al.*, 1979). This has been used for example in urinary steroid determinations (Millner and Taber, 1976). Hydroxyl and other exchangeable protons can be observed on a microscale using deuterated GC column material, for example a Carbowax column fully deuterated by pretreatment with excess deuterium oxide, and exchange noted by monitoring the effluent with MS (Talman *et al.*, 1977). Also subtraction tubes can sometimes be usefully incorporated into the chromatograph. These contain reagents which remove components containing certain functional groups from the mixture (Regnier and Huang, 1970).

However, perhaps the most useful technique entails using hydrogen as a carrier gas and effecting catalytic hydrogenation/hydrogenolysis of the components injected. This is brought about by incorporating a heated platinum or palladium catalyst before the analytical column itself. In the process unsaturation is removed and functional groups stripped from the molecules so that the resulting simple saturated hydrocarbons formed reflect in their geometries the carbon skeletons of the initial components.

Other chromatographic methods

Gas chromatography is, however, but one among many chromatographic methods. All chromatography involves the distribution of the sample under study between two phases one of which is immobile (the stationary phase) and one of which moves through or over this stationary phase (the mobile phase). The stationary phase may pack or coat the inside of a tube (column), or it may take the form of a sheet or layer of material. Separation of a mixture of components is achieved because different components distribute differently between these two phases and so move at different rates through the system. A sensitive detector scans the effluent from the system for components of the mixture and a trace of detector signal with time (chromatogram) is recorded. Chromatographic methods in general are well reviewed by Heftmann (1975) and Johnson and Stevenson (1978).

Of the liquid chromatographic methods, 'flash chromatography' is proving popular (Still et al., 1978) while thin layer chromatography (TLC) (Stahl, 1969; Touchstone and Dobbins, 1978) is an extremely valuable adjunct to gas chromatography in providing a simple means of obtaining a general characterization of the main compound classes present in a natural lipid mixture (Christie, 1973), and also in enabling a preliminary fractionation of samples to be obtained. The power of the technique has more recently been enhanced by the development of much higher resolution TLC methods following improvements in application, separation and detection (high performance TLC or HPTLC) (Zlatkis and Kaiser, 1977; Ericsson and Blomberg, 1980; Lee et al., 1980).

Of particular importance has been the development of high performance liquid chromatography (HPLC) (Johnson and Stevenson, 1978; Engelhardt, 1979; Hamilton and Sewell, 1982). This complements GC particularly in those areas where thermally labile or involatile substances such as porphyrins (Englert et al., 1979) are involved and can be employed either analytically or preparatively (Verzele and Geeraert, 1980). High column efficiencies are obtained using small particle diameter packings ($5-10$ μm) of high uniformity (small particle size range) and rapid throughput. Elevated pressures, short columns and sensitive detectors are employed. Considerable problems are associated in coupling HPLC with mass spectrometry, not least because many of the components for which HPLC is selected are of limited volatility and also because of the proportionately large volume of liquid solvent eluting from the column with the compounds of interest which must be removed before introduction into the mass spectrometer which operates under high vacuum. Even so, HPLC-MS have been coupled satisfactorily in a number of cases (Kenndler and Schmid, 1978; McFadden, 1979, 1980) and this offers promise for extending the study to semiochemical substances. Efforts have been made to couple HPLC with other techniques, such as NMR (Buddrus et al., 1981). HPLC has very great versatility. It can, for example, be used to separate enantiomers (Oelrich et al., 1980) and has been used in association with silver nitrate to separate geometrical isomers of unsaturated compounds encountered in insect pheromone research (Tumlinson and Heath, 1976; Heath et al., 1977; Phelan and Miller, 1981).

A.3 METHODS OF IDENTIFICATION

When assessing what can be achieved in the analysis of a complex mixture, it is important to distinguish between, at the one extreme, the identification of well known substances possessing distinctive spectral properties and at the other extreme new substances of elusive structure which may in some cases require months of investigation by natural product chemists to permit their structural elucidation. The ability of identifying compounds (and even to elucidate the structure of totally new substances) encountered at very low levels in natural mixtures has been revolutionized by the development of sensitive spectrometric methods of various kinds which are now in routine use in many chemical laboratories. Of these mass spectrometry is of central importance to semiochemical research. However, other techniques play an important part and may be coupled with GC. These include nuclear magnetic resonance (NMR), infrared (IR) and ultraviolet spectrometry (UV) (Novotny *et al.*, 1980b), and although for a full description of these the reader should turn to chemistry texts, some few words will be in order.

Infrared spectroscopy (IR)

The infrared spectrum of a compound results from the absorption of infrared radiation by a molecule at levels related to the molecule's various vibrational states, and incorporates characteristics both of the individual functional groups present in the molecule and of the particular compound in question. The technique can be adapted to a microscale (down to 5 μg) and with Fourier transform methods to an order of magnitude lower. The method now has been adapted to direct computerized coupling with GC (GC-FTIR) and offers a very powerful adjunct to GCMS, being sensitive down to the level of tens of nanograms of certain compounds (Hanna *et al.*, 1979).

Nuclear magnetic resonance spectroscopy (NMR)

(James, 1975; Müllen and Pregosin, 1976; Abraham and Loftus, 1978).

NMR is one of the most valuable aids to structural elucidation available to organic chemistry and is amenable to the investigation of suitable solutions of compounds containing atomic nuclei which possess spin (and so have magnetic moments) and can orient themselves to adopt a limited number of discrete energy levels when placed in an external magnetic field. NMR seeks to investigate the small differences which exist between such energy levels by supplying radiofrequency electromagnetic radiation and noting those combinations of frequency and field strength at which resonance occurs. Most attention has focused on the proton, ^{1}H, which possesses a nuclear spin quantum number of $\frac{1}{2}$ and so can adopt two energy states with respect to the magnetic field. Fortunately for the simple applications of the method ^{12}C and ^{16}O nuclei lack spin. The value of the method depends on the fact that the resonance energy differences depend not only on the external field but also on local magnetic effects within the molecule. These are of two types. One (chemical shift effect) concerns electronic shielding of the proton and this is influenced, for example, by the electronegativity of neighbouring atoms. The other (spin–spin coupling) relates to coupling with other protons in the molecule leading to further splitting of energy levels. From a detailed study of the proton NMR spectrum of a molecule a great deal of structural information can be deduced.

However, a major limitation is that the method is relatively insensitive, even using modern Fourier transform methods and other modern aids to increase sensitivity. This is related to the very small energy transitions under investigation. Routine FT-proton NMR requires of the order of 500 μg of a *pure* compound while the limit of sensitivity for many workers is 50 μg and the absolute limit obtainable is in the region of 1 μg. Practical limitations are imposed by the difficulty of removing traces of protonated contaminants from the solvents used ($CDCl_3$, D_2O, CCl_4, etc.) and by factors bearing on the introduction of very small volumes of sample solution (as low as 15 μl) into the spectrometer.

Although the ^{12}C nucleus lacks spin, this is not the case with ^{13}C which occurs naturally at about the 1% level. For this reason ^{13}C-NMR spectra can be of immense value for structural determinations. However, very large quantities of pure compound are required even using FT method, some 50 mg for routine analysis and between 5 and 0.5 mg as the limit of most instruments although spectra on less than 100 μg are reported.

Mass spectrometry (MS)

(Hill, 1966; Burlingame, 1970; Shrader, 1971; Waller, 1972; Williams and Howe, 1972; McFadden, 1973; Frigerio and Castagnoli, 1976; Politzer *et al.*, 1976; Frigerio, 1977; Horning *et al.*, 1977; Daly, 1978; Gross, 1978; Middleditch, 1979; McLafferty, 1980; Waller and Dermer, 1980.)

Mass spectrometry is of such central importance to semiochemical research that it is appropriate to provide a broad description of the scope of the technique and to include mention of developments which are likely to find increasing application in the future. The reader will find directories to much current information on mass spectrometry in the *Specialist Periodical Reports on Mass Spectrometry* published by the Royal Society of Chemistry, London, and in the fundamental and application reviews published each year by the American Chemical Society journal, *Analytical Chemistry*.

Mass spectrometry provides information concerning the identity of a compound in the form of its mass spectrum, the list of ions of varying mass to charge ratio (m/e) and abundance (ion intensity), which results from its gas phase ionization and subsequent decomposition in the mass spectrometer. Mass spectrometry is particularly appropriate for the identification of components present in complex mixtures of volatile organic compounds as indicated in the very readable article by Kolor (1979) because its characteristics of sensitivity and speed of operation permit direct coupling of the mass spectrometer to the effluent of the gas chromatograph in order to obtain mass spectra of the separated components as they emerge from the GC column. To obtain satisfactory mass spectra commonly requires 1–200 ng quantities of substance in the ion source although mass spectrometry may yield valuable information at the picogram level using such techniques as selected ion monitoring. Even where substances initially lack volatility or are too unstable to submit to gas chromatography, they can frequently be rendered suitable for combined gas chromatography-mass spectrometry (GCMS) by prior derivatization. The mass spectrum of a given compound obtained under defined conditions is characteristic of that compound, so that extensive computerized libraries of mass spectra are now available to aid the identification of components present in natural mixtures. However, identification based on such mass spectra alone are always to some degree provisional. It is not unusual for more than one compound to possess closely similar or identical mass spectra. Sometimes also the degree of fragmentation which occurs in the ion source can be so great that little of value can be gained from the examination of the mass spectrum although this can often be overcome by varying the ionization method employed or by suitable derivatization. And when a totally new compound is encountered, the mass spectrum alone is insufficient to provide an identification, although it may prove to be very informative. In such cases other kinds of chemical data are needed to supplement the mass spectral data in order to reach an identification.

The mass spectrometer consists of the following parts.

(1) An *ion source* in which the sample is ionized. A large number of different ionization techniques have been employed in mass spectrometry (Milberg and Cook, 1979a). Of these, by far the most commonly used in organic analysis is electron impact. Chemical ionization, field ionization and field desorption sources are also of growing interest.

(2) A *device for separating ions* according to their mass to charge ratio. The different mass analysers are characterized by the mass resolution of the spectrum they eventually produce (although this also depends on other features of the mass spectrometer operating conditions).

(3) A *device for detecting* and recording the relative numbers of *ions* at each mass to charge ratio value. The current carried through the mass spectrometer from source to detector, the total ion current, is a measure of the intensity of the spectrum. The ion current is usually detected by means of an electron multiplier system.

The problems of interfacing the GC and the MS are substantial. Even so, the two techniques are remarkably compatible, for both operate up to about 300 °C and operate predominantly with materials which are volatile under those conditions. The main problem is that the mass spectrometer generally operates under high vacuum 10^{-4}–10^{-6} torr (higher pressures are appropriate to the chemical ionization mode), while the GC functions at atmospheric pressure. With packed columns at high carrier gas flow rates this requires the selection of one of a number of different types of separator (McFadden, 1979) between the column outlet and the ionization chamber to remove the bulk of the carrier gas but to allow the components of interest to pass. All such separations have their problems so that it is pleasing to note that with the high capacity pumping systems now in use in mass spectrometers, the low gas flow rates (< 4 ml/min) used with capillary columns and the introduction of flexible fused quartz glass columns, the entire GC effluent can be led directly into the mass spectrometer ionizing chamber avoiding contact with any metal surfaces (Koller and Tressl, 1980).

Ionization by electron impact

A number of different processes have been used to ionize substances introduced into the ion source (Fales *et al.*, 1975; Milberg and Cook, 1979a). The most commonly used of these is electron impact. Sample vapour is bombarded under high vacuum (10^{-6}–10^{-8} torr) with electrons accelerated from an incandescent filament by an electric field. Seventy electron-volt electrons are most commonly used, and although the energies of these electrons are substantially in excess of the ionization potentials of organic compounds (7–15 eV, 1 ev = 96.5 kJ/mol), the excited molecular ions initially formed have energies no greater than 8 eV above their ground states, the bulk of the excess energy being lost in the form of the kinetic energy of the resulting electrons:

$$M + e^- = M^{+*} + 2e^-$$

The mass spectrum varies with the energy of the bombarding electrons, particularly at low voltages, but at 70 eV, the mass spectrum is relatively insensitive to small changes in electron energy.

The ions first formed remain in the ion source region for about 10^{-6} sec before being accelerated by electric fields into the mass analyzer region of the mass spectrometer. Because of the high vacuum conditions, ion–molecule and ion–ion collisions do not occur to any extent. However, the excited molecular ion usually has sufficient time to undergo a series of unimolecular fragmentation and rearrangement reactions while in the ion source, yielding an array of charged and neutral fragments. These charged fragments account for the whole of the mass spectrum apart from the molecular ion and their occurrence can be rationalized in terms of the structure of the initial molecule on the basis of known physicochemical principles. The neutral fragments do not register in the mass spectrum. Slightly slower ion fragmentations (ion half-life 10^{-5}–10^{-4} sec) result in decomposition after the ions have left the source but before they can be collected. The resulting fragments have lower energies than expected because neutral fragments formed en route carry away some of the original acceleration energy given to the ion on leaving the ion source. The resulting metastable ions which may be observed occur as small diffuse peaks at non-integral m/e values, the observed mass being related to those of the parent and the daughter ions by the equation

$$M \text{ (observed)} = \frac{[M(\text{daughter ion})]^2}{M(\text{parent ion})}$$

where parent ion $\rightarrow$ daughter ion plus neutral fragment.

Such metastable ions can be a valuable aid to the interpretation of the total mass spectrum of a new compound. Metastable ions are not noted with quadrupole or time-of-flight mass analyzers.

Most ions encountered in mass spectrometry have charge +1, although multiply charged ions are common in certains types of compound. Negative ions are also formed, particularly in compounds containing strong electron-withdrawing groups, although, when examined, the negative ion mass spectrum is generally found to be orders of magnitude less intense than the corresponding positive ion mass spectrum:

$$M + e^- = M^{-*}$$

The negative molecular ion contributes little to most negative ion electron impact mass spectra.

Chemical ionization

However, electron impact provides too drastic an ionization process for many labile molecules, the excess energy in the excited ion initially formed leading to such complete fragmentation in the ion source that important diagnostic features, such as the molecular ion itself, are totally absent from the recorded mass spectrum. To this end, alternative 'softer' ionization processes have been developed. The most important and widely used of these is chemical ionization.

In chemical ionization, the sample, at a partial pressure of about 10^{-6} torr, is introduced into the ionization chamber where it is subjected to the action of a plasma of ionized reagent gas which is present in massive excess (0.2–2 torr pressure). This gas is maintained ionized by a beam of high energy electrons. The sample is ionized as the result of ion–molecule reactions with the reagent gas plasma, the 'quasimolecular' ions formed initially being much more stable than the molecular ions formed in an electron impact source. This is not only because they are formed with much less excess energy but also because they are even-electron entities, in contrast with the odd-electron radical cations formed on electron impact. These quasimolecular ions are of one atomic unit greater than, or less than, the true molecular ion, and result from hydrogen transfer.

Using chemical ionization, the degree of fragmentation can be varied by the selection of different *reagent gases*. This is because the degree of fragmentation is related to the level of excess energy transferred to the sample molecule in the ionizing reaction (Hunt *et al.*, 1976). For example with methane as reagent gas, the following quasimolecular ions can be formed:

$$\text{CH}_5^+ + \text{RH} \quad \begin{array}{l} \xrightarrow{\text{Route 1}} \quad \text{RH}_2^+ + \text{CH}_4 \\ \\ \xrightarrow{\text{Route 2}} \quad \text{R}^+ \quad + \text{CH}_4 + \text{H}_2 \end{array}$$

Quasimolecular ions

Route 1 is favoured if RH is a good proton acceptor (alcohols, aldehydes, esters, etc.). Because CH_5^+ acts as a strong Brönsted acid (high energy proton donor) in the gas phase, a mixed reagent gas composed of methane plus up to 5% of, for example, acetone, ammonia, water vapour or isobutane efficiently generates such weaker Brönsted acid reagents as $\text{CH}_3\text{COHCH}_3^+$, NH_4^+, H_3O^+ or C_4H_9^+ from the CH_5^+ first formed. Because so little excess energy is transferred in chemical ionization using such mild reagents as isobutane, the mass spectra obtained exhibit very little fragmentation but consist principally of the quasimolecular ion (plus ions derived from the reagent gas itself). These reagent gases are therefore useful for the high sensitivity determination of particular components in a mixture, e.g. by single ion monitoring. For structural investigations, spectra possessing more extensive fragmentation are required.

Reagent gases such as argon or nitrogen act as strong ionizing agents forming molecular (rather than quasimolecular) ions of excess energy similar to that obtained on electron impact:

$$\text{Ar}^+ + \text{RH} = \text{Ar} + \text{RH}^+$$

A mass spectrum obtained using an argon–water mixture can be particularly informative, the water yielding an intense quasimolecular ion while the argon produces an informative fragmentation pattern similar to that obtained on electron impact.

Deuterium oxide (deuterium = heavy isotope of hydrogen, atomic mass 2) is widely used as a reagent gas to determine the number of 'active' hydrogen atoms in a molecule. If the chemical ionization spectra recorded using, in turn, isobutane and deuterium oxide are compared, the number of hydrogen atoms per molecule of the sample which can be easily replaced by deuterium (those attached to oxygen or nitrogen) can be calculated, as the replacement of each hydrogen by deuterium increases the mass of the molecule by one atomic mass unit.

Ammonia has been used as a reagent gas to selectively ionize, for example, amides and amines in a complex mixture. This is because the ammonium ion present in the plasma is able to form $(M + H)^+$ ions only in the cases of those relatively few compound classes which possess a proton affinity greater than ammonia. Quasimolecular ions of the type $(M + NH_4)^+$ are, however, formed from a somewhat wider range of substances (ketones, aldehydes, esters, acids). Minimal fragmentation is observed as the energy transferred is so low.

Although the negative ion mass spectrum of a substance is likely to yield valuable structural information complementary to that which can be obtained from the positive ion mass spectrum (Voigt *et al.*, 1977), the electron impact technique frequently fails to yield useful negative ion mass spectra. The negative molecular ion is usually absent, its production requiring 'resonance capture' of very low energy (~ 0 eV) thermal electrons, the higher energy electrons encountered in the electron impact source leading to almost total rupture of the molecule to low mass fragments.

To overcome this limitation, chemical ionization techniques have been used which generate in the plasma high populations of thermal electrons which can then generate a negative ion mass spectrum of added sample molecules. For example, if nitrogen (or argon) is used as the reagent gas, N_2^+ ions and excited N_2^* molecules are formed in the plasma together with large numbers of thermal electrons. In the presence of sample molecules the former species yields, as described above, positive molecular ions of excess energy in the range comparable with those formed in an electron impact source, while the latter form stable negative ions by low energy resonance capture.

$$R + e^- \rightarrow R^-$$

For certain classes of compounds, such electron capture negative ion mass spectrometry can exhibit very high sensitivity, the total negative ion current being up to a thousand-fold greater than the positive ion current (compare electron capture detection in gas chromatography).

Using nitrogen or argon as the chemical ionization reagent gas in a quadrupole mass spectrometer, it has been possible to record simultaneously the positive and negative ion mass spectra of a sample by pulsing the polarity of the ion source potential (and of the focusing lens potential) at 10 kHz while detecting the positive and the negative beams separately. Pulsed positive negative ion chemical ionization mass spectrometry (PPNICI) (Hunt *et al.*, 1976; Hunt and Sethi, 1978) has also been employed using other reagent gases. With methane, similar electron capture negative ion mass spectra are observed. However, very stable negative quasimolecular ions containing virtually no excess energy can also be generated by adding about 0.1% methyl nitrite to the methane. This efficiently absorbs the thermal electrons to form the strong Brönsted base, CH_3O^- which then reacts with sample molecules, thus

$$CH_3O^- + RH = CH_3OH + R^-$$

(although the methoxy anion is a very strong Brönsted base, virtually all the energy liberated in the reaction is carried in the methanol molecule formed). Selectivity can be improved by generating weaker Brönsted bases by adding to this gas mixture a further 1–5% of an appropriate reagent gas with which the methoxy anion can react. For example, acetone will yield a plasma containing $CH_3COCH_2^-$, acetonitrile CH_2CN^-, acetylene CHC, nitromethane

$CH_2NO_2^-$. Similar negative ion mass spectra can also be obtained by using hydrogen gas (which generates the strong Brönsted base, H^-) together with such additives.

Field ionization and field desorption

(Beckey and Schulten, 1979; Milberg and Cook, 1979b.)

Field ionization is a particularly gentle alternative ionizing technique in which molecules, present as a vapour at 10^{-3}–10^{-6} torr pressure, are subjected to the very high potential gradients (10^7–10^8 volts/cm) which form at sharp points or other conducting surfaces of high curvature (radius of curvature 10^{-4}–10^{-6} cm) on an 'emitter' which is maintained at 8–14 kV relative to a counter electrode (Beckey, 1969, 1977; Robertson, 1972). The high potential gradient so distorts the potential surface of the molecule that an electron is easily lost and a positive ion of very low excess energy (~5 eV) formed. Field ionization mass spectra exhibit intense molecular or quasimolecular ions, with relatively little fragmentation (Games *et al.*, 1974; Fales *et al.*, 1975; Milberg and Cook, 1979b). The method suffers from being relatively insensitive (the total ion current may be a hundred-fold lower than that obtained on electron impact), although this may be overcome in the future when sources are designed which allow better contact between the bulk of the sample vapour under study and the region where ionization occurs very close to the ion emitter surface.

Similar emitters are used for field ionization and also for field desorption, described below. A design currently much used consists of a very fine (5–10 μm) tungsten wire covered with fine carbon microneedles formed as required in a special 'activation process' in which the wire is maintained at high potential and at high temperature in an atmosphere consisting of a low pressure of benzonitrile vapour.

All the ionizing techniques described so far are applicable to combined GCMS and require the substance under study in the gas phase. However, many natural substances of molecular weight appropriate for mass spectral analysis decompose rather than form a vapour even under the high vacuum conditions of the mass spectrometer. For these, an elaboration of the field ionization technique, field desorption, is applicable. Here, a solution of the substance under study is applied to an emitter similar to that used in field ionization. The emitter is introduced into the mass spectrometer, the solvent is evaporated off, and the emitter put at high positive potential, as in field ionization. It is then gently heated by passing a current. As the emitter current increases, ionization occurs at the tips of the microneedles and positive ions of sample are generated and stripped from the emitter, much as in field ionization. As in field ionization also, this is a very gentle ionization method (transferred energy ~ 0.1 eV). Spectra have been recorded on as little as 10^{-8} g of substance. As the emitter current is increased further, spectra characteristically show an increasing degree of fragmentation which can itself be of diagnostic value. The technique is particularly appropriate for the qualitative analysis without derivatization of small quantities of polar, heat-labile materials. Mass spectra have been reported, for example, for such substances as salts, carboxylic and sulphonic acids, amino acids, porphyrins, bile pigments, carotenoids, peptides, sugars, glycosides and glucuronides, nucleosides and nucleotides (Beckey and Schulten, 1975; Beckey, 1977; Fales et al., 1975; Schulten et al., 1977). Field desorption from an emitter impregnated with an appropriate lithium salt before sample application can result in intense $(M + Li)^+$ quasimolecular ions even for substances which are too unstable to give M^+ or $(M + 1)^+$ ions on simple field desorption (lithium attachment spectrum) (Veith, 1976).

Other ionization methods also show potential for future development for use with thermally labile and/or involatile substances. These include *in-beam chemical ionization* methods in which the sample is deposited on a tip surface which is then exposed directly to the ion plasma (Hansen and Munson, 1978; Cotter, 1980), *californium-252 plasma desorption* (Macfarlane and Torgerson, 1976) and *fast atom bombardment and secondary ion mass spectrometry* (Day et al., 1980).

The mass analyser

A high potential, V, of 1–8 kV is used to accelerate the ions from the ion source and the resulting beam is focused. After passing the source slit, ions (of the same charge) possess the same energies, but have different velocities, v, depending on their masses, m, and enter a field-free region on their way to the mass analyser.

$$\text{Kinetic energy of each singly charged ion} = Ve = \tfrac{1}{2}mv^2$$

This forms the basis of the simple 'time of flight mass analyser' in which radiofrequency pulses of ions are accelerated from a source and the subsequent times ions of different mass take to traverse a field-free path to a detector are measured electronically. Flight times of the order of 30 μsec are recorded, the lighter ions travelling faster than the heavier. Time of flight mass spectrometers scan very rapidly ($\sim$ 0.001 sec) but have low mass resolutions (400–600).

Note that the resolutions of a mass spectrum, R is given by

$$R = \frac{M(1)}{M(2) - M(1)}$$

where $M(1)$ is the mass of an ion and $M(2)$ is the mass of the next higher mass ion from which it can just be resolved, which is usually taken to mean when two adjacent ions of similar intensity are separated by a '10% valley'.

Magnetic deflection mass analysers are of more relevance to our interests. In a single focusing magnetic deflection mass spectrometer the ion beam enters a homogenous magnetic field, H, where each ion experiences a constant force acting perpendicularly both to its direction of movement, and also to the direction of the field. As the result, the ions move in circular paths, the radius of which, r, is given by

$$\text{Force on ion} = \frac{Hev}{c} = \frac{mv^2}{r}$$

Eliminating v between these two equations, we obtain

$$\frac{m}{e} = \frac{kH^2r^2}{V}$$

where k is a constant.

Therefore

(1) if the position of the detector is fixed (so that r is fixed), ions at each m/e value can be monitored by scanning either H or V (magnetic and voltage scan modes).

(2) the larger the r, the greater the resolution which becomes possible.

However, the kinetic energies of the ions emerging from the source are not all identical, but exhibit a spread of energies of about 0.05 eV depending on the initial energy distribution in the source itself. The larger this spread in relation to the mean kinetic energy of the ions accelerated from the source, the poorer the mass resolution which is possible. Mass spectrometer resolution is greatly increased using a double focusing mass spectrometer. The mass resolution of the single focusing magnetic deflection mass spectrometers range from 500 to 9000 whereas in the double focusing instruments, this range is from 10,000 to 100,000. The speed of operation is slower the higher the resolution ranging from less than one to more than 10–20 seconds per cycle. The maximum mass is often in the region of 4000–5000 amu. In the double focusing instrument, an electrostatic analyser and a slit are introduced between the source and the magnetic mass analyser. This focuses the ion beam and allows the passage of only a very narrow band of ionic kinetic energies. The electrostatic analyser consists of a radial electrostatic field, E, applied perpendicularly to the direction of the ion beam. Ions then move in circular paths the radius of which depends on their kinetic energies. Only ions moving in trajectories of radius close to a radius r' defined by the analyser geometry pass through the analyser.

$$\text{Force on each ion} = Ee = \frac{mv^2}{r'}$$

From this, only ions of kinetic energy given by

$$\tfrac{1}{2}mv^2 = \tfrac{1}{2}Eer'$$

pass through the analyser.

A disadvantage of the magnetic deflection mass analyser is that scan rates are necessarily relatively slow. This can be an important limitation when the mass spectrometer is coupled to a capillary column gas chromatograph when peaks emerge over short time intervals of 3–4 seconds. As the result, faster acting quadrupole mass analysers have gained popularity (Feser and Kögler, 1979).

A quadrupole analyser selects ions on the basis of their mass to charge ratio. The quadrupole consists of four steel rods (electrodes), circular or ideally hyperbolic in cross-section, precisely aligned in a parallel square array, opposite pairs of electrodes being supplied with related radiofrequency potentials. Ions propelled along the axis of this array experience oscillations perpendicular to their direction of motion which are stable only for ions of a particular narrow m/e value range defined by the electrode voltages and the dimensions of the analyser. Other ions are captured by the electrodes as they pass down the quadrupole filter.

Ions progress through the filter to the detector having been accelerated from the source through a potential of 5–30 V in contrast with the kilovolt accelerations required for magnetic deflection mass analysers. The advantages of the quadrupole mass analyser are its speed of scan and its relative simplicity. It is, however, a low resolution analyser which has tended to lack sensitivity to high mass fragments. Unlike the magnetic deflection mass analyser, the quadrupole exhibits a linear relationship of mass to voltage.

Mass spectrometers are commonly employed to scan repetitively throughout a GCMS run in a cyclic scan mode at a frequency which is limited by the time taken for the instrument to scan the spectrum and to return again to its initial state. In any such GCMS run on a complex mixture, a vast quantity of data is generated which requires computerized facilities both to process the data and also to interpret it. Library search methods are extensively used, as are algorithms to interpret the spectra of unknown substances (Grönneberg et al., 1975; McLafferty and Venkataraghavan, 1979; Gray et al., 1980).

High resolution mass spectrometry

For many purposes, a low resolution operation which assigns a nominal (nearest integral) mass to each ion is adequate. However, for some purposes, higher resolution mass spectrometry is valuable, for it becomes possible to distinguish between ions of the same nominal mass but of different exact mass by virtue of their different elemental composition (the mass of any atom is generally not an exact integral). Conversely, from a knowledge of the exact mass of an ion, it is possible to infer an elemental composition. For example, two ions of nominal mass 182 can be distinguished as $C_{13}H_{10}O^+$ (exact mass, 182.0732) and $C_{12}H_{10}N_2^+$ (exact mass 182.0845) at higher resolution.

The resolution of a mass spectrometer is decreased by using wider source and collector slit widths (although this has the advantage of increasing ion current, and so sensitivity) and by using faster mass scan rates. Provided a given mass spectral line is only due to one ion, even quite low resolution measurements can be used to give an indication of the exact mass as deduced from the centroid of the line. For the resolution and elemental compositions of different ions of different exact mass but of the same nominal mass occurring in a mass spectrum, higher resolutions are required. As resolution increases, speed of operation decreases as does sensitivity and for these reasons very high resolution MS cannot easily be coupled with GC. However, a compromise can be reached so that satisfactory exact masses can be obtained in GCMS runs (Kimble, 1978).

The accurate mass of all ions in a mass spectrum is of considerable importance particularly when such high resolution mass spectrometer is fully computerized and in use in conjunction with high resolution gas chromatography in the examination of complex organic mixtures. This has been demonstrated for example in HRGC/HRMS studies (MS resolution = 10,000) on the intricate mixtures of compounds present in urine and other natural materials (Kimble *et al.*, 1974, 1976; Lewis *et al.*, 1979; Meili *et al.*, 1979). The mass spectra of components which remain chromatographically unresolved can be deduced while identification of components which defy identification by GC-LRMS alone can be attempted. Mass chromatography of different components which have fragment ions at the same nominal mass but different exact mass can be obtained. Even so, problems do arise as the result of the relatively slow operation of high resolution mass spectrometers which have a cycle time in excess of the peak width of 3 or 4 seconds often obtained on high resolution gas chromatography.

Mass fragmentography

In GCMS, mass fragmentography (Björkhem *et al.*, 1976) or specific ion monitoring refers to techniques in which, instead of scanning the entire mass spectrum serially, attention is directed to one or a few selected ions and these are continuously monitored with time. The result is a simplification of a complex chromatogram monitoring components of that class only which yield ions at the m/e ratio of interest (Liebich, 1975) and also a considerable enhancement in sensitivity particularly where, as is usually the case, the ions chosen are very intense in the compound of interest. Isotopically labelled internal standards can be incorporated to correct for procedural losses. The presence of a particular compound can be inferred with greater certainty where a number of intense ions characteristic of that compound are monitored and are found to peak in the appropriate ratio at the appropriate retention time. In spite of a loss of sensitivity, much greater reliance can be placed on high resolution mass fragmentography and this can be used to determine particular steroids in crude sample extracts at levels of 10–200 pg/μl (Adlercreutz and Hunneman, 1973; Millington, 1975; Millington *et al.*, 1975).

A.4 RADIOIMMUNOASSAY

(Sönksen, 1974; Aherne and Marks, 1979; Chapman, 1979.)

Endocrinology has derived great benefit from the application of radioimmunoassay techniques to the measurement of the very low levels (nanogram
to picogram quantities) of hormones present in biological substrates, and now
this same technique is being applied to semiochemical systems, notably those
involving the odorous C_{19}-$\Delta 16$ steroids (Andersen, 1975; Claus, 1975).

The advantage of radioimmunoassay is that it can provide information at
and beyond the detection limits of GC and other techniques and that once a
particular assay is established, the procedures are simple and are well adapted
to routine use. However, the institution of an assay is not without its difficulties and any assay provides only one piece of information on a particular
substrate namely the level of the particular compound (or closely related
group of compounds) for which the assay has been developed. It cannot
provide any wider information and even here there is always concern how
specific the assay may be and whether interfering substances may be affecting
the result.

The assay depends on having available a stock of highly specific antibody
(antiserum) which binds tightly with the substances being assayed. This is
produced by immunizing mammals (usually rabbits or guinea pigs) with the
substance in question (sometimes coupled with a protein carrier) so that it
acts as an antigen, and subsequently harvesting the antiserum produced. In
the assay itself a deficit of antiserum is added to the sample under investigation so that some of the antigen binds to form a tight antigen–antibody
complex but the remaining excess antigen is left free. For a given quantity of
antibody, clearly the proportion of antigen which is left free is a measure of
the level of this substance in the sample. This ratio of free to bound antigen is
conveniently determined by having present a much smaller concentration of
radiolabelled antigen which since it is chemically indistinguishable from the
unlabelled antigen mirrors the distribution of the unlabelled antigen between
free and bound fractions. Monitoring the distribution of radioactivity between
the fractions provides the basis for the method.

So, if compound C, for which a specific antiserum A is available, is to be
assayed in a sample:

(1) a very small quantity of radiolabelled C (C^*) is added to the sample.
(2) the sample may be purified to remove possible interfering substances.
(3) a deficit of A is added. This binds tightly with C (and chemically
identical C^*) but since only a limited quantity of A is present a proportion of
C (and C^*) is left free.
(4) bound and free C (and C^*) fractions are separated commonly as the
result of treatment with active charcoal which removes the free fraction but
leaves the bound fraction in solution.

(5) the distribution of C^* between bound and free fractions is determined by scintillation counting and this mirrors exactly the distribution of chemically identical C between the two fractions.

(6) in parallel with this procedure a calibration curve is simultaneously determined using exactly the same quantities of A and C^* but employing standards containing different quantities of C in the concentration range of interest. This calibration curve is used to obtain from the experimental data a concentration level of C in the substrate under study.

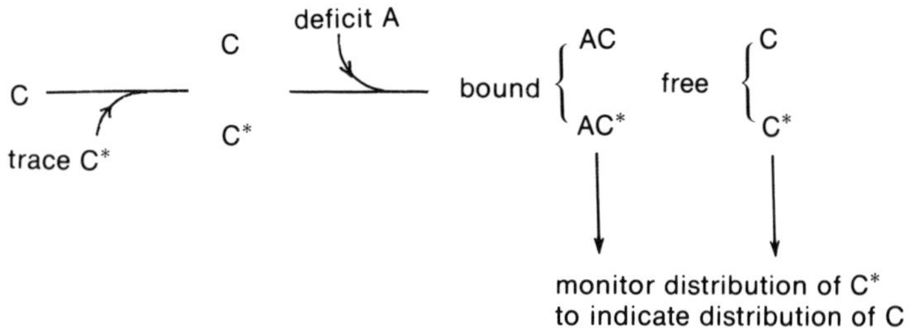

When investigating natural mixtures of very small quantities of very similar biological substances for which the highly specific antisera required are not available, radioimmunoassay may be used with HPLC. This approach has been employed effectively in the study of neuropeptides (McDermott *et al.*, 1981; Spindel *et al.*, 1981).

A.5 INTERRELATION OF THE MAJOR SEX STEROIDS

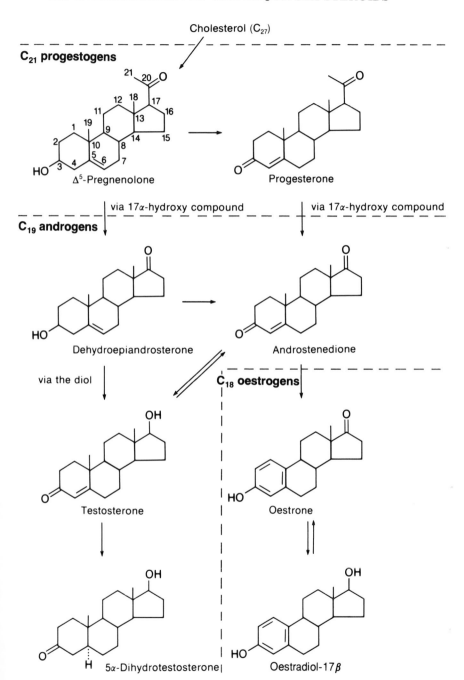

Figure A.2   Biosynthetic relationships between major sex steroids. Many of the transformations involve a number of steps

References

ⓢ Abraham, R. J., and Loftus, P. (1978). *Proton and Carbon-13 NMR spectroscopy; an integrated approach.* Heyden, London.

Acree, F., Turner, R. B., Gouck, H. K., Beroza, M., and Smith, N. (1968). L-Lactic acid; a mosquito attractant isolated from humans', *Science*, **161**, 1346–1347.

Adams, J., Garcia, A., and Foote, C. S. (1978). 'Some chemical constituents of the secretion from the temporal gland of the African elephant, *Loxodonta africana'. J. Chem. Ecol.*, **4**, 17–25.

●Adams, M. G. (1980). 'Odour-producing organs of mammals', *Symp. Zool. Soc. Lond.*, **45**, 57–86.

Adler, A. D. (ed.) (1975). 'The biological role of porphyrins and related structures', *Ann. N.Y. Acad. Sci.*, **244**, 1–694.

Adlercreutz, H., and Hunneman, D. H. (1973). 'Quantitation of up to 12 estrogens in 1–50μl of pregnancy urine', *J. Steroid Biochem.*, **4**, 233–237.

Aeschlimann, A. (1963). 'Observations sur *Philantomba maxwelli* (Hamilton-Smith) une antilope de la forêt éburnéenne', *Acta Tropica*, **20**, 341–368.

Aherne, G. W., and Marks, V. (1979). 'Radioimmunoassay techniques', *Techniques in Metabolic Research*, **B 218**, 1–19.

Aitzetmüller, K., and Koch, J. (1978). 'Liquid chromatographic analysis of sebum lipids, and other lipids of medical interest', *J. Chromatog.*, **145**, 195–202.

Albone, E. S. (1975). 'Dihydroactinidiolide in the supracaudal scent gland secretions of the red fox', *Nature*, **256**, 575.

Albone, E. S. (1977). 'Ecology of mammals — a new focus for chemical research', *Chem. Brit.*, **13**, 92–96, 99, 112.

Albone, E. S., and Fox, M. W. (1971). 'Anal gland secretion of the red fox', *Nature*, **233**, 569–570.

● Albone, E. S., Eglinton, G., Walker, J. M., and Ware, G. C. (1974). 'The anal sac secretion of the red fox *Vulpes vulpes*; its chemistry and microbiology. A comparison with the anal sac secretion of the lion, *Panthera leo'*, *Life Sci.*, **14**, 387–400.

Albone, E. S., and Flood, P. F. (1976). 'The supracaudal scent gland of the red fox, *Vulpes vulpes'*, *J. Chem. Ecol.*, **2**, 167–175.

Albone, E. S., and Perry, G. C. (1976). 'Anal sac secretion of the red fox, *Vulpes vulpes*. Volatile fatty acids and diamines. Implications for a fermentative hypothesis of chemical recognition'. *J. Chem. Ecol.*, **2**, 101–111.

Albone, E. S., Robins, S. P., and Patel, D. (1976). '5-Aminovaleric acid, a major free amino acid component of the anal sac secretion of the red fox, *Vulpes vulpes'*, *Comp. Biochem. Physiol.*, **55B**, 483–486.

● Albone, E. S., Gosden, P. E., and Ware, G. C. (1977). 'Bacteria as a source of chemical signals in mammals', in *Chemical Signals in Vertebrates* (eds D. Müller-Schwarze and M. M. Mozell), Plenum Press, New York, pp. 35–43.

Albone, E. S., and Grönneberg, T. O. (1977). 'Lipids of the anal sac secretions of the red fox, *Vulpes vulpes*, and of the lion, *Panthera leo'*, *J. Lipid Res.*, **18**, 474–479.

● Albone, E. S., Gosden, P. E., Ware, G. C., Macdonald, D. W., and Hough, N. G.

(1978). 'Bacterial action and chemical signalling in the red fox (*Vulpes vulpes*) and other mammals', in *Flavor Chemistry of Animal Foods* (ed. R. W. Bullard), ACS Symposium Series **67**, American Chemical Society, Washington, D.C., pp. 78–91.

Aldrich, T. B. (1896). 'A chemical study of the secretion of the anal glands of *Mephitis mephitis* (common skunk) with remarks on the physiological properties of this secretion', *J. Exp. Med.*, **1**, 323–340.

Aldrich, T. B., and Jones, W. (1897). '*α*-Methylquinoline as a constituent of the secretion of the anal glands of *Mephitis mephitica*', *J. Exp. Med.*, **2**, 439–452.

Allanson, M. (1934). 'The reproductive process of certain mammals, VII Seasonal variation in the reproductive organs of the male hedgehog', *Phil Trans. Roy. Soc.*, *B*, **223**, 277–303.

Allen, G. M. (1967). *Bats*, Dover Publications Inc., New York, pp.138–140.

Allen, T. E., and Bligh, J. (1969). 'A comparative study of the temporal patterns of cutaneous water vapour loss from some domesticated animals with epitrichial sweat glands', *Comp. Biochem. Physiol.*, **31**, 347–363.

Amoore, J. E. (1967). 'Specific anosmia: a clue to the olfactory code', *Nature.*, **214**, 1095–1098.

Amoore, J. E. (1970). *The Molecular Basis of Odour*, C. C. Thomas, Springfield, Illinois, U.S.A.

Amoore, J. E. (1977). 'Specific anosmia and the concept of primary odors', *Chemical Senses and Flavor.*, **2**, 267–281.

Amoore, J. E., Forrester, L. J., and Buttery, R. G. (1975). 'Specific anosmia to 1-pyrroline, the spermous primary odor', *J. Chem. Ecol.*, **1**, 299–310.

Amoore, J. E., and Forrester, L. J. (1976). 'Specific anosmia to trimethylamine; the fishy primary odor', *J. Chem. Ecol.*, **2**, 49–56.

Amoore, J. E., Pelosi, P., and Forrester, L. J. (1977). 'Specific anosmias to 5*α*-androst-16-en-3-one and ω-pentadecalactone; the urinous and musky primary odors', *Chemical Senses and Flavor*, **2**, 401–425.

Andersen, K. K., and Bernstein, D. T. (1975). 'Some chemical constituents of the scent of the striped skunk, *Mephitis mephitis*', *J. Chem. Ecol.*, **1**, 493–499.

Andersen, K. K., Bernstein, D. T., Caret, R. L., and Romanczyk, L. J. (1982). 'Chemical constituents of the defensive secretion of the striped skunk, *Mephitis mephitis*', *Tetrahedron*, **38**, 1965–1970.

Andersson, G. (1979). 'Volatile ketones from the preorbital gland of reindeer (*Rangifer t. tarandus*)', *J. Chem. Ecol.*, **5**, 629–634.

Andersson, G., Andersson, K., Brundin, A., and Rappe, C. (1975). 'Volatile compounds from the tarsal scent gland of reindeer, *Rangifer tarandus*', *J. Chem. Ecol.*, **1**, 275–281.

Andersson, G., Brundin, A., and Andersson, K. (1979). 'Volatile compounds from the interdigital gland of reindeer (*Rangifer tarandus*)', *J. Chem. Ecol.*, **5**, 321–333.

Andersen, Ø. (1975). 'A radioimmunoassay for 5*α*-androst-16-en-3-one in porcine adipose tissue'. *Acta endocrinologica*, **79**, 619–624.

Andrew, R. T., and Klopman, R. B. (1974). 'Urine washing: comparative notes', in *Prosimian Behaviour* (eds R. D. Martin, G. A. Doyle and A. C. Walker), Gerald Duckworth, London, pp. 303–312.

Anema, P. J., Kooiman, W. J., and Geers, J. M. (1973). 'Volatile acid production by *Clostridium sporogenes* under controlled culture conditions', *J. Appl. Bact.*, **36**, 683–687.

Ansari, M. N. A., Fu, H. C., and Nicolaides, N. (1969). 'Fatty acids of the alkane diol diesters of vernix caseosa', *Lipids*, **5**, 279–282.

Aoki, T., and Wada, M. (1951). 'Functional activity of the sweat glands in the hairy skin of the dog', *Science*, **114**, 123–124.

Arao, T., and Perkins, E. (1969). 'Further observations on the Philippine Tarsier, *Tarsius syrichta*', *Amer. J. Phys. Anthrop.*, **31**, 93–96.

304

Ashton, E. H., and Eayrs, J. T. (1970). 'Detection of hidden objects by dogs', in *Taste and Smell in Vertebrates* (eds G. E. W. Wolstenholme and J. Knight), J. and A. Churchill, London, pp. 251–263.

Austad, W., Clamp, J. R., McCarthy, C. F., and Read, A. E. (1967). 'Some observations on the metabolism of L-tryptophan', *Biochim. Biophys. Acta*, **148**, 84–91.

Ayorinde, F., Wheeler, J. W., Wemmer, C., and Murtaugh, J. (1982). 'Volatile components of the occipital gland secretion of the bactrian camel, *Camelus bactrianus*', *J. Chem. Ecol.*, **8**, 177–183.

Bailey, K. (1978). 'Flehmen in the ring-tailed lemur, *Lemur catta*', *Behaviour*, **65**, 309–319.

Bailey, S., Bunyan, P. J., and Page, J. M. J. (1980). 'Variation in the levels of some components of the volatile fraction of urine from captive red fox (*Vulpes vulpes*) and its relationships to the state of the animal' in *Chemical Signals; Vertebrates and Aquatic Invertebrates* (eds D. Müller-Schwarze and R. M. Silverstein), Plenum Press, New York, pp. 391–403.

Baillie, A. H., Ferguson, M. M., and Hart, D. McK. (1966). '*Developments in Steriod Histochemistry*', Academic Press, London.

Baker, E. (1962). 'Diseases and therapy of the anal sacs of the dog', *J. Amer. Vet. Med. Assn.*, **141**, 1347–1350.

Baker, H. G. (1963). 'Evolutionary mechanisms in pollination biology', *Science*, **139**, 877–883.

Baker, R., Bradshaw, J. W. S., Evans, D. A., Higgs, M. D., and Wadhams, L. J. (1976). 'An efficient all-grass splitter and trapping system for gas chromatography', *J. Chromatog. Sci.*, **14**, 425–427.

Bakke, J. M., and Figenschou, E. (1983). 'Volatile compounds from the red deer (*Cervus elaphus*). Secretion from the tail gland', *J. Chem. Ecol.*, **9**, 513–520.

Baldwin, J. D. (1970) 'Reproductive synchronization in squirrel monkeys (Saimiri)', *Primates*, **11**, 317–326.

Bannister, L. H. (1968). 'Fine structure of the sensory endings in the vomeronasal organ of the slow-worm, *Anguis fragilis*', *Nature*, **217**, 275–276.

Bannister, L. H., and Cushieri, A. (1972). 'The fine structure and enzyme histochemistry of vomeronasal receptor cells' in *Olfaction and Taste IV*, (ed. D. Schneider), Wissenschaftliche Verlagsgesellschaft MBH, Stuttgart, pp. 27–33.

Barber, P. C., and Raisman, G. (1978a). 'Cell division in the vomeronasal organ of the adult mouse', *Brain Res.*, **141**, 57–66.

Barber, P. C., and Raisman, G. (1978b). 'Replacement of receptor neurons after section of the vomeronasal nerves in the adult mouse', *Brain Res.*, **147**, 297–314.

Bearder, S. K., and Doyle, G. A. (1974). 'Ecology of bushbabies, *Galago senegalensis* and *Galago crassicaudatus*, with some notes on their behaviour in the field' in *Prosimian Behaviour* (eds R. D. Martin, G. A. Doyle and A. C. Walker), Gerald Duckworth, London, pp. 109–130.

Beauchamp, G. K. (1974). 'The perineal scent gland and social dominance in the male guinea pig', *Physiol. Behav.*, **13**, 669–673.

Beauchamp, G. K. (1976). 'Diet influences attractiveness of urine in guinea pigs', *Nature*, **263**, 587–588.

Beauchamp, G. K., Doty, R. L., Moulton, D. G., and Mugford, R. A. (1976). 'The pheromone concept in mammalian chemical communication: a brief critique', in *Mammalian Olfaction, Reproductive Processes and Behavior* (ed. R. L. Doty), Academic Press, New York, pp. 143–160.

Beauchamp, G. K., Doty, R. L., Moulton, D. G., and Mugford, R. A. (1979). *J. Chem. Ecol.*, **5**, 301–305.

Beauchamp, G. K., Wellington, J. L., Wysocki, C. J., Brand, J. G., Kubie, J. L., and Smith, A. B. (1980). 'Chemical communication in the guinea pig: urinary components of low volatility and their access to the vomeronasal organ', in *Chemical Signals; Vertebrates and Aquatic Invertebrates* (eds D. Müller-Schwarze and R. M. Silverstein), Plenum, New York, pp. 327–339.

Beckey, H. D. (1969). 'Determination of the structures of organic molecules and quantitative analyses with the field ionization mass spectrometer', *Angew. Chem. Int. Ed. Engl.*, **8**, 623–639.

Beckey, H. D. (1977). *Principles of Field Ionization and Field Desorption Mass Spectrometry*, Pergamon Press, Oxford.

Beckey, H. D., and Schulten, H. R. (1975). 'Field desorption mass spectrometry', *Angew. Chem. Int. Ed. Engl.*, **14**, 403–415.

Beckey, H. D., and Schulten, H. R. (1979). 'Field ionization and field desorption mass spectrometry in analytical chemistry', in *Mass Spectrometry, Part A* (eds C. Merritt and C. N. McEwen), Marcel Dekker, New York, pp. 145–266.

Beets, M. G. J. (1978). *Structure–Activity Relationships in Human Chemoreception*, Applied Science Pubishers, London.

Beets, M. G. J., and Theimer, E. T. (1970). 'Odour similarity between structually unrelated odorants', in *Taste and Smell in Vertebrates* (eds G. E. W. Wolstenholme and J. Knight), J. and A. Churchill, London, pp. 313–323.

Bekoff, M., and Wells, M. C. (1980). 'The social ecology of coyotes', *Scientific American*, **242** (4), 112–120.

Bell, D. J. (1980). 'Social olfaction in Lagomorphs', *Symp. Zool. Soc. Lond.*, **45**, 141–164.

Bell, D. J. (1984). 'Lagomorph scents', in *Mammalian Social Odours* (eds R. E. Brown and D. W. Macdonald), Oxford University Press, Oxford.

Bell, F. R., and Sly, J. (1980). 'Cattle can smell salt', *J. Physiol.*, **305**, 68P–69P.

Berberich, J. J., Andrews, R. V., and Folk, G. E. (1977). 'Effects of cold exposure on urinary catecholamines in arctic lemmings', *Comp. Biochem. Physiol.*, **58C**, 133–135.

Berglund, B., Berglund, U., Lindvall, T. and Svensson, L. T. (1973). 'A quantitative principle of perceived intensity summation in odor mixtures', *J. Exp. Psychol.*, **100**, 29–38.

Bergström, G. (1973). 'Studies on natural odoriferous compounds VI. Use of a pre-column tube for the quantitative isolation of natural, volatile compounds for gas chromatography/mass spectrometry', *Chemica Scripta*, **4**, 135–138.

Bergström, S., Carlson, L. A., Weeks, J. R. (1968). 'The prostaglandins; a family of biologically active lipids', *Pharmacol. Rev.*, **20**, 1–48.

Beroza, M. (1962). 'Determination of the chemical structure of microgram amounts of organic compounds by gas chromatography', *Anal. Chem.*, **34**, 1801–1811.

Beroza, M. (1970). 'Determination of the chemical structure of organic compounds at the microgram level by gas chromatography', *Accounts of Chemical Research*, **3**, 33–40.

Beroza, M. (1975). 'Microanalytical methodology relating to the identification of insect sex pheromones and related behavior-control chemicals' , *J. Chromatog. Sci.*, **13**, 314–321.

Beroza, M., and Bierl, B.A. (1967). 'Rapid determination of olefin position in organic compounds in microgram range by ozonolysis and gas chromatography', *Anal. Chem.*, **39**, 1131–1135.

Beroza, M., and Inscoe, M. N. (1969). 'Precolumn reactions for structure determination', in *Ancillary Techniques of Gas Chromatography* (eds L. S. Ettre and W. H. McFadden), Wiley, New York, pp. 89–144.

Bertsch, W., Chang, R. C., and Zlatkis, A. (1974). 'The determination of organic volatiles in air pollution studies; characterization of profiles', *J. Chromatog. Sci.*, **12**, 175–182.

Berüter, J., Beauchamp, G. K., and Muetterties, E. L. (1973). 'Complexity of chemical communication in mammals; urinary components mediating sex discrimination by male guinea pigs', *Biochem. Biophys. Res. Commun.*, **53**, 264–271.

Berüter, J., Beauchamp, G. K., and Muetterties, E. L. (1974). 'Mammalian chemical communication; perineal gland secretion of the guinea pig', *Physiol. Zool.*, **47**, 130–136.

Bibbiani, C., and Viola-Magni, M. P. (1971). 'Catecholamine synthesis and excretion in rats exposed intermittently to cold', *J. Physiol.*, **217**, 533–545.

Bibel, D. J., and Lovell, D. J. (1976). 'Skin flora maps; a tool in the study of cutaneous ecology', *J. Invest. Dermatol.*, **67**, 265–269.

Bigalke, R. C., Novellie, P. A., and le Roux, M. (1980). 'Studies on chemical communication in some African bovids', in *Chemical Signals: Vertebrates and Aquatic Invertebrates* (eds D. Müller-Schwarze and R. M. Silverstein), Plenum, New York, pp. 421–423.

Bird, S., and Gower, D. B. (1980). 'Measurement of 5α-androst-16-en-3-one in human axillary secretions by radioimmunoassay', *J. Endocrinol.*, **85**, 8P–9P.

Björkhem, I., Blomstrand, R., Lantto, O., Svensson, L., and Öhman, G. (1976). 'Toward absolute methods in clinical chemistry: application of mass fragmentography to high-accuracy analyses', *Clin. Chem.*, **22**, 1789–1801.

Blackman, M. W. (1911). 'The anal glands of *Mephitis mephitica*', *Anat. Rec.*, **5**, 491–515.

Blank, M. L., Kasama, K., and Snyder, F. (1972). 'Isolation and identification of an alkyldiacylglycerol containing isovaleric acid', *J. Lipid. Res.*, **13**, 390–395.

Blau, K., and King, G. S. (1978). *Handbook of Derivatives for Chromatography*, Heyden & Son Ltd., London.

Bloodworth, R. A. (1977). 'The squid accessory nidamental gland: ultrastructure and association of bacteria', *Tissue and Cell*, **9**, 197–208.

Blumer, M. (1975). 'Organic compounds in nature: limits of our knowledge', *Angew. Chem. Int. Ed. Engl.*, **14**, 507–514.

Bohlmann, F., Winterfeldt, E., Laurent, H., and Ude, W. (1963). 'Synthese des DL-castoramins', *Tetrahedron*, **19**, 195–208.

Bonsall, R. W., and Michael, R. P. (1980). 'The externalization of vaginal fatty acids by the female Rhesus monkey', *J. Chem. Ecol.*, **6**, 499–509.

Booth, W. D. (1980). 'Endocrine and exocrine factors in the reproductive behaviour of the pig', *Symp. Zool. Soc. Lond.*, **45**, 289–311.

Boudreau, J. C., and White, T. D. (1978). 'Flavor chemistry of carnivore taste systems', in *Flavor Chemistry of Animal Foods* (ed. R. W. Bullard), ACS Symposium Series, 67, American Chemical Society, Washington, D.C., pp. 102–128.

Boulton, A. A., and Milward, L. (1971). 'Separation, detection and qualitative analysis of urinary ß-phenylethylamine', *J. Chromatog.*, **57**, 287–296.

Boulton, A. A., and Marjerrison, G. L. (1972). 'Effect of L-Dopa therapy on urinary *p*-tyramine excretion and EEG changes in Parkinson's disease', *Nature*, **236**, 76–78.

Boulton, A. A., and Dyck, L. E. (1974). 'Biosynthesis and excretion of *meta*- and *para*-tyramine in the rat', *Life Sci.*, **14**, 2497–2506.

Brahmachary, R. L., and Dutta, J. (1979). 'Phenylethylamine as a biochemical marker in the tiger', *Z. Naturforsch.*, **34C**, 632–633.

Brand, J. M., Bracke, J. W., Markovetz, A. J., Wood, D. L., and Browne, L. E. (1975). 'Production of verbenol pheromone by a bacterium isolated from bark beetles', *Nature*, **254**, 136–137.

Brand, J. M., Young, J. C., and Silverstein, R. M. (1979). 'Insect pheromones; a critical review of recent advances in their chemistry, biology and application', *Fortschr. Chem. organ. Naturst.*, **37**, 1–190.

Breen, M. F., and Leshner, A. I. (1977). 'Maternal pheromone: a demonstration of its existence in the mouse (*Mus musculus*)', *Physiol. Behav.*, **18**, 527–529.

Brinck, C., Erlinge, S., and Sandell, M. (1983). 'Anal sac secretion in mustelids. A comparison', *J. Chem. Ecol.*, **9**, 727–745.

Brinck, C., Gerell, R., and Odham, G. (1978). 'Anal pouch secretion in mink, *Mustela vison*', *Oikos*, **30**, 68–75.

Brockie, R. (1976). 'Self-anointing by wild hedgehogs, *Erinaceus europaeus*, in New Zealand', *Anim. Behav.*, **24**, 68–71.

Brody, B. (1975). 'The sexual significance of the axillae', *Psychiatry*, **38**, 278–289.

Brody, S. S., and Chaney, J. E. (1966). 'Flame photometric detector', *J. Gas Chromatography*, **4**, 42–46.

Bronaugh, R. L., Hattox, S. E., Hoehn, M. M., Murphy, R. C., and Rutledge, C. O. (1972). 'The separation and identification of dopamine 3-O-sulphate and dopamine 4-O-sulphate in urine in Parkinsonian patients', *J. Pharmacol. Exp. Ther.*, **195**, 441–452.

•Bronson, F. H. (1971). 'Rodent pheromones', *Biol. Reprod.*, **4**, 344–357.

◦ Bronson, F. H. (1974). 'Pheromonal influences on reproductive activities in rodents', in *Pheromones* (ed. M. C. Birch), North-Holland Publishing Company, Amsterdam/American Elsevier, New York, pp. 344–365.

Bronson, F. H., and Marsden, H. M. (1964). 'Male-induced synchrony of estrus in deermice', *Gen. Comp. Endocrinol.*, **4**, 634–637.

Bronson, F. H., and Whitten, W. K. (1968). 'Oestrus-accelerating pheromone of mice; assay, androgen-dependency and presence in bladder urine', *J. Reprod. Fert.*, **15**, 131–134.

Bronson, F. H., and Caroom, D. (1971). 'Preputial gland of the male mouse; attractant function', *J. Reprod. Fert.*, **25**, 279–282.

Bronson, F. H., and Maruniak, J. A. (1975). 'Male-induced puberty in female mice: evidence for a synergistic action of social cues', *Biol. Reprod.*, **13**, 94–98.

Bronson, F. H., and Coquelin, A. (1980). 'The modulation of reproduction by priming pheromones in house mice: speculations on adaptive function', in *Chemical Signals: Vertebrates and Aquatic Invertebrates* (eds D. Müller-Schwarze and R. M. Silverstein), Plenum, New York, pp. 243–265.

Bronshtein, A. A. (1976). 'Regularities of the ultastructural organization of the olfactory receptors in a comparative series of vertebrates' (in Russian), *Tsitologiya*, **18**, 535–548.

Brooks, J. B., and Moore, W. E. C. (1969). 'Gas chromatographic analysis of amines and other compounds produced by several species of *Clostridium*', *Can. J. Microbiol.*, **15**, 1433–1447.

Brooks, S. C., Lalich, J. J., and Baumann, C. A. (1956). 'Skin sterols. A direct demonstration of fast-acting sterols in sebaceous glands', *Amer. J. Pathol.*, **32**, 1205–1213.

Brooksbank, B. W. L. (1962). 'Urinary excretion of androst-16-en-3α-ol: levels in normal subjects and effects of treatment with trophic hormones', *J. Endocrinol.*, **24**, 435–444.

Brooksbank, B. W. L. (1970). 'Labelling of steroids in axillary sweat after administration of ^{3}H-Δ^5-pregnenolone and ^{14}C-progesterone to a healthy man', *Experientia*, **26**, 1012–1014.

Brooksbank, B. W. L., Wilson, D. A. A., and Clough, G. (1973). 'The in-vivo uptake of ^{3}H-androsta-4,16-dien-3-one in tissues of the adult male rat', *J. Endocrinol.*, **57**, i–ii.

Brooksbank, B. W. L., Brown, R., and Gustafsson, J. A. (1974). 'The detection of 5α-androst-16-en-3-ol in human male axillary sweat', *Experientia*, **30**, 864–865.

Brown, G. K., Stokke, O., and Jellum, E. (1978). 'Chromatographic profile of high boiling organic acids in human urine', *J. Chromatog.*, **145**, 177–184.

Brown, R. E. (1979). 'Mammalian social odors; a critical review', *Adv. Study Behav.*, **10**, 103–162.

Brown, R. E., and Macdonald, D. W. (eds) (1984). *Mammalian Social Odours*, Oxford University Press, Oxford.

Brown, W. J. (1978). 'Microbial ecology of the normal vagina', in *The Human Vagina* (eds E. S. E. Hafez and T. N. Evans), Elsevier North-Holland, Amsterdam, pp. 407–422.

Browne, L. E., Birch, M. C., and Wood, D. L. (1974). 'Novel trapping and delivery systems for airborne insect pheromones', *J. Insect. Physiol.*, **20**, 183–193.

Brownlee, R. G., Silverstein, R. M., Müller-Schwarze, D., and Singer, A. G. (1969). 'Isolation, identification and function of the chief component of the male tarsal scent in black-tailed deer', *Nature*, **221**, 284–285.

Bruce, H. M. (1959). 'An exteroceptive block to pregnancy in the mouse', *Nature*, **184**, 105.

Brüggemann, I. E. M., Ritter, F. J., and Gut, J. (1981). 'Onderzoek naar signaalstoffen voor de muskusrat', *TNO Report CL 81/24*, Netherlands Organization for Applied Scientific Research TNO, Delft, Holland.

Brundin, A., Andersson, G., Andersson, K., Mossing, T., and Källquist, L. (1978). 'Short chain aliphatic acids in the interdigital gland secretions of reindeer (*Rangifer tarandus*) and their discrimination by reindeer', *J. Chem. Ecol.*, **4**, 613–622.

Buddrus, J., Herzog, H., and Cooper, J. W. (1981). 'Coupling of HPLC and NMR. Analysis of flowing liquid chromatographic fractions by proton magnetic resonance in the presence of hydrogen-containing eluting solvents', *J. Mag. Res.*, **42**, 453–459.

Bullard, R. W. (1982). 'Wild canid associations with fermentation products', *Ind. Eng. Chem. Prod. Res. Dev.*, **21**, 646–655.

Bullard, R. W., and Holguin, G. (1977). 'Volatile components of unprocessed rice (*Oryza sativa*)', *J. Agric. Food Chem.*, **25**, 99–103.

Bullard, R. W., Leiker, T. J., Peterson, J. E., and Kilburn, S. R. (1978a). 'Volatile components of fermented egg, an animal attractant and repellent', *J. Agric. Food Chem.*, **26**, 155–159.

Bullard, R. W., Shumake, J. A., Campbell, D. L., and Turkowski, F. J. (1978b). 'Preparation and evaluation of a synthetic fermented egg coyote attractant and deer repellent', *J. Agric. Food Chem.*, **26**, 160–163.

Burger, B. V., le Roux, M., Garbers, C. F., Spies, H. S. C., Bigalke, R. G., Pachler, K. G. R., Wessels, P. L., Christ, V., and Maurer, K. H. (1976). Studies on mammalian pheromones, I, Ketones from the pedal gland of the bontebok, *Damaliscus dorcas dorcas*', *Z. Naturforsch.*, **31C**, 21–28.

Burger, B. V., le Roux, M., Garbers, C. F., Spies, H. S. C., Bigalke, R. C., Pachler, K. G. R., Wessels, P. L., Christ, V., and Maurer, K. H. (1977). 'Studies on mammalian pheromones, II, Further compounds from the pedal gland of the bontebok, *Damaliscus dorcas dorcas*', *Z. Naturforsch.*, **32C**, 49–56.

Burger, B. V., le Roux, M., Spies, H. S. C., Truter, V., and Bigalke, R. C. (1978). 'Mammalian pheromone studies, III, (*E,E*)-7,11,15-trimethyl-3-methylenehexadeca-1,6,10,14-tetraene, a new diterpene analogue of ß-farnesene from the dorsal gland of the springbok, *Antidorcas marsupialis*', *Tet. Lett.*, 5221–5224.

Burger, B. V., le Roux, M., and Bigalke, R. C. (1980). 'Compounds from the dorsal cutaneous exudate of the springbok (*Antidorcas marsupialis*) and from the preorbital gland secretion of the grysbok (*Raphicerus melanotis*)', in *Olfaction and Taste VII* (ed. H. van der Starre), IRL Press Ltd., London, p. 91.

Burger, B. V., le Roux, M., Spies, H. S. C., Truter, V., and Bigalke, R. C. (1981a). 'Mammalian pheromone studies, IV, Terpenoid compounds and hydroxy esters from the dorsal gland of the springbok, *Antidorcas marsupialis*', *Z. Naturforsch.*, **36C**, 340–343.

Burger, B. V., le Roux, M., Spies, H. S. C., Truter, V., Bigalke, R. C., and Novellie, P. A. (1981b). 'Mammalian pheromone studies, V, Compounds from the preorbital gland of the grysbok, *Raphicerus melanotis*', *Z. Naturforsch.*, **36C**, 344–346.

Burlingame, A. L. (ed.) (1970). *Topics in Organic Mass Spectrometry*, Wiley, New York.

Buss, I. O., Rasmussen, L. E., and Smuts, G. L. (1976). 'The role of stress and individual recognition in the function of the African elephant's temporal gland', *Mammalia*, **40**, 437–451.

Butler, L. D., and Burke, M. F. (1976). 'Chromatographic characterization of porous polymers for use as adsorbents in sampling columns', *J. Chromatog. Sci.*, **14**, 117–122.

Buttery, R. G., Seifert, R. M., Guadagni, D. G., and Ling, L. C. (1969). 'Charaterization of some volatile constituents of bell peppers', *J. Agric. Food. Chem.*, **17**, 1322–1327.

Buttery, R. G. Seifert, R. M., Turnbaugh, J. G., Guadagni, D. G., and Ling, L. C. (1981). 'Odor threshold of thiamin odor compound 1-methylbicyclo[3.3.0]-2,4-dithia-8-oxaoctane', *J. Agric. Food Chem.*, **29**, 183–185.

Candland, D. K., Blumer, E. S., and Mumford, M. D. (1980). 'Urine as a communicator in a New World primate, *Saimiri sciureus*', *Animal Learning and Behavior.*, **8**, 468–480.

Cannon, W. B., Shohl, A. T., and Wright, W. S. (1911). 'Emotional glycosuria', *Amer. J. Physiol.*, **29**, 280–287.

Caprioli, R. M., Liehr, J. G., and Seifert, W. E. (1980). 'Applications to clinical medicine', in *Biomedical Applications of Mass Spectrometry. First Supplementary Volume* (eds G. R. Waller and O. C. Dermer), Wiley, New York, pp. 1007–1084.

Care, A. J. E. (1969). 'Hairs and vibrissae in the Rhinocerotidae', *J. Zool.*, **157**, 247–257.

Caroom, D., and Bronson, F. H. (1971). 'Responsiveness of female mice to preputial attractant: Effect of sexual experience and ovarian hormones', *Physiol. Behav.*, **7**, 659–662.

Carter, H. B., and Dowling, D. F. (1954). 'The hair follicle and apocrine gland population of cattle skin', *Aust. J. Agric. Res.*, **5**, 745–754.

Celesk, R. A., Asano, T., and Wagner, M. (1976). 'The size, pH, and redox potential of the cecum in mice associated with various microbial floras', *Proc. Soc. Exp. Biol. Med.*, **151**, 260–263.

Chalmers, R. A., Healy, M. J. R., Lawson, A. M., and Watts, R. W. E. (1976). 'Urinary organic acids in man, II, Effects of individual variations and diet on urinary excretion of acid metabolites', *Clin. Chem.*, **22**, 1288–1291.

Chandler, C. F. (1975). 'Development and function of marking and sexual behavior in the Malagasy prosimian primate, *Lemur fulvus*', *Primates*, **16**, 35–47.

Chapman, D. I. (1979). 'Radioimmunoassay', *Chem. Brit.*, **15**, 439, 441, 443–447.

Charransol, G., Barthelemy, C., and Desgrez, P. (1978). 'Rapid determination of urinary oxalic acid by gas-liquid chromatography without extraction', *J. Chromatog.*, **145**, 452–455.

Chauhan, J., and Darbre, A. (1981). 'Glass capillary SCOT columns by a single step coating method', *HRC&CC.*, **4**, 11–16.

Christensen, N. J. (1973). 'A sensitive assay for the determination of dopamine in plasma', *Scand. J. Clin. Lab. Invest.*, **31**, 343–346.

Christie, W. W. (1973). *Lipid Analysis*, Pergamon Press, Oxford.

Christie, W. W. (1978). 'The composition, structure and function of lipids in the tissues of ruminant animals', *Prog. Lipid Res.*, **17**, 111–205.

Claesson, A., and Silverstein, R. M. (1977). 'Chemical methodology in the study of mammalian communications', in *Chemical Signals in Vertebrates* (eds D. Müller-Schwarze and M. M. Mozell), Plenum, New York, pp. 71–93.

Clark, A. B. (1982a). 'Scent marks as social signals in *Galago crassicaudatus*. I. Sex and reproductive status as factors in signals and responses', *J. Chem. Ecol.*, **8**, 1133–1151.

310

Clark, A. B. (1982b). Scent marks as social signals in *Galago crassicaudatus*. II. Discrimination between individuals by scents', *J. Chem. Ecol.*, **8**, 1153–1165.

Claus, R. (1975). 'Neutralization of pheromones by antisera in pigs', in *Immunization with Hormones in Reproduction Research* (ed. E. Nieschlag), North-Holland Publishing Company, Amsterdam, pp. 189–197.

Claus, R. (1976). 'Messung des Ebergeruchsstoffes im Fett von Schweinen mittels eines Radioimmunotests', *Z. Tierzüchtung un Züchtungsbiologie*, **93**, 38–47.

Claus, R. (1979). 'Mammalian pheromones with special reference to the boar taint steroid and its relationship to other testicular steroids' (German), *Fortschritte in der Tierphysiologie und Tiernaehrung*, **10**, 1–136.

Claus, R., and Alsing, W. (1976a). 'Occurrence of 5α-androst-16-en-3-one, a boar pheromone, in man and its relationship to testosterone', *J. Endocrinol.*, **68**, 483–484.

Claus, R., and Alsing, W. (1976b). 'Einfluss von Choriongonadotropin, Haltungsänderung und sexueller Stimulierung auf die Konzentrationen von Testosteron im Plasma sowie des Ebergeruchsstoffes im Plasma und Fett eines Ebers', *Berl. Muench. Tierarztl. Wochenschr.*, **89**, 354–358.

Claus, R., and Gimenez, T. (1977). 'Diurnal Rhythm of 5α-androst-16-en-3-one and testosterone in peripheral plasma of boars', *Acta endocrinologica.*, **84**, 200–206.

Clayton, R. B., Nelson, A. N., and Frantz, I. D. (1963). 'The skin sterols of normal and triparanol-treated rats', *J. Lipid Res.*, **4**, 166–178.

Clevedon Brown, J., and Williams, J. D. (1972). 'The rodent preputial gland', *Mammal Review.*, **2**, 105–147.

Clulow, F. V., and Langford, P. E. (1971). 'Pregnancy block in the meadow vole, *Microtus pennsylvanicus*', *J. Reprod. Fert.*, **24**, 275–277.

Coblentz, B. E. (1976). 'Functions of scent-urination in ungulates with special reference to feral goats (*Capra hircus*), *Amer. Nat.*, **110**, 549–557.

Cohen, G., Holland, B., Sha, J., and Goldenberg, M. (1959). 'Plasma concentrations of epinephrine and norepinephrine during intravenous infusions in man', *J. Clin. Invest.*, **38**, 1935–1941.

Colborne, J., Norval, R. A. I., and Spickett, A. M. (1981). 'Ecological studies on *Ixodes (Afrixodes) matopi*', *Onderstepoort J. Vet. Res.*, **48**, 31–35.

Colby, D. R., and Vandenbergh, J. G. (1974). 'Regulatory effects of urinary pheromones on puberty in the mouse', *Biol. Reprod.*, **11**, 268–279.

Collins, R. P. (1976). 'Terpenes and odoriferous materials from Microorganisms', *Lloydia*, **39**, 20–24.

Comfort, A. (1971). 'The likelihood of human pheromones', *Nature*, **230**, 432–434.

Conner, W. E., Eisner, T., VanderMeer, R. K., Guerrero, A., Ghiringelli, D., and Meinwald, J. (1980). 'Sex attractant of an arctiid moth (*Utetheisa ornatrix*); a pulsed chemical signal', *Behav. Ecol. Sociobiol.*, **7**, 55–63.

Cooper, J. G., and Bhatnagar, K. P. (1976). 'Comparative anatomy of the vomeronasal organ complex in bats', *J. Anat.*, **122**, 571–601.

Corry, J. E. L. (1978). 'Possible sources of ethanol ante- and post-mortem; its relationship to the biochemistry and microbiology of decomposition', *J. Appl. Bact.*, **44**, 1–56.

Cotter, R. J. (1980). 'Mass spectrometry of non-volatile compounds. Desorption from extended probes', *Anal. Chem.*, **52**, 1589A–1606A.

Cove, J. H., Holland, K. T., and Cunliffe, W. J. (1980). 'An analysis of sebum excretion rate, bacterial population and the production rate of free fatty acids on human skin', *Br. J. Dermatol.*, **103**, 383–386.

Cowley, J. J. (1978). 'Olfaction and the development of sexual behaviour', in *Biological Determinants of Sexual Behaviour* (ed. J. B. Hutchison), Wiley, Chichester, pp. 87–125.

Cowley, J. J. (1980). 'Growth and maturation in mice, *Mus musculus*', *Symp. Zool. Soc. Lond.*, **45**, 213–250.

Cowley, J. J., Johnson, A. L., and Brooksbank, B. W. L. (1977). 'The effect of two odorous compounds on performance in an assessment-of-people test', *Psychoneuroendocrinology*, **2**, 159–172.

Crewe, R. M., Burger, B. V., le Roux, M., and Katsir, Z. (1979). 'Chemical constituents of the chest gland secretion of the thick-tailed galago (*G. crassicaudatus*)', *J. Chem. Ecol.*, **5**, 861–868.

Cribb, A. E., and Flood, P. F. 'Exfoliative cytology of the major vestibular glands of cattle', in preparation.

Cross, J. H. (1980). 'A vapour collection and thermal desorption method to measure semiochemical release rates from controlled release formulations', *J. Chem. Ecol.*, **6**, 781–795.

Cross, J. H., Byler, R. C., Cassidy, R. F., Silverstein, R. M., Greenblatt, R. E., Burkholder, R. F., Levinson, A. R., and Levinson, H. Z. (1976). 'Porapak Q collection of pheromone components and isolation of (*Z*)-and (*E*)-14-methyl-8-hexadecenal, sex pheromone components, from the females of four species of Trogoderma (Coleoptera: Dermestidae)', *J. Chem. Ecol.*, **2**, 457–468.

Cross, J. H., Byler, R. C., Silverstein, R. M., Greenblatt, R. E., Gorman, J. E., and Burkholder, W. E. (1977). 'Sex pheromone components and calling behaviour in the female dermestid beetle, *Trogoderma variabile*', *J. Chem. Ecol.*, **3**, 115–125.

Cross, J. H., Tumlinson, J. H., Heath, R. E., and Burnett, D. E. (1980). 'Applications and procedure for measuring release rates from formulations of lepidopteran semiochemicals', *J. Chem. Ecol.*, **6**, 759–770.

Crump, D. R. (1980a). 'Thietanes and dithiolanes from the anal gland of the stoat, *Mustela erminea*', *J. Chem. Ecol.*, **6**, 341–347.

Crump, D. R. (1980b). 'Anal gland secretion of the ferret (*Mustela putorius* forma *furo*)', *J. Chem. Ecol.*, **6**, 837–844.

Curtis, R. F., Ballantine, J. A., Keverne, E. B., Bonsall, R. W., and Michael, R. P. (1971). 'Identification of primate sexual pheromones and the properties of synthetic attractants', *Nature*, **232**, 396–398.

Cymerman, A., and Franesconi, R. F. (1975). 'Alteration of circadian rhythmicities of urinary 3-methoxy-4-hydroxyphenylglycol and vanilmandelic acid in man during cold exposure', *Life Sci.*, **16**, 225–236.

Daly, N. R. (ed.) (1978). *Advances in Mass Spectrometry*. Vol. 7B, Heyden Press/Institute of Petroleum, London.

Dandeneau, R. D., and Zerenner, E. H. (1979). 'An investigation of glasses for capillary chromatography', *HRC&CC*, **2**, 351–356.

Davies, V. J., and Bellamy, D. (1974). 'Effects of female urine on social investigation in male mice', *Anim. Behav.*, **22**, 239–241.

Day, R. T., Unger, S. E., and Cooks, R. G. (1980). 'Molecular secondary ion mass spectrometry', *Anal. Chem.*, **52**, 557A–572A.

deEds, F. (1968). 'Flavonoid metabolism', in *Comparative Biochemistry*, Vol. 20, (eds M. Florkin and E. H. Stotz), Academic Press, New York, pp. 127–171.

Dekirmenjian, H., and Maas, J. W. (1970). 'An improved procedure for 3-methoxy-4-hydroxyphenylethylene glycol determination by gas-liquid chromatography', *Anal. Biochem.*, **35**, 113–122.

Desjardins, C., Maruniak, J. A., and Bronson, F. H. (1973). 'Social rank in house mice: differentiation revealed by ultraviolet visualization of urinary marking patterns', *Science*, **182**, 939–941.

Detling, J. K., Ross, C. W., Walmsley, M. H., Hilbert, D. W., Bonilla, C. A., and Dyer, M. I. (1981). 'Examination of North American bison saliva for potential plant growth regulators', *J. Chem. Ecol.*, **7**, 239–246.

312

Dey, A. C., Abbott, E. C., Rusted, I. E., and Senciall, I. R. (1972). 'Excretion of conjugated 11-deoxy-17-ketosteroids in "essential" hypertension', *Can. J. Biochem.*, **50**, 1273–1281.

Dirren, H., Robinson, A. B., and Pauling, L. (1975). 'Sex-related patterns in the profiles of human urinary amino acids', *Clin. Chem.*, **21**, 1970–1975.

Dmi'el, R., Robertshaw, D., and Choshniak, I. (1979). 'Sweat gland secretion in the black bedouin goat', *Physiol. Zool.*, **52**, 558–564.

Do, J. C., Kitatsuji, E., and Yoshii, E. (1975). 'Study on the components of musk. I. Ether soluble components', *Chem. Pharm. Bull.*, **23**, 629–635.

Donovan, C. A. (1967). 'Some clinical observations on sexual attraction and deterrence in dogs and cattle', *Vet. Med./Small Animal Clinician*, **62**, 1047–1048, 1051.

Donovan, C. A. (1969). 'Canine anal glands and chemical signals (pheromones)', *J. Amer. Vet. Med. Assn.*, **155**, 1995–1996.

Doty, R. L. (1975). 'Intranasal trigeminal detection of chemical vapors by humans', *Physiol. Behav.*, **14**, 855–859.

Doty, R. L. (1977). 'A review of recent psychophysical studies examining the possibility of chemical communication of sex and reproductive state in humans', in *Chemical Signals in Vertebrates* (eds D. Müller-Schwarze and M. M. Mozell), Plenum, New York, pp. 273–286.

Doty, R. L., and Kart, R. (1972). 'A comparative and developmental analysis of the midventral sebaceous glands in 18 taxa of *Peromyscus*, with an examination of gonadal steroid influences in *Peromyscus maniculatus bairdii*', *J. Mammal.*, **53**, 83–99.

Doty, R. L., and Dunbar, I. (1974). 'Attraction of beagles to conspecific urine, vaginal and anal sac secretion odors', *Physiol. Behav.*, **12**, 825–833.

Doty, R. L., Ford, M., Preti, G., and Huggins, G. R. (1975). 'Changes in the intensity and pleasantness of human vaginal odors during the menstrual cycle', *Science*, **190**, 1316–1318.

Doty, R. L., Orndorff, M. M., Leyden, J., Kligman, A. (1978). 'Communication of gender from human axillary odors; relationship to perceived intensity and hedonicity', *Behav. Biol.*, **23**, 373–380.

Døving, K. B. (1966). 'An electrophysiological study of odour similarities of homologous substances', *J. Physiol.*, **186**, 97–109.

Downing, D. T., Kranz, Z. H., and Murray, K. E. (1960). 'An investigation of the aliphatic constituents of hydrolysed wool wax by gas chromatography', *Aust. J. Chem.*, **13**, 80–94.

Downing, D. T., Strauss, J. S., and Pochi, P. E. (1969). 'Variability in the chemical composition of human skin surface lipids', *J. Investig. Dermatol.*, **53**, 322–327.

Downing, D. T., and Strauss, J. S. (1974). 'Synthesis and composition of surface lipids of human skin', *J. Investig. Dermatol.*, **62**, 228–244.

Doyle, G. A. (1974). 'The behaviour of the lesser bushbaby', in *Prosimian Behaviour* (eds R. D. Martin, G. A. Doyle, A. C. Walker), Gerald Duckworth and Company Ltd., London, pp. 213–231.

Draize, J. H. (1942). 'The determination of the pH of the skin of man and common laboratory animals', *J. Investig. Dermatol.*, **5**, 77–85.

Dravnieks, A. (1975). 'Evaluation of human body odors: methods and interpretations', *J. Soc. Cosmet. Chem.*, **26**, 551–571.

Dreyfus, P. M., Levy, H. L., and Efron, M. L. (1968). 'Concerning amino acids in human saliva', *Experientia*, **24**, 447–448.

Drickamer, L. C. (1977). 'Delay of sexual maturation in female house mice by exposure to grouped females or urine from grouped females', *J. Reprod. Fert.*, **51**, 77–81.

Drickamer, L. C. (1979). 'Acceleration and delay of first estrus in wild *Mus musculus*', *J. Mammal.*, **60**, 215–216.

313

Drickamer, L. C. (1982a). 'Delay and acceleration of puberty in female mice by urinary chemosignals from other females', *Dev. Psychobiol.*, **15**, 433–445.

Drickamer, L. C. (1982b). 'Acceleration and delay of first vaginal oestrus in female mice by urinary chemosignals: dose levels and mixing urine treatment sources', *Anim. Behav.*, **21**, 456–460.

Drickamer, L. C. (1982c). 'Acceleration and delay of sexual maturation in female mice via chemosignals: circadian rhythm effects', *Biol. Reprod.*, **27**, 596–601.

Drickamer, L. C., and Hoover, J. E. (1979). 'Effects of urine from pregnant and lactating female house mice on sexual maturation of juvenile females', *Dev. Psychobiol.*, **12**, 545–551.

? Drickamer, L. C., and McIntosh, T. K. (1980). 'Effects of adrenalectomy on the presence of a maturation-delaying pheromone in the urine of female mice', *Horm. Behav.*, **14**, 146–152.

Drozd, J. (1975). 'Chemical derivatization in gas chromatography', *J. Chromatog.*, **113**, 303–356.

Dryden, G. L., and Conaway, C. H. (1967). 'The origin and hormonal control of scent production in *Suncus murinus*', *J. Mammal.*, **48**, 420–428.

Durden, D. A., Philips, S. R., and Boulton, A. A. (1973). 'Identification and distribution of ß-phenylethylamine in the rat', *Can. J. Biochem.*, **51**, 995–1002.

Dyer, M. I. (1980). 'Mammalian epidermal growth factor promotes plant growth', *Proc. Natl. Acad. Sci. USA*, **77**, 4836–4837.

▸ Dziedzic, S. W., Dziedzic, L. B., and Gitlow, S. E. (1973). 'Separation and determination of urinary homovanillic and iso-homovanillic acid by gas-liquid chromatography and electron capture detection', *J. Lab. Clin. Med.*, **82**, 829–835.

Eberhard, I. H., McNamara, J., Pearse, R. J., and Southwell, I. A. (1975). 'Ingestion and excretion of *Eucalyptus punctata* and its essential oil by the koala, *Phascolarctos cinereus*', *Aust. J. Zool.*, **23**, 169–179.

Eberhard, W. G. (1977). 'Aggressive chemical mimicry by a bolas spider', *Science*, **198**, 1173–1176.

Ebling, F. J. (1974). 'Hormonal control and methods of measuring sebaceous gland activity', *J. Investig. Dermatol.*, **62**, 161–171.

Ebling, F. J. (1977a). 'Hormonal control of mammalian skin glands', in *Chemical Signals in Vertebrates* (eds D. Müller-Schwarze and M. M. Mozell), Plenum, New York, pp. 17–33.

Ebling, F. J. (1977b). 'Sebaceous glands', in *Dermatotoxicity and Pharmacology* (eds F. N. Marzolli and H. I. Maibach), Hemisphere Pub. Corp., Washington, D.C., pp. 55–92.

Ebling, F. J., Ebling, E., McCaffery, V., and Skinner, J. (1973). 'The responses of the sebaceous glands of hypophysectomized-castrated male rat to 5α-androstanedione and 5α-androstane-3ß,17ß-diol', *J. Investig. Dermatol.*, **60**, 183–187.

Ebling, F. J., Ebling, E., Randall, V., and Skinner, J. (1975). 'The synergistic action of α-melanocyte-stimulating hormone on the sebaceous, prostate, preputial, Harderian and lachrymal glands, seminal vesicles and brown adipose tissue in the hypophysectomized-castrated rat', *J. Endocrinol.*, **66**, 407–412.

Eckstein, P., and Zuckermann, S. (1956). 'Morphology of the reproductive tract', in *Marshall's Physiology of Reproduction*, Vol. 1, Part 1 (ed. A. S. Parkes), Longmans, Green and Company, London, 3rd edn., pp. 43–155.

Eisenberg, J. F., McKay, G. M., and Jainudeen, M. R. (1971). 'Reproductive behaviour of the asiatic elephant, *Elephas maximus maximus*', *Behaviour*, **38**, 193–225.

Eisenberg, J. F., and Kleiman, D. G. (1972). 'Olfactory communication in mammals', *Ann. Rev. Ecol. Syst.*, **3**, 1–32.

314

Eisenbraun, E. J., Browne, C. E., Irvin-Willis, R. L., McGurk, D. J., Eliel, E. L., and Harris, D. L. (1980). 'Structure and stereochemistry of 4aß,7α,7aß-nepetalactone from *Nepeta mussini* and its relationship to the 4aα,7α,7aα- and 4aα,7α, 7aß-nepetalactones from *N. cataria*', *J. Org. Chem.*, **45**, 3811–3814.

Eisner, T., Conner, W. E., Hicks, K., Dodge, K. R., Rosenberg, H. I., Jones, T. H., Cohen, M., and Meinwald, J. (1977). 'Stink of stinkpot turtle identified: ω-phenylalkanoic acids', *Science*, **196**, 1347–1349.

Ellenberger, W., and Baum, H. (1943). *'Handbuch der vergleichenden Anatomie der Haustiere'* (eds O. Zietzschmann, E. Ackerknecht, and H. Grau), Springer-Verlag, Berlin,18th edn.

Ellin, R. I., Farrand, R. L., Oberst, F. W., Crouse, C. L., Billups, N. B., Koon, W. S., Musselman, N. P., and Sidell, F. E. (1974). 'An apparatus for the detection and quantitation of volatile human effluents', *J. Chromatog.*, **100**, 137–152.

Ellis, H. (1905). *Studies in the Psychology of Sex*, Vol. 4, Sexual selection in man, Pt II, 'Smell', F. A. Davies and Co., Philadelphia.

Ellis, R. A., and Montagna, W. (1962). 'The skin of primates, VI. The skin of the gorilla *(Gorilla gorilla)*', *Amer. J. Phys. Anthropol.*, **20**, 79–93.

Elstein, M. (1978). 'Functions and physical properties of mucus in the female genital tract', *Br. Med. Bull.*, **34**, 83–88.

Engelhardt, H. (1979). *High Performance Liquid Chromatography: Chemical Laboratory Practice*, Springer-Verlag, Berlin.

Englert, E., Wayne, A. W., Wales, E. E., and Straight, R. C. (1979). 'A rapid, new and direct method for isolation and measurement of porphyrins in biological samples by high performance liquid chromatography', *HRC&CC*, **2**, 570–574.

Epple, G. (1970). 'Quantitative studies on scent marking in the marmoset, *Callithrix jacchus*', *Folia Primat.*, **13**, 48–62.

Epple, G. (1974). 'Primate pheromones', in *Pheromones* (ed. M. C. Birch), North-Holland, Amsterdam/American Elsevier, New York, pp. 366–385.

Epple, G. (1976). 'Chemical communication and reproductive processes in non-human primates', in *Mammalian Olfaction, Reproductive Processes and Behavior* (ed. R. L. Doty), Academic Press, New York, pp. 257–282.

Epple, G. (1978). 'Studies on the nature of chemical signals in scent marks and urine of *Saguinus fuscicollis*', *J. Chem. Ecol.*, **4**, 383–394.

Epple, G. (1979). 'Gonadal control of male scent in the tamarin, *Saguinus fuscicollis*', *Chemical Senses and Flavour*, **4**, 15–20.

Epple, G. (1980). 'Relationship between aggression, scent marking and gonadal state in a primate, the tamarin *Saguinus fuscicollis*', in *Chemical Signals: Vertebrates and Aquatic Invertebrates* (eds D. Müller-Schwarze and R. M. Silverstein), Plenum, New York, pp. 87–105.

Epple, G., and Lorenz, R. (1967). 'Vorkommen, Morphologie und Funktion der Sternal-drüse bei den Platyrrhini', *Folia Primatol.*, **7**, 98–126.

Epple, G., Golob, N. F., and Smith, A. B. (1979). 'Odor communication in the tamarin *Saguinus fuscicollis*: behavioural and chemical studies', in *Chemical Ecology: Odour Communication in Animals* (ed. F. J. Ritter), Elsevier/North-Holland Biomedical Press, Amsterdam, pp. 117–130.

Epps, J. (1974). 'Social interactions of *Perodicticus potto* kept in captivity in Kampala, Uganda', in *Prosimian Behaviour* (eds R. D. Martin, G. A. Doyle and A. C. Walker), Gerald Duckworth and Co., Ltd., London, pp. 233–244.

Ericsson, M., and Blomberg, L. G. (1980). 'Modification of HPTLC-plates by in situ chemical bonding with non-polar and polar organosilanes', *HRC&CC*, **3**, 345–350.

Erlinge, S., Sandell, M., and Brinck, C. (1982). 'Scent marking and its territorial significance in stoats, *Mustela erminea*', *Anim. Behav.*, **30**, 811–812.

Eskelson, C. D., Chvapil, M., Chang, S. Y., and Chvapil, T. (1978). 'Identification of ejaculate-derived propylamine found in collagen sponge contraceptives', *Biomed. Mass. Spectrom.*, **5**, 238–242.

Eskelson, C. D., Chang, S. Y., Chvapil, T., and Chvapil M. (1979). 'The identification of odoriferous compounds formed by coital ejaculate in collagen sponge contraceptives', *NBS Spec. Publ., (US)*, **519**, 469–476.

Espmark, Y. (1964). 'Rutting behaviour in reindeer (*Rangifer tarandus*)', *Anim. Behav.*, **12**, 159–163.

Estes, R. D. (1972). 'The role of the vomeronasal organ in mammalian reproduction', *Mammalia*, **36**, 315–341.

Estes, J. A., and Buss, I. O. (1976). 'Microanatomical structure and development of the African elephant's temporal gland', *Mammalia*, **40**, 429–436.

Evans, C. A., and Mattern, K. L. (1979). 'The aerobic growth of *Propionibacterium acnes* in primary cultures from skin', *J. Investig. Dermatol.*, **72**, 103–106.

Evans, C. M., Mackintosh, J. H., Kennedy, J. F., and Robertson, S. M. (1978). 'Attempts to characterise and isolate aggression-reducing olfactory signals from the urine of female mice, *Mus musculus*', *Physiol. Behav.*, **20**, 129–134.

Evans, C. S., and Goy, R. W. (1968). 'Social behaviour and reproductive cycles in captive ring-tailed lemurs, *Lemur catta*', *J. Zool.*, **156**, 181–197.

Evered, D. F. (1956). 'The excretion of amino acids by the human', *Biochem. J.*, **62**, 416–427.

Ewer, R. F. (1973). *The Carnivores*, Weidenfeld and Nicolson, London.

Ewer, R. F., and Wemmer, C. (1974). 'The behaviour in captivity of the African civet, *Civettictis civetta*', *Z. Tierpsychol.*, **34**, 359–394.

Fagre, D. B., Howard, W. E., and Teranishi, R. (1982). 'Development of coyote attractants for reduction of livestock losses', in *Wildlife-Livestock Symp. Proc. 10*, (eds J. M. Peek and P. D. Dalke), University of Idaho, Moscow, Idaho, pp. 319–326.

Fales, H. M., Milne, G. W. A., Winkler, H. U., Beckey, H. D., Damico, J. N., and Barron, R. (1975). 'Comparison of mass spectra of some biologically important compounds as obtained by various ionization techniques', *Anal. Chem.*, **47**, 207–219.

Feser, K., and Kögler, W. (1979). 'The quadrupole mass filter in GC/MS applications', *J. Chromatog. Sci.*, **17**, 57–63.

Findlay, J. D., and Jenkinson, D. McE. (1960). 'The morphology of bovine sweat glands and the effect of heat on the sweat glands of the Ayrshire calf', *J. Agric. Sci.*, **55**, 247–249.

Fischer, E., Spatz, H., Heller, B., and Reggiani, H. (1972). 'Phenethylamine content of human urine and rat brain: its alteration in pathological conditions after drug administration', *Experientia*, **28**, 307–308.

Fleming, A., Vaccarino, F., Tambosso, L., and Chee, P. (1970). 'Vomeronasal and olfactory system modulation of maternal behavior in the rat', *Science*, **203**, 372–374.

Flood, P. F. (1973). 'Histochemical localization of hydroxysteroid dehydrogenases in maxillary glands of pigs', *J. Reprod. Fert.*, **32**, 125–127.

Flux, J. E. C. (1970). 'Life history of the mountain hare (*Lepus timidus scoticus*) in north-east Scotland', *J. Zool.*, **160**, 75–123.

Fox, M. W. (1971). 'Ontogeny of a social display in *Hyaena hyaena*; anal protrusion', *J. Mammal.*, **52**, 467–469.

Frankel, E. N. (1980). 'Lipid oxidation', *Prog. Lipid. Res.*, **19**, 1–22.

Franklin, W. L. (1980). 'Territorial marking behavior by the South American Vicuna', in *Chemical Signals: Vertebrates and Aquatic Invertebrates* (eds D. Müller-Schwarze and R. M. Silverstein), Plenum Press, New York, pp. 53–66.

Freeman, R. R. (ed.) (1981). *High Resolution Gas Chromatography*, Hewlett-Packard, Palo-Alto, California (2nd edn).

Freinkel, R. K., and Aso, K. (1969). 'Esterification of cholesterol in the skin', *J. Investig. Dermatol.*, **52**, 148–154.

Freinkel, R. K., and Shen, Y. (1969). 'The origin of free fatty acids in sebum'. *J. Investig. Dermatol.*, **53**, 422–427.

Frigerio, A. (ed.) (1977). *Advances in Mass Spectrometry in Biochemistry and Medicine*, Spectrum Publications Inc., New York, Vol II.

Frigerio, A., and Castagnoli, N. (eds) (1976). *Advances in Mass Spectrometry in Biochemistry and Medicine*, Spectrum Publications Inc., New York, Vol I.

Fu, H. C., and Nicolaides, N. (1969). 'The structure of alkane diols of diesters in vernix caseosa lipids', *Lipids*, **4**, 170–172.

Games, D. E., Jackson, A. H., Millington, D. S., and Rossiter, M. (1974). 'Field ionization with GC/MS in structural studies of natural products', *Advances in Mass Spectrometry*, **6**, 137–142.

Gangwal, S. K., and Wagoner, D. E. (1979). 'Response correlation of low molecular weight sulfur compounds using a novel flame photometric detector', *J. Chromatog. Sci.*, **17**, 196–201.

Gautier, J. P., and Gautier, A. (1977). 'Communication in old world monkeys', in *How Animals Communicate* (ed. T. A. Sebeok), Indiana University Press, Bloomington, pp. 890–964.

Gawienowski, A. M. (1977). 'Chemical attractants of the rat preputial gland', in *Chemical Signals in Vertebrates* (eds D. Müller-Schwarze and M. M. Mozell), Plenum, New York, pp. 45–59.

Gawienowski, A. W., Orsulak, P. J., Stacewicz-Sapuntzakis, M., and Joseph, B. M. (1975). 'Presence of sex pheromone in preputial glands of male rats', *J. Endocrinol.*, **67**, 283–288.

Gawienowski, A. M., Orsulak, P. J., Stacewicz-Sapuntzakis, M., and Pratt, J. J. (1976). 'Attractant effect of female preputial gland extracts in the male rat', *Psychoneuroendocrinology*, **1**, 411–418.

Gesteland, R. C., Lettvin, J. Y., Pitts, W. H., and Rojas, A. (1963). 'Odor specificities of the frog's olfactory receptors', in *Olfaction and Taste I* (ed. Y. Zotterman), Pergamon, Oxford, pp. 19–34.

Getchell, M. L., and Gesteland, R. C. (1972). 'The chemistry of olfactory reception; stimulus-specific protection from sulphydryl reagent inhibition', *Proc. Natl. Acad. Sci., U.S.A.*, **69**, 1494–1498.

Getchell, T. V., and Shepherd, G. M. (1978a). 'Responses of olfactory receptor cells to step pulses of odour at different concentrations in the salamander', *J. Physiol.*, **282**, 521–540.

Getchell, T. V., and Shepherd, G. M. (1978b). 'Adaptive properties of olfactory receptors analysed with odour pulses of varying duration', *J. Physiol.*, **282**, 541–560.

Gibbons, R. A. (1978). 'Mucus of the mammalian genital tract', *Br. Med. Bull.*, **34**, 34–38.

Gibbons, R. A., and Sellwood, R. (1973). 'The macromolecular biochemistry of cervical secretions', in *The Biology of the Cervix* (eds R. J. Blandau and K. Moghissi), University of Chicago Press, pp. 251–265.

Goldberg, V. J., and Ramwell, P. W. (1975). 'Role of prostaglandins in reproduction', *Physiol. Rev.*, **55**, 325–351.

Goldfoot, D. A., Goy, R. W., Kravetz, M. A., and Freeman, S. K. (1976a). 'Reply to Michael, Bonsall, and Zumpe', 'Reply to Keverne', *Horm. Behav.*, **7**, 373–378.

Goldfoot, D. A., Kravetz, M. A., Goy, R. W., and Freeman, S. K. (1976b). 'Lack of effect of vaginal lavages and aliphatic acids on ejaculatory responses in rhesus monkeys: behavioural and chemical analysis', *Horm. Behav.*, **7**, 1–27.

Golob, N. F., Yarger, R. G., and Smith, A. B. (1979). 'Primate chemical communication, Part III. Synthesis of the major volatile constituents of the marmoset (*Saguinus fuscicollis*) scent mark, *J. Chem. Ecol.*, **5**, 543–555.

317

Gompertz, D. (1971). 'Crotonic acid, an artefact in screening for organic acidemias', *Clin. Chim. Acta.*, **33**, 457.

Goodrich, B. S., and Mykytowycz, R. (1972). 'Individual and sex differences in the chemical composition of pheromone-like substances from the skin glands of the rabbit, *Oryctolagus cuniculus*', *J. Mammal.*, **53**, 540–548.

Goodrich, B. S., Hesterman, E. R., Murray, K. E., Mykytowycz, R., Stanley, G., and Sugowdz, G. (1978). 'Identification of behaviourally significant volatile compounds in the anal gland of the rabbit, *Oryctolagus cuniculus*', *J. Chem. Ecol.*, **4**, 581–594.

Goodrich, B. S., Hesterman, E. R., Shaw, K. S., and Mykytowycz, R. (1981a). 'Identification of some volatile compounds in the odour of faecal pellets of the rabbit, *Oryctolagus cuniculus*', *J. Chem. Ecol.*, **7**, 817–827.

Goodrich, B. S., Hesterman, E. R., and Mykytowycz, R. (1981b). 'The effect of the volatiles collected from above faecal pellets on the behaviour of the rabbit, *Oryctolagus cuniculus*, tested in an experimental chamber. II. Gas chromatographic fractionation of trapped volatiles', *J. Chem. Ecol.*, **7**, 947–959.

Goodwin, M., Gooding, K. M., and Regnier, F. (1979). 'Sex pheromone in the dog', *Science*, **203**, 559–561.

Gordon, S. G., Smith, K., Rabinowitz, J. L., and Vagelos, P. R. (1973). 'Studies of trans-3-methyl-2-hexenoic acid in normal and schizophrenic humans', *J. Lipid Res.*, **14**, 495–503.

Gorman, M. L. (1976). 'A mechanism for individual recognition by odour in *Herpestes auropunctatus* (Carnivora: Viverridae)', *Anim. Behav.*, **24**, 141–145.

Gorman, M. L. (1980). 'Smelly carnivores', *Symp. Zool. Soc. Lond.*, **45**, 87–105.

Gorman, M. L., Nedwell, D. B., and Smith, R. M. (1974). 'An analysis of the contents of the anal scent pockets of *Herpestes auropunctatus*', *J. Zool.*, **172**, 389–399.

Gorman, M. L., Jenkins, D., and Harper, R. J. (1978). 'The anal scent sacs of the otter, *Lutra lutra*', *J. Zool.*, **186**, 463–474.

Gosden, P. E., and Ware, G. C. (1976). 'The aerobic flora of the anal sac of the red fox (*Vulpes vulpes*)', *J. Appl. Bact.*, **41**, 271–275.

Gower, D. B. (1972). '16-Unsaturated C_{19} steroids. A review of their chemistry, biochemistry and possible physiological role', *J. Steroid Biochem.*, **3**, 45–103.

Graham, C. A., and McGrew, W. C. (1980). 'Menstrual synchrony in female undergraduates living on a coeducational campus', *Psychoneuroendocrinology*, **5**, 245–252.

Graham, D. J. M. (1978). 'Quantitative determination of 3-ethyl-5-hydroxy-4,5-dimethyl-Δ^3-pyrrolin-2-one in urine using gas-liquid chromatography', *Clin. Chim. Acta*, **85**, 205–210.

Grau, G. A. (1976). 'Olfaction and reproduction in ungulates', in *Mammalian Olfaction, Reproductive Processes and Behavior* (ed. R. L. Doty), Academic Press, New York, pp. 219–242.

Gray, N. A. B., Smith, D. H., Varkony, T. H., Carhart, R. E., and Buchanan, B.G. (1980). 'Use of a computer to identify unknown compounds: the automation of scientific inference', in *Biochemical Applications of Mass Spectrometry. First Supplementary Volume* (eds G. R. Waller and O. C. Dermer), Wiley, New York, pp. 125–149.

Graziadei, P., and Tucker, D. (1968). 'Vomeronasal receptor's ultrastructure', *Fed. Proc.*, **27**, 583.

Greaves, J. P., and Scott, P. P. (1960). 'Urinary amino acid pattern of cats on diets of varying protein content', *Nature*, **187**, 242.

Greene, R. S., Downing, D. T., Pochi, R. E., and Strauss, J. S. (1970). 'Anatomical variation in the amount and composition of human skin surface lipid', *J. Investig. Dermatol.*, **54**, 240–247.

Greer, M. B., and Calhoun, M. L. (1966). 'Anal sacs of the cat, *Felix domesticus*', *Amer. J. Vet. Res.*, **27**, 773–781.

318

Gremli, H. A. (1974). 'Interactions of flavor compounds with soy protein', *J. Amer. Oil Chem. Soc.*, **51**, 95A–97A.

Griffin, D. R. (1976). *The Question of Animal Awareness*, Rockefeller University Press, New York.

Griffiths, N. M., and Patterson, R. L. S. (1970). 'Human olfactory responses to 5α-androst-16-en-3-one, principal component of boar taint', *J. Sci. Food Agric.*, **21**, 4–6.

Grob, K. (1979). 'Twenty years of glass capillary columns, *HRC&CC*, **2**, 599–604.

Grob, K., Grob, G., and Grob, K. (1978). 'Preparation of apolar glass capillary columns by the barium carbonate procedure', *HRC&CC*, **1**, 149–155.

Grob, K., Grob, G., and Grob, K. (1979). 'Deactivation of glass capillary columns by silylation', *HRC&CC*, **2**, 31–35.

Grönneberg, T. O. (1978/79). 'Analysis of a wax ester fraction from the anal gland secretion of beaver (*Castor fiber*) by chemical ionization mass spectrometry', *Chemica Scripta*, **13**, 56–58.

Grönneberg, T. O., Gray, N. A. B., and Eglinton, G. (1975). 'Computer-based search and retrieval system for rapid mass spectral screening of samples, *Anal. Chem.*, **47**, 415–418.

Grönneberg, T. O., and Albone, E. S. (1976). 'Computer-aided classification of mass spectra: GC-MS studies of lower molecular weight lipids in the anal sac secretion of the lion, *Panthera leo*', *Chemica Scripta*, **10**, 8–15.

Grosch, W., Laskawy, G., and Fischer, K. H. (1975). 'Aroma compounds formed by enzymatic co-oxidation', in *Aroma Research* (eds H. Maarse and P. J. Groenen), Pudoc, Wageningen, Netherlands.

Gross, M. L. (ed.) (1978). *High Performance Mass Spectrometry: Chemical Applications*, ACS Symposium Series **70**, American Chemical Society, Washington D.C.

Guadagni, D. G., Buttery, R. G., and Turnbaugh, J. G. (1972). 'Odour thresholds and similarity ratings of some potato chip components', *J. Sci. Food Agric.*, **23**, 1435–1444.

Haahti, E. O., and Fales, H. M. (1967). 'The uropygiols: identification of the unsaponifiable constituents of a diester wax from chicken preen glands', *J. Lipid Res.*, **8**, 131–137.

Hachenberg, H., and Schmidt, A. P. (1977). *Gas Chromatographic Headspace Analysis*, Heyden and Son, London.

Hafez, E. S. E. (1968). *Reproduction in Farm Animals*, Lea and Febiger, Philadelphia, 2nd edn.

Hafez, E. S. E. (ed.) (1980). *Human Reproduction: Conception and Contraception*, Harper and Row, Hagerstown, Maryland, USA. 2nd edn.

Hafez, E. S. E., and Evans, T. N. (eds) (1978). *Human Vagina*, Elsevier, New York.

Hais, I. M., Štrych, A., and Chmelař, V. (1968). 'Preliminary thin-layer chromatographic characterization of the rat Harderian gland lipid', *J. Chromatog.*, **35**, 179–191.

Halnan, C. R. E. (1973). 'Anal sacs of the dog', Diploma of the Fellowship of the Royal College of Veterinary Surgeons Thesis, London.

Halpin, Z. T. (1976). 'The role of individual recognition by odors in the social interactions of the Mongolian gerbil (*Meriones unguiculatus*)', *Behaviour*, **58**, 117–130.

Halpin, Z. T. (1980). 'Individual odors and individual recognition. a review and commentary', *Biol. Behav.*, **5**, 233–248.

Hamilton, R. J., and Sewell, P. A. (1982). *Introduction to High Performance Liquid Chromatography*, Chapman and Hall, London, 2nd edn.

Hanna, A., Marshall, J. C., and Isenhour, T. L. (1979). 'A GC/FT-IR Compound Identification System', *J. Chromatog. Sci.*, **17**, 434–440.

Hansen, G., and Munson, B. (1978). 'Surface chemical ionization mass spectrometry', *Anal. Chem.*, **50**, 1130–1134.

Hansen, I. A., Tang, B. K., and Edkins, E. (1969). '*Erythro*-diols of wax from the uropygial gland of the turkey', *J. Lipid Res.*, **10**, 267–270.

Hanson, G., and Montagna, W. (1962). The skin of primates. XII. The skin of the owl monkey (*Aötus trivirgatus*), *Amer. J. Phys. Anthropol.*, **20**, 421–429.

Hanson, S. E. (1969). 'Studies on urinary pigments', *Acta. Chem. Scand.*, **23**, 3461–3465, 3466–3472.

Hargrove, J. W., and Vale, G. A. (1979). 'Aspects of the feasibility of employing odour-baited traps for controlling tsetse flies (Diptera: Glossinidae), *Bull. Ent. Res.*, **69**, 283–290.

Harkness, R. A. (1974). 'Variations in testosterone excretion by man', in *Biorhythms and Human Reproduction* (eds M. Ferin, F. Halberg, R. M. Richart and R. L. van de Wiele), Wiley, New York, pp. 469–478.

Harney, J. W., Barofsky, I. M., and Leary, J. D. (1978). 'Behavioral and toxicological studies on cyclopentanoid monoterpenes from *Nepeta cataria*', *Lloydia*, **41**, 367–374.

Harrington, J. (1974). 'Olfactory communication in *Lemur fulvus*', in *Prosimian Behaviour* (eds R. D. Martin, G. A. Doyle and A. C. Walker), Gerald Duckworth and Co., Ltd., London, pp. 331–346.

Hartigan, M. J., Purcell, J. E., Novotny, M., McConnell, M. L., and Lee, M. L. (1974). 'Analytical performance of a novel nitrogen-sensitive detector and its applications with glass open tubular columns', *J. Chromatog.*, **99**, 339–348.

Harvey, E. B., and Rosenberg, L. E. (1960). 'An apocrine gland complex in the pika', *J. Mammal.*, **41**, 213–219.

Hashimoto, Y., Eguchi, Y., and Arakawa, A. (1963). 'Histological observation of the anal sac and its glands of a tiger', *Jap. J. Vet. Sci.*, **25**, 29–32.

Heath, R. R., Tumlinson, J. H., and Doolittle, R. E. (1977). 'Analytical and preparative separation of geometrical isomers by high efficiency silver nitrate liquid chromatography', *J. Chromatog. Sci.*, **15**, 10–13.

Heftmann, E. (Ed.) (1975). *Chromatography: a Laboratory Handbook of Chromatographic and Electrophoretic Methods*, Van Nostrand Reinhold, New York, 3rd edn.

Hendricks, A. P. J., and Punter, P. H. (1980). 'Specific anosmia to exaltolide: selection criteria', in *Olfaction and Taste, VII* (ed. H. van der Starre), IRL Press, London, p. 431.

Henningsson, S. S. G., and Rosengren, E. (1972). 'Alterations of histamine metabolism after injection of sex hormones in mice', *Br. J. Pharacol.*, **44**, 517–526.

Henningsson, S., and Rosengren, E. (1975). 'Biosynthesis of histamine and putrescine in mice during post-natal development and its hormone dependence', *J. Physiol.*, **245**, 467–479.

Henry, J. D. (1977). 'The use of urine marking in the scavenging behaviour of the red fox, *Vulpes vulpes*', *Behaviour*, **61**, 82–106.

Henry, J. D. (1980). 'The urine marking behaviour and movement patterns of red foxes, *Vulpes vulpes*, during a breeding and post-breeding period', in *Chemical Signal: Vertebrates and Aquatic Invertebrates* (eds D. Müller-Schwarze and R. M. Silverstein), Plenum, New York, pp. 11–27.

Hesterman, E. R., and Mykytowycz, R. (1968). 'Some observations on the odours of anal gland secretions from the rabbit, *Oryctolagus cuniculus*', *CSIRO Wildl. Res.*, **13**, 71–81.

Hesterman, E. R., Goodrich, B. S., and Mykytowycz, R. (1976). 'Behavioural and cardiac responses of the rabbit, *Oryctolagus cuniculus*, to chemical fractions from anal glands', *J. Chem. Ecol.*, **2**, 25–37.

Hesterman, E. R., Goodrich, B. S., and Mykytowycz, R. (1981). 'The effect of the volatiles collected from above faecal pellets on the behaviour of the rabbit, *Oryctolagus cuniculus*, tested in an experimental chamber. I. Total volatiles and some chemically prepared fractions', *J. Chem. Ecol.*, **7**, 799–827.

Hesterman, E. R., and Mykytowycz, R. (1982a). 'Misidentification by wild rabbits,

Oryctolagus cuniculus, of group members carrying the odor of foreign inguinal gland secretion. I. Experiments with all-male groups', *J. Chem. Ecol.*, **8**, 419–427.

Hesterman, E. R., and Mykytowycz, R. (1982b). 'Misidentification by wild rabbits, *Oryctolagus cuniculus*, of group members carrying the odor of foreign inguinal gland secretion. II. Experiments with all-female groups', *J. Chem. Ecol.*, **8**, 723–729.

Hildebrand, M. (1952). 'The integument in Canidae', *J. Mammal.*, **33**, 419–428.

Hill, H. C. (1966). *Introduction to Mass Spectrometry*, Heyden and Son Ltd, London.

Hill, J. O., Pavlik, E. J., Smith, G. L., Burghardt, G. M., and Coulson, P. B. (1976). 'Species-characteristic responses to catnip by undomesticated felids', *J. Chem. Ecol.*, **2**, 239–253.

Hill, W. C. O. (1944). 'An undescribed feature in the drill (*Mandrillus leucophaeus*)', *Nature*, **153**, 199.

Holt, W. V., and Tam, W. H. (1973). 'Steroid metabolism by the chin gland of the male cuis, *Galea musteloides*', *J. Reprod. Fert.*, **33**, 53–59.

Hoppe, P. C. (1975). 'Genetic and endocrine studies of the pregnancy-blocking pheromone of mice', *J. Reprod. Fert.*, **45**, 109–115

Horning, E. C., Horning, M. G., Szafranek, J., van Hout, P., German, A. L., Thenot, J. P., and Pfaffenberger, C. D. (1974). 'Gas phase analytical methods for the study of human metabolites. Metabolic profiles obtained by open tubular capillary chromatography', *J. Chromatog.*, **91**, 367–368.

Horning, E. C., Carroll, D. I., Dzidic, I., Haegele, K. D., Lin, S. N., Oertli, C. U., and Stillwell, R. N. (1977). 'Development and use of analytical systems based on mass spectrometry', *Clin. Chem.*, **23**, 13–21.

Horton, E. W., Main, I. H. M., and Thompson, C. J. (1963). 'The action of intravaginal prostaglandin E on the female reproductive tract', *J. Physiol.*, **168**, 54P–55P.

Hoskins, J. A., Pollitt, R. J., and Evans, S. (1978). 'The determination of 5-methoxyindole-3-acetic acid in human urine by mass fragmentography', *J. Chromatog.*, **145**, 285–289.

Hradecký, P. (1978). 'The relationship between the contents of volatile fatty acids in urine and vaginal secretion and the phase of the reproductive cycle in the cow', Dissertation, Veterinary University, Brno, Czechoslovakia.

Huggins, C. (1945). 'The physiology of the prostate gland', *Physiol. Rev.*, **25**, 281–295.

Huggins, G. R., and Preti, G. (1976). 'Volatile constituents of human vaginal secretions', *Amer. J. Obstet. Gynecol.*, **126**, 129–136.

Hughes, P. E., and Cole, D. J. A. (1976). 'Reproduction in the gilt', *Anim. Prod.*, **23**, 89–94.

Hunt, D. F., Stafford, G. C., Crow, F. W., and Russell, J. W. (1976). 'Pulsed positive negative ion chemical ionization mass spectrometry', *Anal. Chem.*, **48**, 2098–2105.

Hunt, D. F., and Sethi, S. K. (1978). 'Analytical applications of positive and negative ion chemical ionization mass spectrometry', in *High Performance Mass Spectrometry: Chemical Applications* (ed. M. L. Gross), ACS Symposium Series, **70**, American Chemical Society, Washington, D.C., pp. 150–178.

Hurley, R., Stanley, V. C., Leask, B. G. S., and de Louvois, J. (1974). 'Microflora of the vagina during pregnancy', in *The Normal Microbial Flora of Man* (eds F. A. Skinner and J. G. Carr), Academic Press, London, pp. 155–185.

Hyatt, A. T., and Hayes, M. L. (1975). 'Free amino acids and amines in human dental plaque', *Arch. Oral Biol.*, **20**, 203–210.

Inscoe, M. N., King, G. S., and Blau, K. (1978), 'Microreactions', in *Handbook of Derivatives for Chromatography* (eds K. Blau and G. S. King), Heyden and Son Ltd., London, pp. 317–345.

Inwang, E. E., Madubuike, P. U., and Mosnaim, A. D. (1973). 'Evidence for the excretion of 2-phenylethylamine glucuronide in human urine', *Experientia*, **29**, 1080–1081.

Izard, M. K., and Vandenbergh, J. G. (1982a). 'Priming pheromones from oestrous cows increase synchronization of oestrus in dairy heifers after PGF-2α injection', *J. Reprod. Fert.*, **66**, 189–196.

Izard, M. K., and Vandenbergh, J. G. (1982b). 'The effects of bull urine on puberty and calving date in crossbred beef heifers', *J. Anim. Sci.*, **55**, 1160–1168.

Jacob, J., and Poltz, J. (1974). 'Chemical composition of uropygial gland secretions of owls', *J. Lipid. Res.*, **15**, 243–248

Jacob, J., and von Lehmann, E. (1976). 'Chemical composition of the nasal gland secretion from the marsh deer, *Odocoileus dichotomus*', *Z. Naturforsch.*, **31C**, 496–498.

Jacob, J., and Green, U. (1977). 'Composition of the ventral gland-pad from the Mongolian gerbil', *Z. Naturforsch.*, **32C**, 735–738.

James, M. O., Smith, R. L., Williams, R. T., and Reidenberg, M. (1972). 'The conjugates of phenylacetic acid in man, subhuman primates and some non-primate species', *Proc. Roy. Soc. Lond.*, *B*, **182**, 25–35.

James, T. L. (1975). *Nuclear Magnetic Resonance in Biochemistry*, Academic Press, New York.

Jänne, J., Pösö, H., and Raina, A. (1977). 'Polyamines in rapid growth and cancer', *Biochim. Biophys. Acta.*, **473**, 241–293.

Jellum, E., Stokke, O., and Eldjarn, L. (1972). 'Combined use of gas chromatography, mass spectrometry and computer methods in diagnosis and studies of metabolic disorders', *Clin. Chem.*, **18**, 800–809.

Jellum, E., Stokke, O., and Eldjarn, L. (1973). 'Application of gas chromatography, mass spectrometry and computer methods in clinical biochemistry', *Anal. Chem.*, **45**, 1099–1106.

Jenkinson, D.McE., Blackburn, P. S., and Proudfoot, R. (1967). 'Seasonal changes in the skin glands of the goat', *Br. Vet. J.*, **123**, 541–549.

Jenkinson, D.McE., and Robertshaw, D. (1971). 'Studies on the nature of sweat gland "fatigue" in the goat', *J. Physiol.*, **212**, 455–465.

Jennings, W. (1980). 'Evolution and application of the fused silica column', *HRC&CC*, **3**, 601–608.

Jennings, W. G., and Filsoof, M. (1977). 'Comparison of sample preparation techniques for gas chromatographic analysis', *J. Agric. Food Chem.*, **25**, 440–445.

Johns, M. A. (1980). 'The role of the vomeronasal system in mammalian reproductive physiology', in *Chemical Signals: Vertebrates and Aquatic Invertebrates* (eds D. Müller-Schwarze and R. M. Silverstein), Plenum, New York, pp. 341–364.

Johnson, E. (1977). 'Seasonal changes in the skin of mammals', *Symp. Zool. Soc. Lond.*, **39**, 373–404.

Johnson, E. L., and Stevenson, R. (1978). *Basic Liquid Chromatography*, Varian, Palo Alto, California.

Johnson, R. P. (1973). 'Scent marking in mammals', *Anim. Behav.*, **21**, 521–535.

Johnston, R. E. (1974). 'Sexual attraction function of golden-hamster vaginal secretion', *Behav. Biol.*, **12**, 111–117.

Johnston, R. E. (1975). 'Sexual excitation function of hamster vaginal secretion', *Animal Learning and Behavior*, **3**, 161–166.

Johnston, R. E. (1977). 'Sex pheromones in golden hamsters', in *Chemical Signals in Vertebrates* (eds D. Müller-Schwarze and M. M. Mozell), Plenum, New York, pp. 225–249.

Johnston, R. E. (1980). 'Responses of male hamsters to odors of females in different reproductive states', *J. Comp. Physiol. Psychol.*, **94**, 894–904.

Johnston, R. E., and Zahorik, D. M. (1975). 'Taste aversions to sexual attractants', *Science*, **189**, 893–894.

Jones, R. B., and Nowell, N. W. (1973a). 'The coagulating glands as a source of aversive and aggression inhibiting pheromone(s) in the male albino mouse', *Physiol. Behav.*, **11**, 455–462.

Jones, R. B., and Nowell, N. W. (1973b). 'The effects of preputial and coagulating gland secretions upon aggressive behaviour in male mice: a confirmation', *J. Endocrinol.*, **59**, 203–204.

Jones, R. B., and Nowell, N. W. (1973c). 'The effect of urine on the investigatory behaviour of male albino mice', *Physiol. Behav.*, **11**, 35–38.

Jones, T. R., and Plakke, R. K. (1981). 'The histology and histochemistry of the perianal scent gland of the reproductively quiescent black tailed prairie dog, *Cynomys ludovicianus*', *J. Mammal.*, **62**, 362–368.

Jonsson, H. T., Middleditch, B. S., and Desiderio, D. M. (1975), 'Prostaglandins in human seminal fluid: two novel compounds', *Science*, **187**, 1093–1094.

Jorgenson, J. W., Novotny, M., Carmack, M., Copland, G. B., Wilson, S. R., Katona, S., and Whitten, W. K. (1978). 'Chemical scent constituents in the urine of the red fox (*Vulpes vulpes*) during the winter season', *Science*, **199**, 796–798.

Jost, U. (1974). '1-Alkyl-2,3-diacyl-sn-glycerol, the major liquid in the Harderian gland of rabbits', *Hoppe-Seyler's Z. Physiol. Chem.*, **355**, 422–426.

Jüttner, F. (1979). 'Nor-carotenoids as the major volatile excretion products of Cyanidium', *Z. Naturforsch.*, **34C**, 186–191.

Kagan, R., Ikan, R., and Haber, O. (1983). 'Characterization of a sex pheromone in the jird (*Meriones tristrami*)', *J. Chem. Ecol.*, **9**, 775–783.

Kakimoto, Y., and Akazawa, S. (1970). 'Isolation and identification of N^g, N^g- and N^g, $N'g$-dimethylarginine, N^ϵ-mono-, di-, and trimethyllysine, and glucosylgalactosyl- and galactosyl-δ-hydroxylysine from human urine', *J. Biol. Chem.*, **245**, 5751–5758.

Källquist, L., and Mossing, T. (1982). 'Olfactory recognition between mother and calf in reindeer (*Rangifer tarandus*)', *Applied Animal Ethology*, **8**, 561–565.

Kalmus, H. (1955). 'The discrimination by the nose of individual human odours and in particular of the odours of twins', *Br. J. Anim. Behav.*, **3**, 25–31.

Kaplan, J., and Russell, M. (1974). 'Olfactory recognition in the infant squirrel monkey', *Dev. Psychobiol.*, **7**, 15–19.

Kare, M. R., and Beauchamp, G. K. (1977). 'Smell, taste and hearing', in *Dukes' Physiology of Domestic Animals* (ed. M. J. Swenson), Cornell University Press, Ithaca and London, 9th edn, pp. 713–730.

Kärkkäinen, J., Nikkari, T., Ruponen, S., and Haahti, E. (1965). 'Lipids of vernix caseosa', *J. Investig. Dermatol.*, **44**, 333–338.

Karlsen, J. (1972). 'Microanalysis of volatile compounds in biological material by means of gas liquid chromatography', *J. Chromatog. Sci.*, **10**, 642–643.

Karlson, P., and Luscher, M. (1959). '"Pheromones" a new term for a class of biologically active substances', *Nature*, **183**, 155–156.

Kasama, K., Uezumi, N., and Itoh, K. (1970). 'Characterization and identification of glyceryl ether diesters in Harderian gland tumors of mice', *Biochim. Biophys. Acta*, **202**, 56–66.

Kasama, K., Rainey, W. T., and Snyder, F. (1973). 'Chemical identification and enzymatic synthesis of a newly discovered lipid class — hydroxyalkylglycerols', *Arch. Biochem. Biophys.*, **154**, 648–658.

Katsir, Z., and Crewe, R. M. (1980). 'Chemical communication in *Galago crassicaudatus*: investigation of the chest gland secretion', *S. African J. Zool.*, **15**, 249–254.

Katz, R. A., and Shorey, H. H. (1979). 'In defense of the term "pheromone"', *J. Chem. Ecol.*, **5**, 299–301. (Reply by Beauchamp *et al.*, *ibid.*, pp. 301–305).

Kauer, J. S. (1974). 'Response patterns of amphibian olfactory bulb neurons to odour stimulation', *J. Physiol.*, **243**, 695–715.

Kauer, J. S., and Moulton, D. G. (1970). 'Ultrastructure of vomeronasal and olfactory sensory epithelia in the rat and rabbit', *Fed. Proc.*, **29**, 521.

Kauer, J. S., and Moulton, D. G. (1974). 'Responses of olfactory bulb neurones to odour stimulation of small nasal areas in the salamander', *J. Physiol.*, **243**, 717–737.

Keith, L., Stromberg, P., Krotoszynski, B. K., Shah, J., and Dravnieks, A. (1975). 'The odours of the human vagina', *Arch. Gynäk.*, **220**, 1–10.

Keller, A., and Margolis, F. L. (1975). 'Immunological studies of the rat olfactory marker protein', *J. Neurochem.*, **24**, 1101–1106.

Kelly, R. W., Taylor, P. L., Hearn, J. P., Short, R. V., Martin, D. E., and Marston, J. H. (1976). '19-Hydroxyprostaglandin-E as a major component of the semen of primates', *Nature*, **260**, 544–545.

Kenndler, E., and Schmid, E. R. (1978). 'Combination of liquid chromatography and mass spectrometry', in *Instrumentation for High Performance Liquid Chromatography* (ed. J. F. K. Huber), Elsevier, Amsterdam, pp. 163–177.

Kennedy, G. Y. (1970). 'Harderoporphyrin: a new porphyrin from the Harderian glands of the rat', *Comp. Biochem. Physiol.*, **36**, 21–36.

Kennedy, G. Y., Jackson, A. H., Kenner, G. W., and Suckling, C. J. (1970). 'Isolation, structure and synthesis of a tricarboxylic porphyrin from the Harderian glands of the rat', *FEBS Letters*, **6**, 9–12 and 205 (errata).

Kerjaschki, D., and Hörandner, H. (1976). 'The development of mouse olfactory vesicles and their cell contacts: A freeze-etching study', *J. Ultrastruct. Res.*, **54**, 420–444.

Keverne, E. B. (1976a). 'Sex attractants in primates', *J. Soc. Cosmet. Chem.*, **27**, 257–269.

Keverne, E. B. (1976b). 'Reply to Goldfoot *et al.*', *Horm. Behav.*, **7**, 396–372.

Keverne, E. B. (1978). 'Olfactory cues in mammalian sexual behaviour', in *Biological Determinants of Sexual Behaviour* (ed. J. B. Hutchison), Wiley, Chichester, pp. 727–763.

Keverne, E. B. (1979). 'The dual olfactory projections and their significance for behaviour', in *Chemical Ecology: Odour Communication in Animals* (ed. F. J. Ritter), Elsevier/North-Holland Biomedical Press, Amsterdam, pp. 75–83.

Kiddy, C. A., Mitchell, D. S., Bolt, D. J., and Hawk, H. W. (1978). 'Detection of estrus-related odors in cows by trained dogs', *Biol. Reprod.*, **19**, 389–395.

Kimble, B. J. (1978). 'Introduction to gas chromatography/high resolution mass spectrometry', in *High Performance Mass Spectrometry: Chemical Applications* (ed. M. L. Gross), ACS Symposium Series, **70**, American Chemical Society, Washington, D.C., pp. 120–149.

Kimble, B. J., Cox, R. E., McPherron, R. V., Olsen, R. W., Roitman, E., Walls, F. C., and Burlingame, A. L. (1974). 'Real-time gas chromatography/high resolution mass spectrometry and its application to the analysis of physiological fluids', *J. Chromatog. Sci.*, **12**, 647–657.

Kimble, B. J., Cox, R. E., McPherron, R. V., Olsen, R. W., Roitman, E., Walls, F. C., and Burlingame, A. L. (1976). 'Real-time gas chromatography/high resolution mass spectrometry and its application to the analysis of physiological fluids', in *Advances in Mass Spectrometry in Biochemistry and Medicine*, Vol. 1 (eds A. Frigerio and N. Castagnoli), Spectrum Publications Inc., New York, pp. 565–576.

King, G. S., Goodwin, B. L., Ruthven, C. R. J., and Sandler, M. (1974). 'Urinary excretion of *o*-tyramine', *Clin. Chim. Acta*, **51**. 105–107.

Kirk-Smith, M., Booth, D. A., Carroll, D., and Davies, P. (1978). 'Human social attitudes affected by androstenol', *Res. Commun. Psychol. Psychiat. Behav.*, **3**, 379–384.

Kivett, V. K. (1978). 'Integumentary glands of Columbian ground squirrels (*Spermophilus columbianus*), Sciuridae', *Can. J. Zool.*, **56**, 374–381.

Kivett, V. K., Murie, J. O., and Steiner, A. L. (1976). 'A comparative study of scent gland location and related behaviour in some northwestern nearctic ground squirrel species (Sciuridae): an evolutionary approach', *Can. J. Zool.*, **54**, 1294–1306.

Kjaer, A. (1977). 'Low molecular weight sulphur-containing compounds in nature: A survey', *Pure Appl. Chem.*, **49**, 137–152.

324

Kleiman, D. G. (1974). 'Patterns of behaviour in hystricomorph rodents', *Symp. Zool. Soc. Lond.*, **34**, 171–209.

Kligman, A, M., and Shelley, W. B. (1958). 'An investigation of the biology of the human sebaceous gland', *J. Investig. Dermatol.*, **30**, 99–124.

Kloek, J. (1961). 'The smell of some steroid sex-hormones and their metabolites. Reflections and experiments concerning the significance of smell for the mutual relation of the sexes', *Psychiatria Neurologia Neurochirurgia*, **64**, 309–344.

Kloek, J. (1967). 'Psychological and neurophysiological aspects of flavour compounds', *Bibl. "Nutritio et Dieta"*, **9**, 105–119.

Knight, T. W., and Lynch, P. R. (1980). 'Source of ram pheromones that stimulate ovulation in the ewe', *Anim. Reprod. Sci.*, **3**, 133–136.

Knights, B. A., Legendre, M., Laseter, J. L., and Storer, J. S. (1975). 'Use of high-resolution open tubular glass capillary columns to separate acidic metabolites in urine', *Clin. Chem.*, **21**, 888–891.

Koelega, H. S., and Köster, E. P. (1974). 'Some experiments on sex differences in odor perception', *Ann. N.Y. Acad. Sci.*, **237**, 234–246.

Kolb, B., and Bischoff, J. (1974). 'A new design of a thermionic nitrogen and phosphorus detector for GC', *J. Chromatog. Sci.*, **12**, 625–629.

Koller, W. D., and Tressl, G. (1980). 'Simple GC/MS interface with transferline of fused silica for open or direct coupling, *HRC&CC*, **3**, 359–360.

Kolor, M. G. (1979). 'Food and Flavor Industry', in *Mass Spectometry*, Part A, (eds C. Merritt and C. N. McEwen), Marcel Dekker, Inc., New York, pp. 67–117.

Koshland, D. E. (1980). 'Biochemistry of sensing and adaptation', *Trends in Biochemical Sciences*, **5**, 297–302.

Kossa, W. C., MacGee, J., Ramachandran, S., and Webber, A. J. (1979). 'Pyrolytic methylation/gas chromatography. A short review', *J. Chromatog. Sci.*, **17**, 177–187.

Kostelc, J. G., Preti, G., Zelson, P. R., Stoller, N. H., and Tonzetich, J. (1980). 'Salivary volatiles as indicators of periodontitis', *J. Periodont. Res.*, **15**, 185–192.

Köster, E. P. (1969). 'Intensity in mixtures of odorous substances', in *Olfaction and Taste III* (ed. C. Pfaffmann), Rockefeller University Press, New York, pp. 142–149.

Köster, E. P., and Koelega, H. S. (1976). 'Sex differences in odour perception', *J. Soc. Cosmet. Chem.*, **27**, 319–327.

Kowalski, B. R. (1980). 'Chemometrics', *Anal. Chem.*, **52**, 112R–122R.

Krotoszynski, B., Gabriel, G., O'Neill, H., and Claudio, M. P. A. (1977). 'Characterization of human expired air: a promising investigative and diagnostic technique', *J. Chromatog. Sci.*, **15**, 239–244.

Kruse, S. McK., and Howard, W. E. (1983). 'Canid sex attractant studies', *J. Chem. Ecol.*, **9**, 1503–1510.

Kruuk, H. (1972). *The Spotted Hyaena*, Univ. Chicago Press, Chicago and London.

Kruuk, H. (1976). 'Feeding and social behaviour in the striped hyaena', *E. Afr. Wildl. J.*, **14**, 91–111.

Kühnel, W. (1971). 'Struktur und Cytochemie der Harderschen Drüse von Kaninchen', *Z. Zellforsch.*, **119**, 384–404.

Kundrotas, L. W., and Gregg, R. (1977). 'Urinary excretion of 5-hydroxyindoleacetic acid in the rat after immobilization stress', *Physiol. Behav.*, **19**, 739–741.

Kuno, Y. (1956). *Human Perspiration*, Charles C. Thomas, Springfield, Illinois.'

Labows, J. N. (1979). 'Human odors — what can they tell us?' *Perfumer and Flavorist*, **4**, 12–17.

Labows, J. N. (1981). 'Odorants as chemical messengers', in *Odor Quality and Chemical Structure* (eds H. R. Moskowitz and C. B. Warren), ACS Symposium Series **148**, American Chemical Society, Washington, D.C., pp. 195–210.

Labows, J. N., McGinley, K. J., Leyden, J. J., and Webster, G. F. (1979a). 'Characteristic γ-lactone odor production of the genus Pityrosporum', *Appl. Environ. Microbiol.*, **38**, 412–415.

Labows, J. N., Preti, G., Hoelzle, E., Leyden, J., and Kligman, A. (1979b). 'Steroid analysis of human apocrine secretion', *Steroids*, **34**, 249–258.

Labows, J., Preti, G., Hoelzle, E., Leyden, J., and Kligman, A. (1979c). 'Analysis of human axillary volatiles: compounds of exogenous origin', *J. Chromatog.*, **163**, 294–299.

Labows, J. N., McGinley, K. J., Webster, G. F., and Leyden, J. J. (1980). 'Headspace analysis of volatile metabolites of *Pseudomonas aeruginosa* and related species by gas chromatography-mass spectrometry', *J. Clin. Microbiol.*, **12**, 521–526.

Labows, J. N., McGinley, K. J., and Kligman, A. M. (1982). 'Perspectives in axillary odor', *J. Soc. Cosmet. Chem.*, **34**, 193–202.

Landman, A. D., Sanford, L. M., Howland, B. E., Dawes, C., and Pritchard, E. T. (1976). 'Testosterone in human saliva', *Experientia*, **32**, 940–941.

Ladewig, J., Price, E. O., and Hart, B. L. (1980). 'Flehmen in male goats: role in sexual behavior', *Behavioral and Neural Biology*, **30**, 312–322.

Larsen, B., Markovetz, A. J., and Galask, R. P. (1976). 'The bacterial flora of the female rat genital tract', *Proc. Soc. Exp. Biol. Med.*, **151**, 571–574.

Larsen, B., Markovetz, A. J., and Galask, R. P. (1977). 'Role of estrogen in controlling the genital microflora of female rats', *Appl. Env. Microbiol.*, **34**, 534–540.

Lawson, A. D., and Whitsett, J. M. (1979). 'Inhibition of sexual maturation by a urinary pheromone in male prairie deer mice', *Horm. Behav.*, **13**, 128–138.

Lawson, A. M., Chalmers, R. A., and Watts, R. W. E. (1976). 'Urinary organic acids in man. I. Normal patterns', *Clin. Chem.*, **22**, 1283–1287.

Lederer, E. (1946). 'Chemistry and biochemistry of the scent glands of the beaver, *Castor fiber*', *Nature*, **157**, 231–232.

Lederer, E. (1949). 'Chemistry and biochemistry of some mammalian secretions and excretions', *J. Chem. Soc.*, 2115–2125.

Lederer, E., (1950). 'Odeurs et parfums des animaux', *Fortschr. Chem. organ. Naturst.*, **6**, 87–153.

Leduc, J. (1961). 'Catecholamine production and release in exposure and acclimation to cold', *Acta Physiol. Scand.*, **53**, *Suppl. 183*, 1–101.

Lee, C. T. (1976). 'Agonistic behavior, sexual attraction and olfaction in mice', in *Mammalian Olfaction, Reproductive Processes and Behavior* (ed. R. L. Doty), Academic Press, New York, pp. 161–180.

Lee C. T., Lukton, A., Bobotas, G., and Ingersoll, D. W. (1980). 'Partial purification of male *Mus musculus* urinary aggression-promoting chemosignal', *Aggressive Behavior*, **6**, 149–160.

Lee, K. Y., Nurok, D., and Zlatkis, A. (1978). 'Combined headspace and extraction technique for profile analysis by capillary gas chromatography', *J. Chromatog.*, **158**, 377–386.

Lee, K. Y., Poole, C. F., and Zlatkis, A. (1980). 'Simultaneous multi-mycotoxin determination by high performance thin-layer chromatography', *Anal. Chem.*, **52**, 837–842.

Lee, S. van der., and Boot, L. M. (1956). 'Spontaneous pseudopregnancy in mice', *Acta Physiol. Pharmacol. Neer.*, **5**, 213–215.

Leon, M. (1974). 'Maternal pheromone', *Physiol. Behav.*, **13**, 441–453.

Leon, M. (1978). 'Emission of maternal pheromone', *Science*, **201**, 938–939.

Leon, M. (1980). 'Development of olfactory attraction by young norway rats', in *Chemical Signals: Vertebrates and Aquatic Invertebrates* (eds D. Müller-Schwarze and R. M. Silverstein), Plenum Press, New York, pp. 193–209.

Leonhardt, B. A., and Beroza, M. (eds) (1982). *Insect Pheromone Technology: Chemistry and Applications*, ACS Symp. Series, 190, American Chemical Society, Washington D.C.

Levi, L. (ed) (1972). 'Stress and distress in response to psychosocial stimuli', *Acta Med. Scand. Suppl.*, **528**, 1–166.

Lewis, S., Kenyon, C. N., Meili, J., and Burlingame, A. L. (1979). 'High resolution gas chromatographic/real-time high resolution mass spectrometric identification of organic acids in human urine', *Anal. Chem.*, **51**, 1275–1285.

Lewis, V. J., Moss, C. W., and Jones, W. L. (1967). 'Determination of volatile acid production of Clostridium by gas chromatography', *Can. J. Microbiol.*, **13**, 1033–1040.

Leyden, J. J., McGinley, K. J., Hölzle, E., Labows, J. N., and Kligman, A. M. (1981). 'The microbiology of the human axilla and its relationship to axillary odor', *J. Investig. Dermatol.*, **77**, 413–416.

Liappis, N. (1973). 'Sex specific differences in the free amino acids in adult urine', *Z. Klin. Chem. Klin. Biochem.*, **11**, 279–285.

Liardon, R., and Kühn, U. (1978). 'Analysis of volatile fatty acids by gas chromatography separation of benzyl esters', *HRC&CC*, **2**, 47–53.

Liddell, K. (1976). 'Smell as a diagnostic marker', *Postgrad. Med. J.*, **52**, 136–138.

Liebich, H. M. (1975). 'Specific detection of volatile metabolites in urines of normal subjects and patients with *Diabetes mellitus* using computerized mass fragmentography', *J. Chromatog.*, **112**, 551–557.

Liebich, H. M., and Al-Babbili, O. (1975). 'Gas chromatographic-mass spectrometric study of volatile organic metabolites in urines of patients with *Diabetes mellitus*', *J. Chromatog.*, **112**, 539–550.

Liebich, H. M., and Wöll, J. (1977). 'Volatile substances in blood serum: profile analysis and quantitative determination', *J. Chromatog.*, **142**, 505–526.

Liebich, H. M., Zlatkis, A., Bertsch, W., Van Dahm, R., and Whitten, W. K. (1977). 'Identification of dihydrothiazoles in urine of male mice', *Biomed. Mass Spectom.*, **4**, 69–72.

Lindstedt, S., Steen, G., and Wahl, E. (1974). '3,4-Methylenehexanedioic acid — a previously unknown compound in human urine', *Clin. Chim. Acta*, **53**, 143–144.

Lindstedt, S., and Steen, G. (1975). '5-Decynedioic acid, an acetylenic acid compound in human urine', *Clin. Chem.*, **21**, 1964–1969.

Ling, G. V., and Ruby, A. L. (1978). 'Aerobic bacterial flora of the prepuce, urethra and vagina of normal adult dogs', *Amer. J. Vet. Res.*, **39**, 695–698.

Linhart, S. B., and Knowlton, F. F. (1975). 'Determining the relative abundance of coyotes by scent station lines', *Wildl. Soc. Bull.*, **3**, 119–124.

Lison, M., Blondheim, S. H., and Melmed, R. N. (1980). 'A polymorphism of the ability to smell urinary metabolites of asparagus', *Br. Med. J.*, **281**, 1676–1678.

Lombardi, J. R., and Vandenbergh, J. G. (1977). 'Pheromonally induced sexual maturation in females: regulation by the social environment of the male', *Science*, **196**, 545–546.

Lombardi, J. R., and Whitsett, J. M. (1980). 'Effects of urine from conspecifics on sexual maturation in female prairie deermice, *Peromyscus maniculatus bairdii*', *J. Mammal.*, **61**, 766–768.

Lovell, J. E., and Getty, R. (1957). 'The hair follicle, epidermis, dermis, and skin glands of the dog', *Amer. J. Vet. Res.*, **18**, 873–885.

Lowell, W. R., and Flanigan, W. F. (1980). 'Marine mammal chemoreception', *Mammal Rev.*, **10**, 53–59.

McClintock, M. K. (1971). 'Menstrual synchrony and suppression', *Nature*, **229**, 244–245.

McClintock, M. K. (1978). 'Estrous synchrony and its mediation by airborne chemical communication (*Rattus norvegicus*)', *Horm. Behav.*, **10**, 264–276.

McConnell, M. L., and Novotny, M. (1975). 'Automated high-resolution gas chromatographic system for recording and evaluation of metabolic profiles', *J. Chromatog.*, **112**, 559–571.

McConnell, M. L., Rhodes, G., Watson, U., and Novotny, M. (1979). 'Application of pattern recognition and feature extraction techniques to volatile constituent metabolic profiles obtained by capillary gas chromatography', *J. Chromatog.*, **162**, 494–506.

McDermott, J. R., Smith, A. I., Biggins, J. A., Al-Noaemi, M. C., and Edwardson, J. A. (1981). 'Characterization and determination of neuropeptides by high performance liquid chromatography and radioimmunoassay', *J. Chromatog.*, **222**, 371–379.

Macdonald, D. W. (1977). 'The behavioural ecology of the red fox *Vulpes vulpes*', D.Phil. Thesis, Oxford University, UK.

Macdonald, D. W. (1980). 'Patterns of scent marking with urine and faeces amongst carnivore communities', *Symp. Zool. Soc. Lond.*, **45**, 107–139.

Macdonald, D. W. (1981). 'Dwindling resources and the social behaviour of capybaras (*Hydrochoerus hydrochaeris*)', *J. Zool.*, **194**, 371–391.

Macdonald, D. W., Kranz, K., and Aplin, R. T. (1984). 'Behavioural, anatomical and chemical aspects of scent marking amongst capybaras, *Hydrochoerus hydrochaeris* (Rodentia: Caviomorpha)', *J. Zool.* (in press).

Macdonald, R. R., and Lumley, I. B. (1970). 'Endocervical pH measured in vivo through the normal menstrual cycle', *Obstet. Gynecol.*, **35**, 202–206.

McFadden, W. H. (1973). *Techniques of Combined Gas Chromatography/Mass Spectrometry*, John Wiley, Chichester.

McFadden, W. H. (1979). 'Interfacing chromatography and mass specrometry', *J. Chromatog. Sci.*, **17**, 2–16.

McFadden, W. H. (1980). 'Liquid chromatography-mass spectrometry systems and applications', *J. Chromatog. Sci.*, **18**, 97–115.

McFadden, W. H., Bradford, D. C., Eglinton, G., Hajlbrahim, S. K., and Nicolaides, N. (1979). 'Application of combined liquid chromatography/mass spectrometry (LC/MS): Analysis of petroporphyrins and meibomian gland waxes', *J. Chromatog. Sci.*, **17**, 518–522.

Macfarlane, R. D., and Torgerson, D. F. (1976). 'Californium-252 plasma desorption mass spectroscopy', *Science*, **191**, 920–925.

McGarry, E., Sehon, A. H., and Rose, B. (1955). 'The isolation and electrophoretic characterization of the proteins in the urine of normal subjects', *J. Clin. Invest.*, **34**, 832–844.

McLafferty, F. W. (1980). *Interpretation of Mass Spectra*, University Science Books, Mill Valley, California, 3rd edn.

McLafferty, F. W., and Venkataraghavan, R. (1979). 'Computer techniques for mass spectral identification', *J. Chromatog. Sci.*, **17**, 24–29.

MacLeod, P. (1971). 'An experimental approach to the peripheral mechanisms of olfactory discrimination', in *Gustation and Olfaction* (eds G. Ohloff and A. F. Thomas), Academic Press, London, pp. 28–44.

McNair, H. M., and Bonelli, E. J. (1969). *Basic Gas Chromatography*, Varian Aerograph, Palo Alto, California, 5th edn.

Ma, T. S., and Ladas, A. S. (1976). *Organic Functional Group Analysis by Gas Chromatography*, Academic Press, London.

Maas, J. W., and Landis, D. H. (1971). 'The metabolism of circulating norepinephrine by human subjects', *J. Pharmacol. Exp. Ther.*, **177**, 600–612.

Mace, J. W., Goodman, S. I., Centrewall, W. R., and Chinnock, R. F. (1976). 'The child with an unusual odor', *Clin. Pediatrics*, **15**, 57–62.

Machida, H., and Montagna, W. (1964). 'The skin of primates, XXII. The skin of the lutong (*Presbytis pyrrus*)', *Amer. J. Phys. Anthropol.*, **22**, 443–451.

Machida, H., Perkins, E., and Montagna, W. (1964). 'The skin of primates, XXIII. A comparative study of the skin of the green monkeys, *Cercopithecus aethiops*, and Syke's monkey, *C. mitis*', *Amer. J. Phys. Anthropol.*, **22**, 453–465.

Machida, H., Perkins, E., and Hu, F. (1967). 'The skin of primates, XXXV. The skin of the squirrel monkey, *Saimiri sciureus*', *Amer. J. Phys. Anthropol.*, **26**, 45–54.

Mahon, M. E., and Mattok, G. L. (1967). 'The differential determination of conjugated hydroxyskatoles in human urine', *Can. J. Biochem.*, **45**, 1317–1322.

Maier, H. G. (1970). 'Volatile flavoring substances in foodstuffs', *Angew. Chem. Int. Ed. Eng.*, **9**, 917–926.

Maier, H. G. (1975). 'Binding of volatile aroma substances to nutrients and foodstuffs', in *Aroma Research* (eds H. Maarse and P. J. Groenen), Pudoc, Wageningen, Netherlands, pp. 143–157.

Mair, R. G., Gesteland, R. C., and Adamek, G. D. (1980). 'Morphological variation among frog olfactory cilia', in *Olfaction and Taste VII* (ed. H. van der Starre), IRL Press, London, p. 206.

Mamer, O. A., and Tjoa, S. S. (1974). '2-Ethylhydracrylic acid: a newly discovered urinary organic acid', *Clin. Chim. Acta*, **55**, 199–204.

Manley, G. H. (1974). 'Functions of the external genital glands of Perodicticus and Arctocebus', in *Prosimian Braviour* (eds R. D. Martin, G. A. Doyle and A. C. Walker), Gerald Duckworth and Co., Ltd., London, pp. 313–329.

Mann, T. (1964). *The Biochemistry of Semen and of the Male Reproductive Tract*, Methuen and Co., London.

March, K. S. (1980). 'Deer, bears and blood: a note on non-human animal response to menstrual odor', *Amer. Anthropol.*, **82**, 125–127.

̌ Marchlewska-Koj, A. (1977). 'Pregnancy block elicited by urinary proteins of male mice', *Biol. Reprod.*, **17**, 729–732.

Marchlewska-Koj, A. (1980). 'Partial isolation of pregnancy block pheromone in mice', in *Chemical Signals: Vertebrates and Aquatic Invertebrates* (eds D. Müller-Schwarze and R. M. Silverstein), Plenum Press, New York, pp. 413–414.

Margolis, F. L. (1971). 'Regulation of porphyrin biosynthesis in the Harderian gland of inbred mouse strains', *Arch. Biochem. Biophys.*, **145**, 373–381.

Margolis, F. L. (1972). 'A brain protein unique to the olfactory bulb', *Proc. Natl. Acad. Sci. U.S.A.*, **69**, 1221–1224.

Marler, P. (1977). 'The evolution of communication', in *How Animals Communicate* (ed. T. A. Sebeok), Indiana University Press, Bloomington and London, pp. 45–70.

Marples, M. J. (1965). *The Ecology of Human Skin*, CC Thomas, Springfield, Illinois.

Marples, R. R. (1974). 'The microflora of the face and acne lesions', *J. Investig. Dermatol.*, **62**, 326–331.

Marples, R. R., Downing, D. T., and Kligman, A. M. (1972). 'Influence of Pityrosporum species in the generation of free fatty acids in human surface lipids', *J. Investig. Dermatol.*, **58**, 155–159.

Marques, D. M. (1979). 'Roles of the main olfactory and vomeronasal systems in the response of the female hamster', *Behav. Neural. Biol.*, **26**, 298–329.

Marsden, H. M., and Bronson, F. H. (1964). 'Estrous synchrony in mice: alteration by exposure to male urine', *Science*, **144**, 1469.

Martan, J., and Price, D. (1967). 'Comparative responsiveness of supracaudal and other sebaceous glands in male and female guinea pigs to hormones', *J. Morphol.*, **121**, 209–221.

Martin, I. G. (1980). '"Homeochemic", intraspecific chemical signal', *J. Chem. Ecol.*, **6**, 517–519.

Martin, R. D. (1968). 'Reproduction and ontogeny in tree shrews (*Tupaia belangeri*) with reference to their general behaviour and taxonomic relationships', *Z. Tierpsychol.*, **25**, 409–495, 505–532.

Martin, R. D. (1972). 'A preliminary field study of the lesser mouse lemur, *Microcebus murinus*', *Z. Tierpsychol. Suppl.*, **9**, 43–89.

Mason, J. W. (1968a). 'A review of psychoendocrine research on the pituitary-adrenal cortical system', *Psychosomatic Med.*, **30**, 576–607.

Mason, J. W. (1968b). 'A review of psychoendocrine research on the sympathetic-adrenal medullary system', *Psychosomatic Med.*, **30**, 631–653.

Mason, J. W., Brady, J. V., and Tolliver, G. A. (1968a). 'Plasma and urinary 17-hydroxycorticosteroid responses to 72 hr avoidance sessions in the monkey', *Psychosomatic Med.*, **30**, 608–630.

Mason, J. W., Tolson, W. W., Brady, J. V., Tolliver, G. A. and Gilmore, L. I. (1968b). 'Urinary epinephrine and norepinephrine responses to 72 hr avoidance sessions in the monkey', *Psychosomatic Med.*, **30**, 654–681.

Mason, J. W., Mongey, E. H., and Kenion, C. C. (1973). 'Urinary epinephrine and norepinephrine responses to chair restraint in the monkey', *Physiol. Behav.*, **10**, 801–804.

Massey, A., and Vandenbergh, J. G. (1980). 'Puberty delay by a urinary cue from female house mice in feral populations', *Science*, **209**, 821–822.

Massey, A., and Vandenbergh, J. G. (1981). 'Puberty acceleration by a urinary cue from male mice in feral populations', *Biol. Reprod.*, **24**, 523–527.

Matsumoto, K. E., Partridge, D. H., Robinson, A. B., Pauling, L., Flath, R. A., Mon, T. R., and Taranishi, R. (1973). 'The identification of volatile compounds in human urine', *J. Chromatog.*, **85**, 31–34.

Mead, G. C. (1971). 'The amino acid-fermenting Clostridia', *J. Gen. Microbiol.*, **67**, 47–56.

Meili, J., Walls, F. C., McPherron, R., and Burlingame, A. L. (1979). 'Design, implementation and performance of high resolution gas chromatography/high resolution mass spectrometry/real-time computer system for the analysis of complex organic mixtures', *J. Chromatog. Sci.*, **17**, 29–42.

Menco, B. P. M. (1980a). 'Qualitative and quantitative freeze-fracture studies on olfactory and respiratory structures of frog, ox, rat and dog. I. A general survey', *Cell and Tissue Res.*, **207**, 183–209.

Menco, B. P. M. (1980b). 'Qualitative and quantitative freeze-fracture studies on olfactory and nasal respiratory epithelial surfaces of frog, ox, rat and dog. II. Cell apices, cilia and microvilli', *Cell and Tissue Res.*, **211**, 5–29.

Menco, B. P. M., Dodd, G. H., Davey, M., and Bannister, L. H. (1976). 'Presence of membrane particles in freeze-etched bovine olfactory cilia', *Nature*, **263**, 597–599.

Menevse, A., Dodd, G. H., Poynder, T. M., and Squirrell, D. (1977). 'A chemical modification approach to the olfactory code', *Biochem. Soc. Trans.*, **5**, 191–194.

Menevse, A., Dodd, G. H., and Poynder, T. M. (1978). 'A chemical modification approach to the olfactory code', *Biochem. J.*, **176**, 845–854.

Meredith, M. (1980). 'The vomeronasal organ and accessory olfactory system in the hamster', in *Chemical Signals: Vertebrates and Aquatic Invertebrates* (eds D. Müller-Schwarze and R. M. Silverstein), Plenum Press, New York, pp. 303–326.

Meredith, M., and O'Connell, R. J. (1979). 'Efferent control of stimulus access to the hamster vomeronasal organ', *J. Physiol.*, **286**, 301–316.

Meredith, M., Marques, D. M., O'Connell, R. J., and Stern, F. L. (1980). 'Vomeronasal pump: significance for male hamster sexual behaviour', *Science*, **207**, 1224–1226.

Merritt, G. C., Goodrich, B. S., Hesterman, E. R., and Mykytowycz, R. (1982). 'Microflora and volatile fatty acids present in the inguinal pouches of the wild rabbit, *Oryctolagus cuniculus*, in Australia', *J. Chem. Ecol.*, **8**, 1217–1225.

Michael, R. P., Keverne, E. B., and Bonsall, R. W. (1971). 'Pheromones: isolation of male sex attractants from a female primate', *Science*, **172**, 964–966.

Michael, R. P., Zumpe, D., Keverne, E. B., and Bonsall, R. W. (1972). 'Neuroendocrine factors in the control of primate behavior', *Recent Progr. Horm. Res.*, **28**, 665–704.

Michael, R. P., Bonsall, R. W., and Warner, P. (1974). 'Human vaginal secretions: volatile fatty acid content', *Science*, **186**, 1217–1219.

Michael, R. P., Bonsall, R. W., and Kutner, M. (1975). 'Volatile fatty acids, "copulins", in human vaginal secretions', *Psychoneuroendocrinology*, **1**, 153–163.

Michael, R. P., Bonsall, R. W., and Zumpe, D. (1976a). 'Evidence for chemical communication in primates', *Vitamins and Hormones*, **34**, 137–186.

Michael, R. P., Bonsall, R. W., and Zumpe, D. (1976b). 'A reply to Goldfoot *et al.*', *Horm Behav.*, **7**, 365–367.

Middleditch, B. S. (ed.) (1979). *Practical Mass Spectrometry*, Plenum, New York.

Milberg, R. M., and Cook, J. C. (1979a). 'Design considerations of MS sources: EI,CI,FI,FD, and API', *J. Chromatog. Sci.*, **17**, 17–23.

Milberg, R. M., and Cook, J. C. (1979b). 'Some applications of high sensitivity combined field ionization gas chromatography-mass spectrometry', *J. Chromatog. Sci.*, **17**, 43–47.

Miller, S., Ruttinger, V., and Macy, I. C. (1954). 'Urinary excretion of ten amino acids by women during the reproductive cycle', *J. Biol. Chem.*, **209**, 795–801.

Milligan, S. R. (1976). 'Pregnancy blocking in the vole, *Microtus agrestis*', *J. Reprod. Fert.*, **46**, 91–100.

Milligan, S. R. (1980). 'Pheromones and rodent reproductive physiology', *Symp. Zool. Soc. Lond.*, **45**, 251–275.

Millington, D. S. (1975). 'Determination of hormonal steroid concentrations in biological extracts by high resolution mass fragmentography', *J. Steroid Biochem.*, **6**, 239–245.

Millington, D. S., Buoy, M. E., Brooks, G., Harper, M. E., and Griffiths, K. (1975). 'Thin-layer chromatography and high resolution selected ion monitoring for the analysis of C_{19} steroids in human hyperplastic prostate tissue', *Biomed. Mass Spectrom.*, **2**, 219–224.

Millner, S. N., and Taber, C. A. (1976). 'Rapid determination of urine estriol by an on-column derivitazation procedure', *Clin. Chim. Acta*, **71**, 67–74.

Mills, M. G. L., Gorman, M. L., and Mills, M. E. J. (1980). 'The scent marking behaviour of the brown hyaena, *Hyaena brunnea*', *S. Afr. J. Zool.*, **15**, 240–248.

Moghissi, K. S., Syner, F. N., and Evans, T. N. (1972). 'A composite picture of the menstrual cycle', *Amer. J. Obst. Gynecol.*, **114**, 405–416.

Moir, M., Gallacher, I. M., Seaton, J. C., and Suggett, A. (1980). 'Terpene methyl sulphides in the essential oil of hops', *Chem. Ind.*, 624–625.

Moltz, H. (1978). 'Emission of maternal pheromone', *Science*, **201**, 939.

Moltz, H., and Leidahl, L. C. (1977). 'Bile, prolactin, and the maternal pheromone', *Science*, **196**, 81–83.

Moltz, H., and Lee, T. M. (1981). 'The maternal pheromone of the rat: identity and functional significance', *Physiol. Behav.*, **26**, 301–306.

Montagna, W. (1950). 'The brown inguinal glands of the rabbit', *Amer. J. Anat.*, **87**, 213–237.

Montagna, W. (1962). 'The skin of lemurs', *Ann. N.Y. Acad. Sci.*, **102**, 190–209.

Montagna, W. (1964). 'Histology and histochemistry of human skin. XXIV. Further observations on the axillary organ', *J. Investig. Dermatol.*, **42**, 119–129.

Montagna, W. (1970). 'Histology and histochemistry of human skin. XXXV. The nipple and areola', *Br. J. Dermatol.*, **83**, 2–13.

Montagna, W. (1972). 'The skin of non-human primates', *Amer. Zoologist*, **12**, 109–124.

Montagna, W., and Parks, H. F. (1948). 'A histochemical study of the glands of the anal sac of the dog', *Anat. Rec.*, **100**, 297–315.

Montagna, W., and Ellis, R. A. (1959). 'The skin of primates. I. The skin of the potto (*Perodicticus potto*)', *Amer. J. Phys. Anthropol.*, **17**, 137–161.

Montagna, W., and Ellis, R. A. (1960). 'The skin of primates. II. The skin of the slender loris (*Loris tardigradus*), *Amer. J. Phys. Anthropol.*, **18**, 19–43.

Montagna, W., Yasuda, K., and Ellis, R. A. (1961). 'The skin of primates. III. The skin of the slow loris (*Nycticebus coucang*)', *Amer. J. Phys. Anthropol.*, **19**, 1–21.

Montagna, W., and Yun, J. S. (1962a). 'The skin of primates. X. The skin of the ring-tailed lemur (*Lemur catta*)', *Amer. J. Phys. Anthropol.*, **20**, 95–117.

Montagna, W., and Yun, J. S. (1962b). 'The skin of primates. VIII. The skin of the anubis baboon (*Papio doguera*)', *Amer. J. Phys. Anthropol.*, **20**, 131–141.

Montagna, W., and Yun, J. S. (1962c). 'The skin of primates. XIV. Further observations on *Perodicticus potto*', *Amer. J. Phys. Anthropol.*, **20**, 441–449.

Montagna, W., Yun, J. S., Silver, A. F., and Quevedo, W. C. (1962). 'The skin of the tree shrew (*Tupaia glis*)', *Amer. J. Phys. Anthropol.*, **20**, 431–440.

Montagna, W., and Yun, J. S. (1963a). 'The skin of primates. XV. The skin of the chimpanzee (*Pan satyrus*)', *Amer. J. Phys. Anthropol.*, **21**, 189–203.

Montagna, W., and Yun, J. S. (1963b). 'The skin of primates. XVI. The skin of Lemur mongoz', *Amer. J. Phys. Anthropol.*, **21**, 371–381.

Montagna, W., and Yun, J. S. (1964). 'The skin of the domestic pig', *J. Investig. Dermatol.*, **43**, 11–21.

Montagna, W., Yun, J. S., and Machida, H. (1964). 'The skin of primates. XVIII. The skin of the rhesus monkey (*Macaca mulatta*)', *Amer. J. Phys. Anthropol.*, **22**, 307–319.

Montagna, W., and Machida, H. (1966). 'The skin of primates. XXXII. The Philippine tarsier, *Tarsius syrichta*', *Amer. J. Phys. Anthropol.*, **25**, 71–84.

Montagna, W., Machida, H., and Perkins, E. (1966a). 'The skin of primates. XXVIII. The stump-tailed macaque (*Macaca speciosa*)', *Amer. J. Phys. Anthropol.*, **24**, 71–85.

Montagna, W., Machida, H., and Perkins, E. M. (1966b). 'The skin of primates. XXXIII. The skin of the angwantibo, *Arctocebus calabarensis*', *Amer. J. Phys. Anthropol.*, **25**, 277–290.

Montagna, W., Yasuda, K., and Ellis, R. A. (1967). 'The skin of primates. V. The skin of the black Lemur (*Lemur macaco*)', *Amer. J. Phys. Anthropol.*, **19**, 115–129.

Montagna, W., and Ford, D. M. (1969). 'Histology and Histochemistry of Human Skin. XXXIII. The eyelid', *Arch. Dermatol.*, **100**, 328–336.

Montagna, W., and Parakkal, P. F. (1974). *The Structure and Function of Skin*, Academic Press, New York, 3rd edn.

Monti Graziadei, G. A., Margolis, F. L., Harding, J. W., and Graziadei, P. P. C. (1977). 'Immunocytochemistry of the olfactory marker protein', *J. Histochem. Cytochem.*, **25**, 1311–1316.

Moore, W. E. C., Cato, E. P., and Holdeman, L. V. (1966). 'Fermentation patterns of some Clostridium species', *Int. J. Syst. Bacteriol.*, **16**, 383–415.

Morello, A. M., and Downing, D. T. (1976). 'Trans-unsaturated fatty acids in human skin surface lipids', *J. Investig. Dermatol.*, **67**, 270–272.

Morgan, E. D., and Wadhams, L. J. (1972a). 'Gas chromatography of volatile compounds in small samples of biological materials', *J. Chromatog. Sci.*, **10**, 528–529.

Morgan, E. D., and Wadhams, L. J. (1972b). 'Chemical constituents of Dufour's gland in the ant, *Myrmica rubra*', *J. Insect Physiol.*, **18**, 1125–1135.

Morgan, E. D., Evershed, R. P., and Tyler, R. C. (1979). 'Gas chromatographic detection and structure analysis of volatile pheromones in insects', *J. Chromatog.*, **186**, 605–610.

Mori, A., Yasaka, Y., Masamoto, K., and Hiramatsu, M. (1978). 'Gas chromatography of 5-hydroxy-3-methylindole in human urine', *Clin. Chim. Acta*, **84**, 63–68.

Morris, J. G. (1975). 'The physiology of obligate anaerobiosis', *Adv. Microbiol. Physiol.*, **12**, 169–246.

Morris, J. G. (1977). 'Obligately anaerobic bacteria', *Trends in Biochemical Sciences*, 81–84.

Moskowitz, H. R., and Warren, C. B. (eds) (1981). *Odor Quality and Chemical Structure*, ACS Symposium Series, **148**, American Chemical Society, Washington, D.C.

Moss, C. W., Howell, R. T., Farshy, D. C., Dowell, V. R., and Brooks, J. B. (1970). 'Volatile acid production by *Clostridium botulinum* type F', *Can. J. Microbiol.*, **16**, 421–425.

Mossing, T., and Damber, J. E. (1981). 'Rutting behaviour and androgen variation in reindeer, *Rangifer tarandus*', *J. Chem. Ecol.*, **7**, 377–389.

Mossing, T., and Källquist, L. (1981). 'Variations in cutaneous glandular structures in reindeer, *Rangifer tarandus*', *J. Mammal.*, **62**, 606–612.

Moulton, D. G. (1974). 'Dynamics of cell populations in the olfactory epithelium', *Ann. N. Y. Acad. Sci.*, **237**, 52–61.

Moulton, D. G., Ashton, E. H., and Eayrs, J. T. (1960). 'Studies in olfactory acuity. Relative detectability of n-aliphatic acids by the dog', *Anim. Behav.*, **8**, 117–128.

Mugford, R. A., and Nowell, N. W. (1971). 'The preputial glands as a source of aggression-promoting odors in mice', *Physiol. Behav.*, **6**, 247–249.

Müllen, K., and Pregosin, P. S. (1976). *Fourier Transform NMR Techniques: a Practical Approach*, Academic Press, London.

Müller-Schwarze, D. (1969). 'Complexity and relative specificity in a mammalian pheromone', *Nature*, **223**, 525–526.

Müller-Schwarze, D. (1971). 'Pheromones in black-tailed deer, *Odocoileus hemionus columbianus*', *Anim. Behav.*, **19**, 141–152.

Müller-Schwarze, D. (1972). 'Responses of young black-tailed deer to predator odors', *J. Mammal.*, **53**, 392–394.

Müller-Schwarze, D. (1977). 'Complex mammalian behavior and pheromone bioassay in the field', in *Chemical Signals in Vertebrates* (eds D. Müller-Schwarze and M. M. Mozell), Plenum, New York, pp. 413–433.

Müller-Schwarze, D. (1979). 'Flehmen in the context of mammalian urine communication', in *Chemical Ecology: Odour Communication in Animals* (ed. F. J. Ritter), Elsevier/North-Holland Biomedical Press, Amsterdam, pp. 85–96.

Müller-Schwarze, D., and Müller-Schwarze, C. (1972). 'Social scents in hand-reared pronghorn, *Antilocapra americana*', *Zoologia Africana*, **7**, 257–271.

Müller-Schwarze, D., Müller-Schwarze, C., Singer, A. G., and Silverstein, R. M. (1974). 'Mammalian pheromone: identification of active component in the subauricular scent of the male pronghorn', *Science*, **183**, 860–862.

Müller-Schwarze, D., and Müller-Schwarze, C. (1975). 'Subspecies specifity of response to a mammalian social odor', *J. Chem. Ecol.*, **1**, 125–131.

Müller-Schwarze, D., Silverstein, R. M., Müller-Schwarze, C., Singer, A. G., and Volkman, N. J. (1976). 'Response to a mammalian pheromone and its geometric isomer', *J. Chem. Ecol.*, **2**, 389–398.

Müller-Schwarze, D., Quay, W. B., and Brundin, A. (1977a). 'The caudal gland in reindeer (*Rangifer tarandus*): its behavioral role, histology and chemistry', *J. Chem. Ecol.*, **3**, 591–601.

Müller-Schwarze, D., Volkman, N. J., and Zemanek, K. F. (1977b). 'Osmetrichia: specialized scent hair in black-tailed deer', *J. Ultrastructure Res.*, **59**, 223–230.

Müller-Schwarze, D., David, U., Claesson, A., Singer, A. G., Silverstein, R. M., Müller-Schwarze, C., Volkman, N. J., Zemanek, K. F., and Butler, R. G. (1978a). 'The deer-lactone: source, chemical properties and responses of black-tailed deer', *J. Chem. Ecol.*, **4**, 247–256.

Müller-Schwarze, D., Källquist, L., Mossing, T., Brundin, A., and Andersson, G. (1978b). 'Response of reindeer to interdigital secretions of conspecifics', *J. Chem. Ecol.*, **4**, 325–335.

Müller-Schwarze, D., Källquist, L., and Mossing, T. (1979). 'Social behavior and chemical communication in reindeer (*Rangifer tarandus*), *J. Chem. Ecol.*, **5**, 483–517.

Müller-Schwarze, D., and Heckman, S. (1980). 'The social role of scent marking in beaver, *Castor canadensis*', *J. Chem. Ecol.*, **6**, 81–95.

Murawski, U., and Jost, U. (1974). 'Unsaturated wax esters in the Harderian gland of the rat', *Chem. Phys. Lipids*, **13**, 155–158.

Murphy, E. L., Flath, R. A., Black, D. R., Mon, T. R., Teranishi, R., Timm, R. M., and Howard, W. E. (1978). 'Isolation, identification and biological activity assay of chemical fractions from estrus urine attractive to the coyote', in *Flavor Chemistry of Animal Foods* (ed. R. W. Bullard), ACS Symposium Series **67**, American Chemical Society, Washington, D.C., pp. 66–77.

Murphy, M. R., and Schneider, G. E. (1970). 'Olfactory bulb removal eliminates mating behavior in the male golden hamster', *Science*, **167**, 302–304.

Murphy, M. R. (1980). 'Sexual preferences of male hamsters: Importance of pre-weaning and adult experience, vaginal secretion and olfactory or vomeronasal sensation', *Behavioral and Neural Biology*, **30**, 323–340.

Murray, K. E. (1977). 'Concentration of headspace, airborne and aqueous volatiles on Chromosorb 105 for examination by gas chromatography and gas chromatography-mass spectrometry', *J. Chromatog.*, **135**, 49–61.

Mykytowycz, R. (1965). 'Further observations on the territorial function and histology of the submandibular cutaneous (chin) glands in the rabbit, *Oryctolagus cuniculus*', *Anim. Behav.*, **13**, 400–412.

Mykytowycz, R. (1966). 'Observations on odoriferous and other glands in the Australian wild rabbit, *Oryctolagus cuniculus*, and the hare, *Lepus europaeus*: the anal gland', *CSIRO Wildl. Res.*, **11**, 11–29.

Mykytowycz, R. (1970). 'The role of skin glands in mammalian communication', in *Advances in Chemoreception*, Vol. 1 (eds J. W. Johnston, D. G. Moulton and A. Turk), Appleton-Century-Crofts, New York, pp. 327–360.

Mykytowycz, R. (1979). 'Some difficulties in the study of the function and composition of semiochemicals in mammals, particularly wild rabbits, *Oryctolagus cuniculus*', in *Chemical Ecology: Odour Communication in Animals* (ed. F. J. Ritter), Elsevier/North-Holland Biomedical Press, Amsterdam, pp. 105–115.

Mykytowycz, R., and Gambale, S. (1969). 'The distribution of dung-hills and the behaviour of free-living wild rabbits, *Oryctolagus cuniculus* on them', *Forma et Functio*, **1**, 333–349.

Mykytowycz, R., and Goodrich, B. S. (1974). 'Skin glands as organs of communication in mammals', *J. Investig. Dermatol.*, **62**, 124–131.

Mykytowycz, R., Hesterman, E. R., Gambale, S., and Dudzinski, M. L. (1976). 'A comparison of the effectiveness of the odors of rabbits, *Oryctolagus cuniculus*, in enhancing territorial confidence', *J. Chem. Ecol.*, **2**, 13–24.

Naftchi, N. E., Becker, M. A., and Akerkar, A. S. (1975). 'Reactions of biogenic amines with aqueous potassium dichromate. An application to the determination of dopamine', *Anal. Biochem.*, **66**, 423–433.

Nicolaides, N. (1965), 'Skin lipids. II. Lipid class composition of samples from various species and anatomical sites', *J. Amer. Oil Chem. Soc.*, **42**, 691–702.

Nicolaides, N. (1974). 'Skin lipids: their biochemical uniqueness', *Science*, **186**, 19–26.

Nicolaides, N., and Kellum, R. E. (1965). 'Skin lipids. I. Sampling problems of the skin and its appendages', *J. Amer. Oil Chem. Soc.*, **42**, 685–690.

Nicolaides, N. and Ansari, M. N. A. (1968). 'Fatty acids of unusual double-bond positions and chain lengths found in rat skin surface lipids', *Lipids*, **3**, 403–410.

Nicolaides, N., Fu, H. C., and Rice, G. R. (1968). 'The skin surface lipids of man compared with those of eighteen species of animals', *J. Investig. Dermatol.*, **51**, 83–89.

Nicolaides, N., Fu, H. C., and Ansari, M. N. A. (1970). 'Diester waxes in surface lipids of animal skin', *Lipids*, **5**, 299–307.

Nicolaides, N., Fu, H. C., Ansari, M. N. A., and Rice, G. R. (1972). 'The fatty acids of wax esters and sterol esters from vernix caseosa and from human skin surface lipid', *Lipids*, **7**, 506–517.

Nicolaides, N., and Apon, J. M. B. (1977). 'The saturated methyl branched fatty acids of adult human skin surface lipid', *Biomed. Mass Spectrom.*, **4**, 337–347.

Nicolaides, N., Kaitaranta, J. K., Rawdah, T. N., Macy, J. I., Boswell, F. M., and Smith, R. E. (1981). 'Meibomian gland studies: comparison of steer and human lipids', *Invest. Ophthalmol. Vis. Sci.*, **20**, 522–536.

Nicoll, R. A. (1970). 'Identification of tufted cells in the olfactory bulb', *Nature*, **227**, 623–625.

334

Nikkari, T. (1965). 'Composition and secretion of the skin surface lipid of the rat: effects of dietary lipids and hormones', *Scand. J. Clin. Lab. Investig.*, **17 (Suppl. 85)**, 1–140.

Nikkari, T. (1974). 'Comparative chemistry of sebum', *J. Investig. Dermatol.*, **62**, 257–267.

Nikkari, T., and Haahti, E. (1968). 'Isolation and analysis of two types of diester waxes of the skin surface lipids of the rat', *Biochim. Biophys. Acta*, **164**, 294–305.

Nikkari, T., Schreibman, P. H., and Ahrens, E. H. (1974). 'In vivo studies of sterol and squalene secretion by human skin', *J. Lipid Res.*, **15**, 563–573.

Noble, W. C. (1981). *Microbiology of Human Skin* (2nd edn), Lloyd-Duke (Medical Books) Ltd., London.

Nordlund, D. A., and Lewis, W. J. (1976). 'Terminology of chemical releasing stimuli in intraspecific and interspecific interactions', *J. Chem. Ecol.*, **2**, 211–220.

Novotny, M., Lee, M. L., and Bartle, K. D. (1974). 'Some analytical aspects of the chromatographic headspace concentration method using a porous polymer', *Chromatographia*, **7**, 333–338.

Novotny, M., Jorgenson, J. W., Carmack, M., Wilson, S. R., Boyse, E. A., Yamazaki, K., Wilson, M., Beamer, W., and Whitten, W. K. (1980a). 'Chemical studies of the primer mouse pheromones', in *Chemical Signals: Vertebrates and Aquatic Invertebrates* (eds D. Müller-Schwarze and R. M. Silverstein), Plenum, New York, pp. 377–390.

Novotny, M., Schwende, F. J., Hartigan, M. J., and Purcell, J. E. (1980b). 'Capillary gas chromatography with ultraviolet spectrometric detection', *Anal. Chem.*, **52**, 736–740.

Nowell, N. W., Thody, A. J., and Woodley, R. (1980a). 'α-Melanocyte stimulating hormone and aggressive behavior in the male mouse', *Physiol. Behav.*, **24**, 5–9.

Nowell, N. W., Thody, A. J., and Woodley, R. (1980b). 'The source of an aggression promoting olfactory cue, released by α-melanocyte stimulating hormone in the male mouse', *Peptides*, **1**, 69–72.

Nyby, J., Whitney, G., Schmitz, S., and Dizinno, G. (1978). 'Postpubertal experience establishes signal value of mammalian sex odor', *Behav. Biol.*, **22**, 545–552.

Nyby, J., and Whitney, G. (1980). 'Experience affects behavioural responses to sex odors', in *Chemical Signals: Vertebrates and Aquatic Invertebrates* (eds D. Müller-Schwarze and R. M. Silverstein), Plenum Press, New York, pp. 173–192.

Nyby, J., and Zakeski, D. (1980). 'Elicitation of male mouse ultrasounds: bladder urine and aged urine from females', *Physiol. Behav.*, **24**, 737–740.

O'Brien, P. H. (1982). 'Flehmen: its occurrence and possible functions in feral goats', *Anim. Behav.*, **30**, 1015–1019.

O'Connell, R. J., Singer, A. G., Pfaffmann, C., and Agosta, W. C. (1979). 'Pheromones of hamster vaginal discharge. Attraction to femtogram amounts of dimethyl disulphide and to mixtures of volatile components', *J. Chem. Ecol.*, **5**, 575–585.

O'Connell, R. J., Singer, A. G., Stern, F. L., Jesmajian, S., and Agosta, W. C. (1981). 'Cyclic variations in the concentration of sex attractant pheromone in hamster vaginal secretion', *Behavioral and Neural Biology*, **31**, 457–464.

Oelrich, E., Preusch, H., and Wilhelm, E. (1980). 'Separation of enantiomers by high performance liquid chromatography using chiral eluents', *HRC&CC*, **3**, 269–272.

Oertel, G. W., and Treiber, L. (1969). 'Metabolism and excretion of C_{19}- and C_{18}-steroids by human skin', *Europ. J. Biochem.*, **7**, 234–238.

Ohloff, G. (1978). 'Recent developments in the field of naturally occurring aroma compounds', *Fortsch. Chem. organ. Naturst.*, **35**, 431–527.

Oita, K., Oh, J. H., Tiedeman, G. T., and Gauditz, I. (1976). 'Wildlife damage control I: development of deer repellents', *Amer. Chem. Soc. 172nd National Meeting*, San Francisco.

Okamoto, M., Setaishi, C., Horiuchi, Y., Mashimo, K., Moriya, K., and Itoh, S. (1971). 'Urinary excretion of testosterone and epitestosterone and plasma levels of LH and testosterone in the Japanese and Ainu', *J. Clin. Endocrinol. Metab.*, **32**, 673–674.

Oppenheimer, J. R. (1977). 'Communication in New World Monkeys', in *How Animals Communicate* (ed. T. A. Seboek), Indiana University Press, Bloomington and London, pp. 851–889.

Orsulak, P. J., and Gawienowski, A. M. (1972). 'Olfactory preferences for the rat preputial gland', *Biol. Reprod.*, **6**, 219–223.

Paleologou, A. M. (1977). 'Detecting oestrus in cows by a method based on bovine sex pheromones', *Vet. Rec.*, **100**, 319.

Parakkal, P., Montagna, W., and Ellis, E. A. (1962). 'The skin of primates. XI. The skin of the white-browed gibbon, *Hylobates hoolock*', *Anat. Rec.*, **143**, 169–177.

Parkes, A. S., and Bruce, H. M. (1961). 'Olfactory stimuli in mammalian reproduction', *Science*, **134**, 1049–1054.

Parliment, T. H. (1981). 'Concentration and fractionation of aromas on reverse-phase adsorbents', *J. Agric. Food Chem.*, **29**, 836–841.

Parliment, T. H., and Spencer, M. D. (1981). 'Applications of simultaneous FID/ NPD/FPD detectors in the capillary gas chromatographic analysis of flavors', *J. Chromatog. Sci.*, **19**, 435–438.

Parry, R. J. (1977). 'Biosynthesis of lipoic acid. I. Incorporation of specifically tritiated octanoic acid into lipoic acid', *J. Amer. Chem. Soc.*, **99**, 6464–6466.

Parsons, J. S., and Mitzner, S. (1975). 'Gas chromatographic method for concentration and analysis of traces of industrial organic pollutants in environmental air and stacks', *Env. Sci. Technol.*, **9**, 1053–1058.

Pasteels, J. M. (1982). 'Is kairomone a valid and useful term?', *J. Chem. Ecol.*, **8**, 1079–1081.

Patterson, J. F. (1960). 'Biosynthesis of squalene by rat preputial gland', *Proc. Soc. Exp. Biol. Med.*, **105**, 461–463.

Patterson, P. L., and Howe, R. L. (1978), 'Thermionic nitrogen-phosphorus detection with an alkali-ceramic bead', *J. Chromatog. Sci.*, **16**, 275–280.

Patterson, R. L. S. (1967). 'A possible contribution of phenolic components to boar odour', *J. Sci. Food Agric.*, **18**, 8–10.

Patterson, R. L. S. (1968a). '5α-androst-16-ene-3-one; compound responsible for taint in boar fat', *J. Sci. Food Agric.*, **19**, 31–38.

Patterson, R. L. S. (1968b). 'Acidic components of boar preputial fluid', *J. Sci. Food Agric.*, **19**, 38–40.

Patterson, R. L. S. (1968c). 'Identification of 3α-hydroxy-5α-androst-16-ene as the musk odour component of boar submaxillary salivary gland and its relationship to the sex odour taint in pork meat', *J. Sci. Food Agric.*, **19**, 434–438.

Patterson, R. L. S. (1969). 'Boar taint: its chemical nature and estimation', in *Meat Production from Entire Male Animals* (ed. D. H. Rhodes), J. and A. Churchill Ltd., London, pp. 247–260.

Patton, S., and Jenson, R. G. (1975). 'Lipid metabolism and membrane functions of the mammary gland', *Progress in the Chemistry of Fats and Other Lipids*, **14**, 163–277.

Payne, A. P. (1977). 'Pheromonal effects of Harderian gland homogenates on aggressive behaviour in the hamster', *J. Endocrinol.*, **73**, 191–192.

Payne, A. P. (1979). 'The attractiveness of Harderian gland smears to sexually naive and experienced male golden hamsters', *Anim. Behav.*, **27**, 897–904.

Payne, A. P., McGadey, J. M., Moore, M. R., and Thompson, G. (1977). 'Androgenic control of the Harderian gland in the male golden hamster', *J. Endocrinol.*, **75**, 73–82.

Payne, A. P., McGadey, J. M., Moore, M. R., and Thompson, G. G. (1979). 'Changes in Harderian gland activity in the female golden hamster during the oestrous cycle, pregnancy and lactation', *Biochem. J.*, **178**, 597–604.

Pelosi, P., Pisanelli, A. M., Baldaccini, N. E., and Gagliardo, A. (1981). 'Binding of [³H]-2-isobutyl-3-methoxypyrazine to cow olfactory mucosa', *Chemical Senses*, **6**, 77–85.

Pelosi, P., Baldaccini, N. E., and Pisanelli, A. M. (1982). 'Identification of a specific olfactory receptor for 2-isobutyl-3-methoxypyrazine', *Biochem. J.*, **201**, 245–248.

Perkins, E. M. (1966). 'The skin of primates. XXXI. The skin of the black-collared tamarin, *Tamarinus nigricollis*', *Amer. J. Phys. Anthropol.*, **25**, 41–70.

Perkins, E. M. (1968). 'The skin of primates. XXXVI. The skin of the pigmy marmoset, *Callithrix (= Cebuella) pigmaea*', *Amer. J. Phys. Anthropol.*, **29**, 349–364.

Perkins, E. M. (1969a). 'The skin of primates. XL. The skin of the cotton-headed tamarin, *Saguinus oedipus*', *Amer. J. Phys. Anthropol.*, **30**, 13–28.

Perkins, E. M. (1969b). 'The skin of primates. XXIV. The skin of Goeldi's marmoset, *Callimico goeldii*', *Amer. J. Phys. Anthropol.*, **30**, 231–250.

Perkins, E. M. (1975). 'Phylogenetic significance of the skin of New World monkeys', *Amer. J. Phys. Anthropol.*, **42**, 395–424.

Perkins, E. M., Arao, T., and Dolnick, E. H. (1968). 'The skin of primates. XXXVII. The skin of the pig-tail macaque, *Macaca nemastrina*', *Amer. J. Phys. Anthropol.*, **28**, 75–84.

Perry, G. C., Patterson, R. L. S., Macfie, H. J. H., and Stinson, C. G. (1980). 'Pig courtship behaviour: pheromonal property of androstene steroids in male submaxillary secretion', *Anim. Prod.*, **31**, 191–199.

Perry, T. L., and Schroeder, W. A. (1963). 'The occurrence of amines in human urine', *J. Chromatog.*, **12**, 358–373.

Perry, T. L., Hansen, S., Hestrin, M., and Macintyre, L. (1965). 'Exogenous urinary amines of plant origin', *Clin. Chim. Acta*, **11**, 24–34.

Persaud, K. C., Wood, P. H., Squirrell, D. J., and Dodd, G. H. (1981). 'Biochemical studies in olfaction', *Biochem. Soc. Trans.*, **9**, 107–108.

Persaud, K., Wood, P., and Dodd, G. (1980). 'An approach to affinity labelling of rat olfactory receptors', in *Olfaction and Taste VII* (ed. H. van der Starre), IRL Press, London, p. 98.

Peters, R. P., and Mech, L. D. (1975). 'Scent-marking in wolves', *Amer. Scientist*, **63**, 628–637.

Petropavlovskii, V. V., and Rykova, A. I. (1958). The stimulation of sexual functions in cows', *Anim. Br. Abstr.*, **27**, 798.

Pettersen, J. E., and Jellum, E. (1972). 'The identification and metabolic origin of 2-furoylglycine and 2,5-furandicarboxylic acid in human urine', *Clin. Chim. Acta*, **41**, 199–207.

Pettersen, J. E., and Stokke, O. (1973). 'Branched short-chain dicarboxylic acids in human urine', *Biochim. Biophys. Acta*, **304**, 316–325.

Pfaffenberger, C. D., and Horning, E. C. (1977). 'Sex differences in human urinary steroid metabolic profiles determined by gas chromatography', *Anal. Biochem.*, **80**, 329–343.

Phelan, P. L., and Miller, J. R. (1981). 'Separation of isomeric insect pheromonal compounds using reversed-phase HLPC with AgNO₃ in the mobile phase', *J. Chromatog. Sci.*, **19**, 13–17.

Pillsbury, D. M., and Rebell, G. (1952). 'The bacterial flora of the skin', *J. Investig. Dermatol.*, **18**, 173–186.

Pochi, P. E., and Strauss, J. S. (1974). 'Endocrinologic control of the development and activity of the human sebaceous gland', *J. Investig. Dermatol.*, **62**, 191–201.

Politzer, I. R., Githens, S., Dowty, B. J., and Laseter, J. L. (1975). 'Gas chromatographic evaluation of the volatile constituents of lung, brain and liver tissues', *J. Chromatog. Sci.*, **13**, 378–379.

Politzer, I. R., Dowty, B. J., and Laseter, J. L. (1976). 'Use of gas chromatography and mass spectrometry to analyze underivatized volatile human or animal constituents of clinical interest', *Clin. Chem.*, **22**, 1775–1788.

Popper, K. (1976). *Unended Quest; an intellectual autobiography*, Fontana, London.

Powers, J. B., Fields, R. B., and Winans, S. S. (1979). 'Olfactory and vomeronasal system participation in male hamsters' attraction to female vaginal secretions', *Physiol. Behav.*, **22**, 77–84.

Poynder, T. M. (1974). 'Response of the frog olfactory system to controlled odor stimuli', *J. Soc. Cosmet. Chem.*, **25**, 184–202.

Prelog, V., Führer, J., Hagenbach, R., and Schneider, R. (1948). 'Untersuchungen über Organextrakte. Über die Isolierung von Jonon-Derivaten aus dem Harn trächtiger Stuten', *Helv. Chim. Acta*, **31**, 1799–1814.

Preti, G., and Huggins, G. (1975). 'Cyclical changes in volatile acidic metabolites of human vaginal secretions and their relation to ovulation', *J. Chem. Ecol.*, **1**, 361–376.

Preti, G., Muetterties, E. L., Furman, J. M., Kennelly, J. J., and Johns, B. E. (1976). 'Volatile constituents of dog (*Canis familiaris*) and coyote (*Canis latrans*) anal sacs', *J. Chem. Ecol.*, **2**, 177–186.

Preti, G., Smith, A. B., and Beauchamp, G. K. (1977). 'Chemical and behavioral complexity in mammalian chemical communication systems: guinea pigs (*Cavia porcellus*), marmosets (*Saguinas fuscicollis*) and humans (*Homo sapiens*)', in *Chemical Signals in Vertebrates* (eds D. Müller-Schwarze and M. M. Mozell), Plenum Press, New York, pp. 95–114.

Preti, G., Huggins, G. R., and Tonzetich, J. (1978). 'Predicting and determining ovulation by monitoring the concentration of volatile sulfur-containing compounds present in mouth air' (U.S. Patent), *Chemical Abstracts*, **90**, 51087g.

Preti, G., Huggins, G. R., and Silverberg, G. D. (1979). 'Alterations in the organic compounds of vaginal secretions caused by sexual arousal', *Fertil. Steril.*, **32**, 47–54.

Puah, C. M., Kjeld, J. M., and Joplin, G. F. (1978). 'A radioimmuno-chromatographic scanning method for the analysis of testosterone conjugates in urine and serum', *J. Chromatog.*, **145**, 247–255.

Quay, W. B. (1955). 'Histology and cytochemistry of skin gland areas in the caribou, *Rangifer*', *J. Mammal.*, **36**, 187–201.

Quay, W. B. (1968). 'The specialized posteriolateral sebaceous regions in microtine rodents', *J. Mammal.*, **49**, 427–445.

Quay, W. B., and Müller-Schwarze, D. (1970). 'Functional histology of integumentary glandular regions in black-tailed deer, *Odocoileus hemionus columbianus*', *J. Mammal.*, **51**, 675–694.

Quay, W. B., and Müller-Schwarze, D. (1971). 'Relations of age and sex to integumentary glandular regions in Rocky Mountain mule deer, *Odocoileus hemionus hemionus*', *J. Mammal.*, **52**, 670–685.

Raffi, R. O., Moghissi, K. S., and Sacco, A. G. (1977). 'Proteins of human vaginal fluid', *Fertil. Steril.*, **28**, 1345–1348.

Rakoff, A. E., Feo, L. G., and Goldstein, L. (1944). 'The biological characteristics of the normal vagina', *Amer. J. Obst. Gynec.*, **47**, 467–494.

Ralls, K. (1971). 'Mammalian scent marking', *Science*, **171**, 443–449.

Ramasastry, P., Downing, D. T., Pochi, P. E., and Strauss, J. S. (1970). 'Chemical composition of human skin surface lipids from birth to puberty', *J. Investig. Dermatol.*, **54**, 139–144.

Rasmussen, L. E., Schmidt, M. J., Henneous, R., Groves, D., and Daves, G. D.

(1982). 'Asian bull elephants: flehmen-like responses to extractable components in female elephant estrous urine', *Science*, **217**, 159–162.

Rauschkolb, E. W., Davis, H. W., Fenimore, D. C., Black, H. S., and Fabre, L. F. (1969). 'Identification of vitamin D_3 in human skin', *J. Investig. Dermatol.*, **53**, 289–294.

Reed, H. C. B., Melrose, D. R., and Patterson, R. L. S. (1974). 'Androgen steroids as an aid to the detection of oestrus in pig artificial insemination', *Br. Vet. J.*, **130**, 61–67.

Regnier, F. E. (1971). 'Semiochemicals — structure and function', *Biol. Reprod.*, **4**, 309–326.

Regnier, F. E., and Huang, J. C. (1970). 'Identification of some oxygen-containing functional groups by reaction gas chromatography', *J. Chromatog. Sci.*, **8**, 267–272.

Regnier, F. E., and Goodwin, M. (1977). 'On the chemical and environmental modulation of pheromone release from vertebrate scent marks', in *Chemical Signals in Vertebrates* (eds D. Müller-Schwarze and M. M. Mozell), Plenum, New York, pp. 115–133.

Regnier, F. E., Waller, G. R., Eisenbraun, E. J., and Auda, H. (1968). 'The biosynthesis of methylcyclopentane monoterpenoids. II. Nepetalactone', *Phytochemistry*, **7**, 221–230.

Reynolds, G. P., and Gray, D. O. (1976). 'A method for the estimation of 2-phenylethylamine in human urine by gas chromatography', *Clin. Chim. Acta*, **70**, 213–217.

Reynolds, G. P., and Gray, D. O. (1978). 'Gas chromatographic detection of N-methyl-2-phenylethylamine: a new component of human urine', *J. Chromatog.*, **145**, 137–140.

Reynolds, G. P., Seakins, J. W. T., and Gray, D. O. (1978). 'The urinary excretion of 2-phenylethylamine in phenylketonuria', *Clin. Chim. Acta*, **83**, 33–39.

Reynolds, J., and Keverne, E. B. (1979). 'The accessory olfactory system and its role in the pheromonally mediated suppression of estrus in grouped mice', *J. Reprod. Fert.*, **57**, 31–36.

Richard, A. (1974). 'Patterns of mating in *Propithecus verreauxi verreauxi*', in *Prosimian Behaviour* (eds R. D. Martin, G. A. Doyle and A. C. Walker), Gerald Duckworth and Co. Ltd., London, pp. 49–74.

Rijkens, F., and Boelens, H. (1975). 'The future of aroma research', in *Aroma Research* (eds H. Maarse and P. J. Groenen), Pudoc, Wageningen, Netherlands, pp. 203–220.

Ritter, F. J. (ed.) (1979). *Chemical Ecology: Odour Communication in Animals*, Elsevier/North-Holland Biomedical Press, Amsterdam.

Ritter, F. J., Brüggemann, I. E. M., Gut, J., and Persoons, C. J. (1982a). 'Recent pheromone research in the Netherlands on muskrats and some insects', in *Insect Pheromone Technology: Chemistry and Applications* (eds B. A. Leonhardt and M. Beroza), ACS Symposium Series, **190**, American Chemical Society, Washington, D. C., pp. 107–130.

Ritter, F. J., Brüggemann, I. E. M., Gut, J., Persoons, C. J., and Verwiel, P. E. J. (1982b). 'Chemical stimuli of the muskrat', in *Determination of Behaviour by Chemical Stimuli* (eds J. E. Steiner and J. R. Ganchrow), IRL Press, London, pp. 77–89.

Robertson, A. J. B. (1972). 'Field ionisation', *M.T.P. Int. Rev. Sci., Phys. Chem., Ser. One*, **5**, 103–131. Butterworth, London.

Rochelle, J. A., Gauditz, I., Oita, K., and Oh, J. H. K. (1974). 'New developments in big-game repellents', in *Wildlife and Forest Management in the Pacific Northwest* (ed. H. C. Black), Oregon State University, Corvallis, pp. 103–112.

Rock, C. O. (1977). 'Harderian gland', in *Lipid Metabolism in Mammals*, Vol. 2 (ed. F. Snyder), Plenum, New York, pp. 311–321.

Rock, C. O., Fitzgerald, V., Rainey, W. T., and Snyder, F. L. (1976). 'Mass spectral identification of 2-(*O*-acyl)hydroxy fatty acid esters in the white portion of the rabbit Harderian gland', *Chem. Phys. Lipids*, **17**, 207–212.

Rogel, M. J. (1978). 'A critical evaluation of the possibility of higher primate reproductive and sexual pheromones', *Psychol. Bull.*, **85**, 810–830.

Rothe, M. (1975). 'Aroma values — a useful concept?' in *Aroma Research* (eds H. Maarse and P. J. Groenen), Pudoc, Wageningen, Netherlands, pp. 111–119.

Rothman, R. J., and Mech, L. D. (1979). 'Scent marking in lone wolves and newly formed pairs', *Anim. Behav.*, **27**, 750–760.

Rothman, S. (1954). *Physiology and Biochemistry of the Skin*, University of Chicago Press, Chicago.

Rothschild, M., and Ford, B. (1973). 'Factors influencing the breeding of the rabbit flea (*Spilopsyllus cuniculi*). A spring-time accelerator and a kairomone in nestling rabbit urine, with notes on *Cediopsylla simplex*, another "hormone bound" species', *J. Zool.*, **170**, 87–137.

Roughton, R. D. (1982). 'A synthetic alternative to fermented egg as a canid attractant', *J. Wildl. Management*, **46**, 230–234

Rubin, R. T., Miller, R. G., Clark, B. R., Poland, R. E., and Arthur, R. J. (1970). 'The stress of aircraft carrier landings. II. 3-Methoxy-4-hydroxyphenylglycol excretion in naval aviators', *Psychosomatic Med.*, **32**, 589–597.

Russell, D. H. (1977). 'Clinical relevance of polyamines as biochemical markers of tumor kinetics', *Clin. Chem.*, **23**, 22–27.

Russell, J. W. (1975). 'Analysis of air pollutants using sampling tubes and gas chromatography', *Env. Sci. Technol.*, **9**, 1175–1178.

Russell, M. J. (1976). 'Human olfactory communication', *Nature*, **260**, 520–522.

Rutowski, R. L. (1981). 'The function of pheromones', *J. Chem. Ecol.*, **7**, 481–484.

Ruzicka, L. (1926). 'Zur Kenntnis des Kohlenstoffringes. I. Über die Konstitation des Zibetons', *Helv. Chim. Acta*, **9**, 230–248.

Rylance, H. J. (1969). 'The estimation of indoxyl sulphate in urine', *Clin. Chim. Acta*, **26**, 99–103.

Sakai, T., Nakajima, K., Yoshihara, K., and Sakan, T. (1980). 'Revisions of the adsolute configurations of C-8 methyl groups in dehydroiridodiol, neonepetalactone and matatabiether from *Actinidia polygama*', *Tetrahedron*, **36**, 3115–3119.

Salo, P. (1975). 'Use of odour thresholds in sensorial testing and comparisons with instrumental analysis', in *Aroma Research* (eds H. Maarse and P. J. Groenen), Pudoc, Wageningen, Netherlands, pp. 121–130.

Sanderson, G. W., Co, H., and Gonzale, J. G. (1971). 'Biochemistry of tea fermentation. The role of carotenes in black tea aroma formation', *J. Food Sci.*, **36**, 231–236.

Sansone, G., and Hamilton, J. G. (1969). 'Glyceryl ether, wax ester and triglyceride composition of mouse preputial gland', *Lipids*, **4**, 435–440.

Sansone-Bazzano, G., Bazzano, G., Reisner, R. M., and Hamilton, J. G. (1972). 'The hormonal induction of alkylglycerol, wax and alkyl acetate synthesis in the preputial gland of the mouse', *Biochim. Biophys. Acta*, **260**, 35–40.

Sansone-Bazzano, G., and Reisner, R. M. (1974). 'Steroid pathways in sebaceous glands', *J. Investig. Dermatol.*, **62**, 211–216.

Sastry, S. D., Buck, K. T., Janák, J., Dressler, M., and Preti, G. (1980). 'Volatiles emitted by humans', in *Biochemical Applications of Mass Spectrometry* (eds G. R. Waller and O. C. Dermer), Wiley, Chichester, pp. 1085–1129.

Sato, K. (1977). 'The physiology, pharmacology and biochemistry of the eccrine sweat gland', *Rev. Physiol. Biochem. Pharmacol.*, **79**, 51–131.

Schaffer, J. (1940). *Die Hautdrüsenorgane der Säugetiere*, Urban & Schwarzenberg, Berlin and Vienna.

Schalken, A. P. M. (1976). 'Three types of pheromones in the domestic rabbit, *Oryctolagus cuniculus*', *Chemical Senses and Flavor*, **2**, 139–155.

Schaumburg-Lever, G., and Lever, W. F. (1975). 'Secretion from human apocrine glands: an electron microscopic study', *J. Investig. Dermatol.*, **64**, 38–41.

Schiffman, S. S., and Erickson, R. P. (1980). 'The issue of primary tastes versus a taste continuum', *Neurosci. Biobehav. Rev.*, **4**, 109–117.

340

⁄ Schildknecht, H., Wilz, I., Enzmann, F., Grund, N., and Ziegler, M. (1976). 'Mustelan, the malodorous substance from the anal gland of the mink, *Mustela vison*, and the polecat, *M. putorius*', *Angew. Chem. Int. Ed. Engl.*, **15**, 242–243.

Schilling, A. (1974). 'A study of marking behaviour in *Lemur catta*', in *Prosimian Behaviour* (eds R. D. Martin, C. A. Doyle and A. C. Walker), Gerald Duckworth and Co. Ltd., London, pp. 347–362.

Schilling, A. (1979). 'Olfactory communication in prosimians', in *The Study of Prosimian Behaviour* (eds G. A. Doyle and R. D. Martin), Academic Press, New York, pp. 461–542.

Schilling, A. (1980). 'The possible role of urine in territoriality of some nocturnal prosimians', *Symp. Zool. Soc. Lond.*, **45**, 165–193.

Schinckel, P. G. (1954). 'The effect of the ram on the incidence and occurrence of oestrus in ewes', *Aust. Vet. J.*, **30**, 189–195.

Schleidt, M. (1980). 'Personal odor and nonverbal communication', *Ethology and Sociobiology*, **1**, 225–231.

Schleidt, M., Hold, B., and Attili, G. (1981). 'A cross-cultural study on the attitude towards personal odors', *J. Chem. Ecol.*, **7**, 19–31.

Schomburg, G., Dielmann, R., Husmann, H., and Weeke, F. (1976). 'Gas chromatographic analysis with glass capillary columns', *J. Chromatog.*, **122**, 55–72.

Schreck, C. E., Smith, N., Carlson, D. A., Price, G. D., Haile, D., and Godwin, D. R. (1982). 'A material isolated from human hands that attracts female mosquitoes', *J. Chem. Ecol.*, **8**, 429–438.

Schulten, H. R., Komori, T., and Kawasaki, T. (1977). 'Field desorption mass spectrometry of natural products. I. Steroid and triterpene saponins', *Tetrahedron*, **33**, 2595–2602.

Schultz, A. H. (1921). 'The occurrence of a sternal gland in orangutan', *J. Mammal.*, **2**, 194–196.

Schultze-Westrum, T. (1965). 'Innerartliche Verständigung durch Düfte beim Gleitbeutler *Petaurus breviceps papuanus* (Marsupialia, Phalangeridae), *Z. Vergl. Physiol.*, **50**, 151–220.

Sebeok, T. A. (ed.) (1977). *How do Animals Communicate?*, Indiana University Press, Bloomington and London.

Seitz, E. (1969). 'Die Bedeutung geruchlicher Orientierung beim Plumplori *Nycticebus coucang*', *Z. Tierpsychol.*, **26**, 73–103.

Senf, W., Menco, B. P. M., Punter, P. H., and Duyvesteyn, P. (1980). 'Determination of odour affinities based on the dose-response relationships of the frog's electo-olfactogram', *Experientia*, **36**, 213–215.

Seth, S. D. S., Mukhopadhyay, A., Bagchi, N., Prabhakar, M. C., and Arora, R. B. (1973). 'Antihistaminic and spasmolytic effects of musk', *Jap. J. Pharmacol.*, **23**, 673–679.

Shafer, K. H., Bjørseth, A., Tabor, J., and Jakobsen, R. J. (1980). 'Advancing the chromatography of GC/FT-IR to WCOT capillary columns', *HRC&CC*, **3**, 87–88.

Shalita, A. R. (1974). 'Genesis of free fatty acids', *J. Investig. Dermatol.*, **62**, 332–335.

Sharman, D. F. (1973). 'The catabolism of catecholamines', *Br. Med. Bull.*, **29**, 110–115.

Shelley, W. B., Hurley, H. J., and Nichols, A. C. (1953). 'Axillary odor', *Arch. Dermatol. Syphilology*, **68**, 430–446.

Shelton, M. (1960). 'Influence of the presence of a male goat on the initiation of oestrus, cycling and ovulation of Angora does', *J. Anim. Sci.*, **19**, 368–375.

Shepherd, G. M. (1963). 'Neuronal system controlling mitral cell excitability', *J. Physiol.*, **168**, 101–117.

Sheth, M. R., Mugatwala, P. P., Shah, G. V., and Rao, S. S. (1975). 'The occurrence of prolactin in human semen', *Fertil. Steril.*, **26**, 905–907.

Shirley, S. G., Polak, E., and Dodd, G. H. (1981). 'Chemical modification of the olfactory epithelium', *Biochem. Soc. Trans.*, **9**, 108–109.

Short, R. V. (1972). 'Role of hormones in sex cycles', in *Hormones in Reproduction* (eds C. R. Austin and R. V. Short), Cambridge University Press, p. 52.

Shrader, S. R. (1971). *Introductory Mass Spectrometry*, Allyn and Bacon, Inc., Boston.

Shum, A., Johnson, G. E., and Flattery, K. V. (1971). 'Catecholamines and metabolite excretion in cold-stressed immunosympathectomized rats', *Amer. J. Physiol.*, **221**, 64–68.

Shumake, S. A. (1977). 'The search for applications of chemical signals in wildlife management', in *Chemical Signals in Vertebrates* (eds D. Müller-Schwarze and M. M. Mozell), Plenum, New York, pp. 357–376.

Shuster, S., and Thody, A. J. (1974). 'The control and measurement of sebum secretion', *J. Investig. Dermatol.*, **62**, 172–190.

Signoret, J. P. (1976). 'Chemical communication and reproduction in domestic mammals', in *Mammalian Olfaction, Reproductive Processes and Behavior* (ed. R. L. Doty), Academic Press, New York, pp. 244–255.

Singer, A. G., Agosta, W. C., O'Connell, R. J., Pfaffmann, C., Bowen, D. V., and Field, F. H. (1976). 'Dimethyl disulphide; an attractant pheromone in hamster vaginal secretion', *Science*, **191**, 948–950.

Singer, A. G., Macrides, F., and Agosta, W. C. (1980). 'Chemical studies of hamster reproductive pheromones', in *Chemical Signals: Vertebrates and Aquatic Invertebrates* (eds D. Müller-Schwarze and R. M. Silverstein), Plenum Press, New York, pp. 365–375.

Singh, E. J., and Swartwout, J. R. (1972). 'Human cervical mucus lipids', *J. Reprod. Med.*, **8**, 35–40.

Singh, E. J., Swartwout, J. R., and Boss, S. (1972). 'Hydrocarbons in human cervical mucus, and the effect of oral contraception', *J. Reprod. Med.*, **8**, 128–132.

Singh, U. B., and Bharadwaj, M. B. (1978). 'Anatomical, histological and histochemical observations and changes in the poll glands of the camel, *Camelus dromedarius*', *Acta Anat.*, **102**, 74–83.

Sisson, J. K., and Fahrenbach, W. H. (1967). 'Fine structure of steroidogenic cells of a primate cutaneous organ', *Amer. J. Anat.*, **121**, 337–367.

Sisson, S. (Revised Grossman, J. D.) (1955). *The Anatomy of Domestic Animals*, W. B. Saunders and Co, Philadelphia, 4th edn.

Skeen, J. T., and Thiessen, D. D. (1977). 'Scent of gerbil cuisine', *Physiol. Behav.*, **19**, 11–14.

Skinner, F. A., and Carr, J. G. (eds) (1974). *The Normal Microbial Flora of Man*, Academic Press, New York.

Slingsby, J. M., and Boulton, A. A. (1976). 'Separation and quantitation of some urinary arylalkylamines', *J. Chromatog.*, **123**, 51–56.

Smith, A. B., Byrne, K. J., and Beauchamp, G. K. (1977). '*Syn*- and *anti*-phenylacetaldehyde oxime, two novel testosterone-dependent mammalian metabolites', *J. Chem. Ecol.*, **3**, 309–319.

Smith, I., and Lerner, R. P. (1971). 'Comparative biochemistry of the primates. II. The indole compounds of primate blood and urine', *Folia primatol.*, **14**, 110–117.

Smith, J. D., and Hearn, G. W. (1979). 'Ultrastructure of the apocrine-sebaceous anal scent gland of the woodchuck, *Marmota monax*. Evidence for apocrine and merocrine secretion by a single cell type', *Anat. Rec.*, **193**, 269–292.

Smith, L. D., and Holdeman, L. V. (1968). *The Pathogenic Anaerobic Bacteria*, C. C. Thomas, Springfield, Illinois.

Smith, M. E., and Ahmed, S. U. (1976). 'The lipid composition of cattle sebaceous glands; a comparison with skin surface lipid', *Res. Vet. Sci.*, **21**, 250–252.

Smith, R. G., Besch, P. K., Dill, B., and Buttram, V. C. (1979). 'Saliva as a matrix for measuring free androgens; comparison with sebum androgens in polycystic ovarian disease', *Fertil. Steril.*, **31**, 513–517.

Smith, T. A. (1977). 'Phenethylamine and related compounds in plants', *Phytochemistry*, **16**, 9–18.

Snyder, F., and Blank, M. L. (1969). 'Relationship of chain lengths and double bond locations in *O*-alkyl, *O*-alk-1-enyl, acyl and fatty alcohol moieties in preputial glands of mice', *Arch. Biochem. Biophys.*, **130**, 101–110.

Sokolov, V. E., Khorlina, I. M., Golovnya, R. V., and Zhuravleva, I. L. (1974). 'Change in the composition of amines in the volatile substances of the vaginal secretions of American mink (*Mustela vison*) depending on the sexual cycle', *Dokl. Akad. Nauk. SSSR (Biochem.)*, **216**, 220–222.

Sokolov, V. E., and Khorlina, I. M. (1976). 'Pheromones of mammals; study of the composition of volatile acids in vaginal discharges of minks (*Mustela vison*)', *Dokl. Akad. Nauk. SSSR. (Biochem.)*, **228**, 225–227.

Sokolov, V. E., Brundin, A., and Zinkevich, E. P. (1977). 'Differences in the chemical composition of cutaneous gland secretions from the Northern reindeer, *Rangifer tarandus*', *Dokl. Akad. Nauk. SSSR. (Ecology)*, **237**, 1529–1532.

Sokolov, V. E., Albone, E. S., Flood, P. F., Heap, P. F., Kagan, M. Z., Vasilieva, V. S., Roznov, V. V., and Zinkevich, E. P. (1980). 'Secretion and secretory tissues of the anal sac of the mink, *Mustela vison*: chemical and histological studies', *J. Chem. Ecol.*, **6**, 805–825.

Sönksen, P. H. (Ed.) (1974). 'Radioimmunoassay and saturation analysis', *Br. Med. Bull.*, **30**, 1–103.

Southern, H. N. (1948). 'Sexual and aggressive behaviour in the wild rabbit', *Behaviour*, **1**, 173–194.

Southwell, I. A. (1975). 'Essential oil metabolism in the koala. III. Novel urinary monoterpenoid lactones', *Tet. Lett.*, 1885–1888.

Spannhof, I. (1969). 'The histophysiology and function of the anal sac of the red fox, *Vulpes vulpes*', *Forma et functio*, **1**, 26–45.

Spener, F., Mangold, H. K., Sansone, G., and Hamilton, J. G. (1969). 'Long-chain alkyl acetates in the preputial gland of the mouse', *Biochim. Biophys. Acta*, **192**, 516–521.

Spickett, A. M., Keirans, J. E., Norval, R. A. I., and Clifford, C. M. (1981). 'Ixodes (Afrixodes) metopi n. sp. (Acarina: Ixodidae): A tick found aggregating on preorbital gland scent marks of the Klipspringer in Zimbabwe', *Onderstepoort J. Vet. Res.*, **48**, 23–30.

Spindel, E., Pettibone, D., Fisher, L., Fernstrom, J., and Wurtman, R. (1981). 'Characterization of neuropeptides by reversed-phase ion-pair liquid chromatography with post column detection by radioimmunoassay', *J. Chromatog.*, **222**, 381–387.

Stacewicz-Sapuntzakis, M., and Gawienowski, A. M. (1977). 'Rat olfactory response to aliphatic acetates', *J. Chem. Ecol.*, **3**, 411–417.

Stack, M. V. (1979). 'Application of gas chromatography in dental research', *J. Chromatog.*, **165**, 103–116.

Stafford, M., Horning, M. G., and Zlatkis, A. (1976). 'Profiles of volatile metabolites in body fluids', *J. Chromatog.*, **126**, 495–502.

Stahl, E. (ed.) (1969). *Thin Layer Chromatography; a Laboratory Handbook.* Springer-Verlag, Berlin, 2nd edn.

Stahl, W. H. (ed.) (1973). Compilation of odor and taste threshold data values. *ASTM Data Series DS 48*, American Society for Testing and Materials, Philadelphia.

Ställberg-Stenhagen, S. (1972). 'Studies on natural odoriferous compounds. V. Splitter-free all glass intake systems for glass capillary gas chromatography of volatile compounds from biological material', *Chemica Scripta*, **2**, 97–100.

Stanley, G., and Kennett, B. H. (1973). 'Reaction gas chromatography of microgram and sub-microgram samples using sealed glass capillaries', *J. Chromatog.*, **75**, 304–307.

Stehn, R. A., and Richmond, M. E. (1975). 'Male-induced pregnancy termination in the prairie vole, *Microtus ochrogaster*', *Science*, **187**, 1211–1213.

Stein, W. H. (1953). 'A chromatographic investigation of the amino acid constituents of normal urine', *J. Biol. Chem.*, **201**, 45–58.

Stevens, P. G. (1945). 'American musk IV. On the biological origin of animal musk. Two more large ring ketones from the muskrat', *J. Amer. Chem. Soc.*, **67**, 907–909.

Stewart, R. E. A., and Brooks, R. J. (1976). 'The perineal gland of brown lemmings, *Lemmus trimucronatus*', *Can. J. Zool.*, **54**, 1013–1018.

Still, W. C., Kahn, M., and Mitra, A. (1978). 'Rapid chromatographic technique for preparative separations with moderate resolution', *J. Org. Chem.*, **43**, 2923–2925.

Stinson, C. G., and Patterson, R. L. S. (1972). 'C_{19}-Δ^{16} steroids in boar sweat glands', *Br. Vet. J.*, **128**, xli–xliii.

Stoddart, D. M. (1973). 'Preliminary characterization of the caudal organ secretion of *Apodemus flavicollis*', *Nature*, **246**, 501–503.

Stoddart, D. M. (1977). 'Two hypotheses supporting the social function of odor secretions of some old world rodents', in *Chemical Signals in Vertebrates* (eds D. Müller-Schwarze and M. M. Mozell), Plenum, New York, pp. 333–355.

Stoddart, D. M. (1979). 'A specialized scent-releasing hair in the crested rat, *Lophiomys imhausi*', *J. Zool.*, **189**, 551–553.

Stoddart, D. M. (1980). *The Ecology of Vertebrate Olfaction*, Chapman and Hall, London and New York.

Stoddart, D. M., Alpin, R. T., and Wood, M. J. (1975). 'Evidence for social difference in the flank organ secretion of *Arvicola terrestris*', *J. Zool.*, **177**, 529–540.

Stott, A. W., Lindsay Smith, J. R., Hanson, P., and Robinson, R. (1975). 'A simple chromatographic procedure for the concurrent estimation of urinary 4-hydroxy-3-methoxymandelic acid (HMMA) and homovanillic acid (HVA) using a scanning technique', *Clin. Chim. Acta*, **63**, 7–12.

Strauss, J. S., and Pochi, P. E. (1961). 'The quantitative gravimetric determination of sebum production', *J. Investig. Dermatol.*, **36**, 293–298.

Strauss, J. S., and Ebling, F. J. (1970). 'Control and function of skin glands in mammals', in *Hormones and Environment: Mem. Soc. Endocrinol. No. 18* (eds G. K. Benson and J. G. Phillips), Cambridge University Press, pp. 341–371.

Stubbe, M. (1970). 'Zur Evolution der analen Markierungsorgane bei Musteliden', *Biol. Zbl.*, **89**, 213–223.

Suemitsu, R., Yoshikawa, T., Tanaka, M., and Akatsuchi, T. (1974). 'The structures of two new glycine conjugated compounds from cattle urine', *Agr. Biol. Chem.*, **38**, 885–886.

Sugiyama, T., Sasada, H., Masaki, J., and Yamashita, K. (1981). 'Unusual fatty acids with specific odor from mature male goat', *Agr. Biol. Chem.*, **45**, 2655–2658.

Sully, B. D. (1971). 'The analysis of odoriferous vapours, including head space analysis', *J. Soc. Cosmet. Chem.*, **22**, 3–14.

Svendsen, G. E. (1978). 'Castor and anal glands of the beaver, *Castor canadensis*', *J. Mammal.*, **59**, 618–620.

Svendsen, G. E. (1979). 'Territoriality and behavior in a population of pikas (*Ochotona princeps*)', *J. Mammal.*, **60**, 324–330.

Svendsen, G. E. (1980). 'Patterns of scent-mounding in a population of beaver, *Castor canadensis*', *J. Chem. Ecol.*, **6**, 133–148.

Svendsen, G. E., and Jollick, J. D. (1978). 'Bacterial contents of the anal and castor glands of beaver (*Castor canadensis*)', *J. Chem. Ecol.*, **4**, 563–569.

Taber, A. B., and Macdonald, D. W. (1984). 'Scent dispensing papillae and associated behaviour of the mara, *Dolichotis patagonum*', *J. Zool.* (in press).

Tachibana, D. K. (1976). 'Microbiology of the foot', *Ann. Rev. Microbiol.*, **30**, 351–375.

Takagi, K., and Sakurai, T. (1970). 'A sweat reflex due to pressure on the body surface', *Jap. J. Physiol.*, **1**, 22–28.

Takayasu, S., Wakimoto, H., Itami, S., and Sano, S. (1980). 'Activity of testosterone 5α-reductase in various tissues of human skin', *J. Investig. Dermatol.*, **74**, 187–191.

Tallan, H. H., Moore, S., and Stein, W. H. (1954). 'Studies on the free amino acids and related compounds in the tissues of the cat', *J. Biol. Chem.*, **211**, 927–939.

Talman, E., Verwiel, P. E. J., and Lakwijk, A. C. (1977). 'Reaction GC-MS in Biochemistry', *Adv. Mass Spec. Biochem. Med.*, **2**, 215–229.

Tandy, J. M. (1976). 'Communication in *Galago crassicaudatus*', *Primates*, **17**, 513–526.

Taylor, C. R. (1966). 'The vascularity and possible thermoregulatory function of the horns in goats', *Physiol. Zool.*, **39**, 127–139.

Taylor, P. L., and Kelly, R. W. (1974). '19-Hydroxylated E prostaglandins as the major prostaglandins of human semen', *Nature*, **250**, 665–667.

Teague, L. G., and Bradley, E. L. (1978). 'The existence of puberty accelerating pheromone in the urine of the male prairie deermouse (*Peromyscus maniculatus bairdii*)', *Biol. Reprod.*, **19**, 314–317.

Teranishi, R., Issenberg, P., Hornstein, I., and Wick, E. L. (1971). *Flavor Research: Principles and Techniques*, Marcell Dekker, Inc., New York.

Teranishi, R., Mon, T. E., Robinson, A. B., Cary, P., and Pauling, L. (1972). 'Gas chromatography of volatiles from breath and urine', *Anal. Chem.*, **44**, 18–20.

Theimer, E. T. (ed.) (1982). *'Fragrance Chemistry'*, Academic Press, New York.

Thiessen, D. D. (1973). 'Footholds for survival', *Amer. Scientiol.*, **61**, 346–351.

Thiessen, D. D. (1977a). 'Methodology and strategies in the laboratory', in *Chemical Signals in Vertebrates* (eds D. Müller-Schwarze and M. M. Mozell), Plenum, New York, pp. 391–412.

Thiessen, D. D. (1977b). 'Thermoenergetics and the evolution of pheromone communication', in *Progress in Psychobiology and Physiological Psychology*, Vol. 7 (eds J. M. Sprague and A. N. Epstein), Academic Press, New York, pp. 91–191.

Thiessen, D. D., Blum, S. L., and Lindzey, G. (1969). 'The scent-marking response associated with the ventral sebaceous gland of the mongolian gerbil, *Meriones unguiculatus*', *Anim. Behav.*, **18**, 26–30.

Thiessen, D. D., Regnier, F. E., Rice, M., Goodwin, M., Isaacks, N., and Lawson, N. (1974). 'Identification of ventral scent marking pheromone in the male mongolian gerbil, *Meriones unguiculatus*', *Science*, **184**, 83–85.

Thiessen, D. D., Clancy, A., and Goodwin, M. (1976). 'Harderian gland pheromone in the mongolian gerbil, *Meriones unguiculatus*', *J. Chem. Ecol.*, **2**, 231–238.

Thiessen, D., and Rice, M. (1976). 'Mammalian scent marking and social behavior', *Psychol. Rev.*, **83**, 505–539.

Thiessen, D. D., Graham, M., Perkins, J., and Marcks, S. (1977). 'Temperature regulation and social grooming in the mongolian gerbil, *Meriones unguiculatus*', *Behav. Biol.*, **19**, 279–288.

Thiessen, D. D., and Kittrell, E. M. W. (1980). 'The Harderian gland and thermoregulation in the gerbil, *Meriones unguiculatus*', *Physiol. Behav.*, **24**, 417–424.

Thody, A. J., and Shuster, S. (1975). 'Control of sebaceous gland function in the rat by α-MSH', *J. Endocrinol.*, **64**, 503–510.

Thomas, B. S. (1980). 'Steroid analysis by gas chromatography with SCOT and wide bore WCOT columns', *HRC&CC*, **3**, 241–247.

Thompson, J. A., Markey, S. P., and Fennessey, P. W. (1975). 'Gas chromatographic/ mass spectrometic identification and quantitation of tetronic and deoxytetronic acids in urine from normal adults and neonates', *Clin. Chem.*, **21**, 1892–1898.

Tiedeman, G. T., Oh, J. H., Oita, K., and Christoffers, G. V. (1976). 'Wildlife Damage Control II: Partial identification of the active ingredients in big game repellents derived from fresh fish and eggs', *Amer. Chem. Soc. 172nd Natl. Meeting*, San Francisco.

Todd, N. B. (1962). 'Inheritance of the catnip response in domestic cats', *J. Heredity*, **53**, 54–56.

Todd, N. B. (1963). *The Catnip Response*, Ph.D. Thesis, Harvard University.

Tomio, J. M., and Grinstein, M. (1968). 'Porphyrin biosynthesis: Biosynthesis of protoporphyrin IX in Harderian glands', *Europ. J. Biochem.*, **6**, 80–83.

Tonzetich, J. (1977). 'Production and origin of oral malodor: a review of mechanisms and methods of analysis', *J. Periodontology*, **48**, 13–20.

Tonzetich, J., Preti, G., and Huggins, G. R. (1978). 'Changes in concentration of volatile sulfur compounds of mouth air during the menstrual cycle', *J. Int. Med. Res.*, **6**, 245–254.

Touchstone, J. C., and Dobbins, M. F. (1978). *Practice of Thin Layer Chromatography*, Wiley-Interscience, Chichester and New York.

Tressl, R., Holzer, M., and Apetz, M. (1975). 'Biogenesis of volatiles in fruit and vegetables' in *Aroma Research* (eds H. Maarse and P. J. Groenen), Pudoc, Wageningen, Netherlands, pp. 41–62.

Troyer, K. (1982). 'Transfer of fermentative microbes between generations in a herbivorous lizard', *Science*, **216**, 540–542.

Tucker, D. (1962/1963). 'Physical variables in the olfactory stimulation process', *J. Gen. Physiol.*, **46**, 453–489.

Tumlinson, J. H., Moser, J. C., Silverstein, R. M., Brownlee, R. G., and Ruth, J. M. (1972). 'A volatile trail pheromone of a leaf-cutting ant, *Atta texana*', *J. Insect. Physiol.*, **18**, 809–814.

Tumlinson, J. H., and Heath, R. R. (1976). 'Structure elucidation of insect pheromones by microanalytical methods', *J. Chem. Ecol.*, **2**, 87–99.

Udenfriend, S., Lovenberg, W., and Sjoerdsma, A. (1959). 'Physiologically active amines in common plants and vegetables', *Arch. Biochem. Biophys.*, **85**, 487–490.

Ullrich, W. (1954). 'Zur Frage des Sichselbstbespukens bei Säugetieren', *Z. Tierpsychol.*, **11**, 50.

Vale, G. A. (1979). 'Field responses of tsetse flies (Diptera: Glossinidae) to odours of men, lactic acid and carbon dioxide', *Bull. Ent. Res.*, **69**, 459–467.

Vale, G. A., and Hargrove, J. W. (1975). 'Field attraction of tsetse flies (Diptera: Glossinidae) to ox odour; the effects of dose', *Trans. Rhod. Scient. Ass.*, **56**, 46–50.

Valenta, Z., and Khaleque, A. (1959). 'The structure of castoramine', *Tet. Lett.*, 1–5.

Vandenbergh, J. G. (1973). 'Acceleration and inhibition of puberty in female mice by pheromones', *J. Reprod. Fert., Suppl.*, **19**, 411–419.

Vandenbergh, J. G., and Drickamer, L. C. (1974). 'Reproductive coordination among free-ranging rhesus monkeys', *Physiol. Behav.*, **13**, 373–376.

Vandenbergh, J. G., Whitsett, J. M., and Lombardi, J. R. (1975). 'Partial isolation of a pheromone accelerating puberty in female mice', *J. Reprod. Fert.*, **43**, 515–523.

Vandenbergh, J. G., Finlayson, J. S., Dobrogosz, W. J., Dills, S. S., and Kost, T. A. (1976). 'Chromatographic separation of puberty accelerating pheromone from male mouse urine', *Biol. Reprod.*, **15**, 260–265.

van der Pijl, L. (1961). 'Ecological aspects of flower evolution', *Evolution*, **15**, 44–59.

Van Dorp, D. A., Klok, R., and Nugteren, D. H. (1973). 'New macrocyclic compounds from the secretion of the civet cat and the musk rat', *Recl. Trav. Chim., Pays-Bas Belg.*, **92**, 915–924.

van Heerden, J. (1981). 'The role of integumental glands in the social and mating behaviour of the hunting dog, *Lycaon pictus*', *Onderstepoort J. Vet. Res.*, **48**, 19–21.

Vaughan, J. A., and Mead-Briggs, A. R. (1970). 'Host-finding behaviour of the rabbit flea, *Spilopsyllus cuniculi*, with special reference to the significance of urine as an attractant', *Parasitology*, **61**, 397–409.

Veith, H. J. (1976). 'Li$^+$-Attachment — a soft ionization method in field desorption mass spectrometry', *Angew. Chem. Int. Ed. Engl.*, **15**, 696.

Verberne, G. (1976). 'Chemocommunication among domestic cats, mediated by the olfactory and vomeronasal senses. II. The relations between the functions of

Jacobson's organ (vomeronasal organ) and Flehmen behaviour', *Z. Tierpsychol.*, **42**, 113–128.

Verberne, G., and De Boer, J. (1976). 'Chemocommunication among domestic cats mediated by the olfactory and vomeronasal senses. I. Chemocommunication', *Z. Tierpsychol.*, **42**, 86–109.

Vernet-Maury, E. (1980). 'Trimethyl-thiazoline in fox faeces: a natural alarming substance for the rat', in *Olfaction and Taste VII*, (ed. H. van der Starre), IRL Press, London, p. 407.

Verzele, M. (1979). 'Surface modification in glass capillary gas chromatography', *HRC&CC*, **2**, 647–653.

Verzele, M., and Sandra, P. (1979). 'Advances in the preparation and evaluation of GC capillary columns', *HRC&CC*, **2**, 303–311.

Verzele, M., and Geeraert, E. (1980). 'Preparative liquid chromatography', *J. Chromatog. Sci.*, **18**, 559–570.

Voeltler, W., Jung, G., Breitmaier, E., König, W., Gupta, D., and Breitmaier, G. (1971). 'Isolierung von Oxindol aus Urin von Kindern', *Z. Naturforsch.*, **26B**, 1380–1381.

Voigt, D., Schreiber, K., and Adam, G. (1977). 'Comparative anion and cation mass spectrometry by low energy ionization of selected natural products', in *Advances in Mass Spectrometry in Biochemistry and Medicine*, Vol. II, (ed. A. Frigerio), Spectrum Publications, New York, pp. 183–190.

Wada, M. (1950). 'Sudorific action of adrenaline on the human sweat glands and determination of their excitability', *Science*, **111**, 376–377.

Waites, G. M. H., and Voglmayr, J. K. (1963). 'The functional activity and control of apocrine sweat glands of the scrotum of the ram', *Aust. J. Agric. Res.*, **14**, 839–851.

Waller, G. R. (ed.) (1972). *Biochemical applications of Mass Spectrometry*, Wiley Interscience, New York.

Waller, G. R., and Dermer, O. C. (eds) (1980). *Biochemical Applications of Mass Spectrometry, First Supplimentary Volume*, John Wiley and Sons, New York.

Walro, J. M., and Svendsen, G. E. (1982). 'Castor sacs and anal glands of the North American Beaver (*Castor canadensis*). Their histology, development and relationship to scent communication', *J. Chem. Ecol.*, **8**, 809–819.

Ward, M. E., Politzer, I. R., Laseter, J. L., and Alam, S. Q. (1976). 'Gas chromatographic mass sprectrometric evaluation of free organic acids in human saliva', *Biomed. Mass Spectrom.*, **3**, 77–80.

Ward, R. J., and Moore, T. (1953). '7-Dehydrosterol in the sexual organs of the rat', *Biochem. J.*, **55**, 295–298.

Ware, G. C., and Gosden, P. E. (1980). 'Anaerobic microflora of the anal sac of the red fox, *Vulpes vulpes*', *J. Chem. Ecol.*, **6**, 97–102.

Watts, R. W. E., Chalmers, R. A., and Lawson, A. M. (1975). 'Abnormal organic acidureas in mentally retarded parients', *Lancet*, **1**, 368–372.

Weldon, P. J. (1980). 'In defense of "Kairomone" as a class of chemical releasing stimuli', *J. Chem. Ecol.*, **6**, 719–725.

Wellington, J. L., Byrne, K. J., Preti, G., Beauchamp, G. K., and Smith, A. B. (1979). 'Perineal scent gland of wild and domestic guinea pigs. A comparative chemical and behavioural study', *J. Chem. Ecol.*, **5**, 737–751.

Wemmer, C., and Murtaugh, J. (1980). 'Olfactory aspects of rutting behaviour in the bacterian camel, *Camelus bactrianus ferus*', in *Chemical Signals: Vertebrates and Aquatic Invertebrates* (eds D. Müller-Schwarze and R. M. Silverstein), Plenum, New York, pp. 107–124.

Westall, R. G. (1953). 'The amino acids and other ampholytes of urine. 2. The isolation of a new sulphur-containing amino acid from cat urine', *Biochem. J.*, **55**, 244–247.

Wheatley, V. R., and James, A. T. (1957). 'The composition of the sebum of some common rodents', *Biochem. J.*, **65**, 36–42.

Wheeler, J. W., Von Endt, D. W., and Wemmer, C. (1975). '5-Thiomethylpentane-2,3-dione. A unique natural product from the striped hyaena', *J. Amer. Chem. Soc.*, **97**, 441.

Wheeler, J. W., Blum, M. S., and Clark, A. (1977). 'ß-(*p*-Hydroxyphenyl)ethanol in the chest-gland secretion of a galago (*Galago crassicaudatus*)', *Experientia*, **33**, 988–989.

Wheeler, J. W., Rasmussen, L. E., Ayorinde, F., Buss, I. O., and Smuts, G. L. (1982). 'Constituents of the temporal gland secretion of the African elephant, *Loxodonta africana*', *J. Chem. Ecol.*, **8**, 821–835.

Whissell-Buechy, D., and Amoore, J. E. (1973). 'Odour-blindness to musk: simple recessive inheritance', *Nature*, **242**, 271–273.

White, R. H. (1975). 'Occurrence of *S*-methylthioesters in urine of humans after they have eaten asparagus', *Science*, **189**, 810–811.

Whitten, W. K. (1959). 'Occurrence of anoestrus in mice caged in groups', *J. Endocrinol.*, **18**, 102–107.

Whitten, W. F., Bronson, F. H., and Greenstein, J. A. (1968). 'Estrus-inducing pheromone of male mice: transport by movement of air', *Science*, **161**, 584–585.

Wiener, H. (1966/1967). 'External chemical messengers', *N.Y. State J. Med.*, **66**, 3153–3170; **67**, 1144–1165; **67**, 1287–1310.

Wilkinson, D. I. (1970). 'Monounsaturated fatty acids of mouse skin surface lipids', *Lipids*, **5**, 148–149.

Wilkinson, D. I., and Karasek, M. A. (1966). 'Skin lipids of a normal and a mutant (asebic) mouse strain', *J. Investig. Dermatol.*, **47**, 449–455.

Williams, D. H., and Howe, I. (1972). *Principles of Organic Mass Spectrometry*, McGraw-Hill, New York.

Williams, H. J., and Vinson, S. B. (1980). 'All-glass system for preparative gas chromatography using capillary columns', *J. Chem. Ecol.*, **6**, 973–978.

Williams-Ashman, H. G., and Lockwood, D. H. (1970). 'Role of polyamines in reproductive physiology and sex hormone action', *Ann. N.Y. Acad. Sci.*, **171**, 882–894.

Wilson, J. D., and Gloyna, R. E. (1970). 'The intranuclear metabolism of testosterone in the accessory organs of reproduction', *Rec. Progr. Horm. Res.*, **26**, 309–330.

Wilz, I. (1967). 'Die Konstitution der Geruchskomponenten des Analsekretes von Nerzen', Doctoral Thesis, University of Heidelberg.

Winars, S. S., and Powers, J. B. (1977). 'Olfactory and vomeronasal differentiation of male hamsters: histological and behavioural analysis', *Brain Res.*, **126**, 325–344.

Winkel, P., and Slob, A. (1973). 'Catecholamine excretion of normal male adolescents during various periods of the day cycle', *Clin. Chim. Acta*, **45**, 113–118.

Wislocki, G. B. (1930). 'A study of the scent glands in the marmosets, especially *Oedipomidas geoffroyi*', *J. Mammal.*, **11**, 475–483.

Wislocki, G. B., and Schultz, A. H. (1925). 'On the nature of modifications of the skin in the sternal region of certain primates', *J. Mammal.*, **6**, 236–244.

Wood, W. F., Leahy, M. G., Galun, R., Prestwich, G. D., Meinwald, J., Purnell, R. E., and Payne, R. C. (1975). 'Phenols as pheromones of ixodid ticks: a general phenomenon?', *J. Chem. Ecol.*, **1**, 501–509.

Wysocki, C. J., Wellington, J. L., and Beauchamp, G. K. (1980). 'Access of urinary nonvolatiles to the mammalian vomeronasal organ', *Science*, **207**, 781–783.

Yahr, P. (1977a). 'Central control of scent marking' in *Chemical Signals in Vertebrates* (eds D. Müller-Schwarze and M. M. Mozell), Plenum, New York, pp. 547–562.

Yahr, P. (1977b). 'Social subordination and scent-marking in male mongolian gerbils, *Meriones unguiculatus*', *Anim. Behav.*, **25**, 292–297.

348

Yahr, P., Jackson, J. C., Newman, A., Stephens, D. R., and Clancy, A. N. (1980). 'Paradigm for comparing sexual behavior and scent marking in male gerbils', *Physiol. Behav.*, **24**, 263–266.

Yamazaki, K., Yamaguchi, M., Boyse, E. A., and Thomas, L. (1980). 'The major histocompatibility complex as a source of odors imparting individuality among mice', in *Chemical Signals: Vertebrates and Aquatic Invertebrates* (eds D. Müller-Schwarze and R. M. Silverstein), Plenum, New York, pp. 267–273.

Yarger, R. G., Smith, A. B., Preti, G., and Epple, G. (1977). 'The major volatile constituents of the scent mark of a South American primate, *Saguinus fuscicollis*, Callithricidae', *J. Chem. Ecol.*, **3**, 45–56.

Yasuda, K., Aoki, T., and Montagna, W. (1961). 'The skin of primates. IV. The skin of lesser bushbaby (*Galago senegalensis*)', *Amer. J. Phys. Anthropol.*, **19**, 23–33.

Young, J. C., and Silverstein, R. M. (1975). 'Biological and chemical methodology in the study of insect communication', in *Methods in Olfactory Research* (eds D. G. Moulton, A. Turk and J. W. Johnson), Academic Press, New York, pp. 75–161.

Young, J. F. (1973). 'The fluorescence of normal urine', Ph.D. Thesis, University of Florida.

Yuasa, S. (1978). 'Isolation and structure determination of a new amino acid, α-amino-γ,δ-dihydroxyadipic acid, from the hydrolysate of normal human urine', *Biochim. Biophys. Acta*, **540**, 93–100.

Yun, J. S., and Montagna, W. (1964). The skin of primates. XX. Development of appendages in *Lemur catta* and *Lemur fulvus*', *Amer. J. Phys. Anthropol.*, **22**, 399–405.

Zlatkis, A., and Liebich, H. M. (1971). 'Profile of volatile metabolites in human urine', *Clin. Chem.*, **17**, 592–594.

Zlatkis, A., and Pretorius, V. (eds) (1971). *Preparative Gas Chromatography*, Wiley Interscience, New York.

Zlatkis, A., Bertsch, W., Lichtenstein, H. A., Tishbee, A., Shunbo, F., Liebich, H. M., Coscia, A. M., and Fleischer, N. (1973a). 'Profile of volatile metabolites in urine by gas chromatography-mass spectrometry', *Anal. Chem.*, **45**, 763–767.

Zlatkis, A., Lichtenstein, H. A., Tishbee, A., Bertsch, W., Shunbo, F., and Liebich, H. M. (1973b). 'Concentration and analysis of volatile urinary metabolites', *J. Chromatog. Sci.*, **11**, 299–302.

Zlatkis, A., Lichtenstein, H. A., and Tishbee, A. (1973c). 'Concentration and analysis of trace volatile organics in gases and biological fluids with a new solid adsorbent', *Chromatographia*, **6**, 67–70.

Zlatkis, A., Bertsch, W., Bafus, D. A., and Liebich, H. M. (1974). 'Analysis of trace volatile metabolites in serum and plasma', *J. Chromatog.*, **91**, 379–383.

Zlatkis, A., and Andrawes, F. (1975). 'A micromethod for the determination of volatile metabolites in biological samples', *J. Chromatog.*, **112**, 533–538.

Zlatkis, A., and Kim, K. (1976). 'Column elution and concentration of volatile compounds in biological fluids', *J. Chromatog.*, **126**, 475–485.

Zlatkis, A., and Kaiser, R. E. (eds) (1977). *HPTLC, High Performance Thin-layer Chromatography*, Elsevier, Amsterdam/Institute of Chromatography, Bad Dürkheim.

Zlatkis, A., Lee, K. Y., Poole, C. F., and Holzer, G. (1979a). 'Capillary column gas chromatographic profile analysis of volatile compounds in sera of normal and virus-infected patients', *J. Chromatog.*, **163**, 125–133.

Zlatkis, A., Poole, C. F., Brazell, R., Lee, K. Y., and Singhawangcha, S. (1979b). 'Profiles of volatile metabolites in biological fluids using capillary columns', *HRC&CC*, **2**, 423–428.

Zlatkis, A., and Shanfield, H. (1979). 'Concentration techniques for volatile samples', in *Practical Mass Spectrometry* (ed. B. S. Middleditch), Plenum, New York, pp. 151–160.

Zomzely-Neurath, C., and Keller, A. (1977). 'Nervous system-specific proteins of vertebrates. A search for function and physiological roles', *Neurochem. Res.*, **2**, 353–378.

Zweig, J. S., Roman, R., Hagerman, W. B., and Van den Heuvel, W. J. A. (1980). 'Gas chromatographic analysis of estrogens from pregnant mare's urine using glass capillary columns', *HRC&CC*, **3**, 169–171.

Index

The names of compounds have not been uniformly systematized and reflect the diversity of current usage across much of the biological literature. In the Index, compounds are generally referred to by classes and only a relatively few substances are individually listed.

Accessory sex organs, female 226–228
  male 210–211
Acetates, in scent glands 81, 108,
  217–218, 223
Acetonitrile, from man 32, 236
Acid phosphatase, urinary 212
Acids, 5-aminovaleric 145, 148–149,
  235
  aromatic 148–149, 160, 230–231,
    236, 242
    in urine 188–189, 191–194
    (*see also* Acids, phenolic and
        Acids, phenylacetic)
  citric 215
  fatty, biosynthesis 72–73
  fatty, free long-chain 48, 60–63, 82,
    91–93, 95, 133, 149, 218–219,
    236
  fatty, moieties, families in sebum
    65–66
  fatty, in urine 177, 187–191,
    192–193
  fatty, volatile 36, 49–50, 72–73,
    82–84, 87, 91–93, 95–96, 99,
    105–106, 115, 137, 142–148,
    158, 192–193, 228–233, 242, 259
  gymnemic 251, 267
  2-hydroxycarboxylic 63–64, 149
  lactic 32, 49, 92, 190, 215, 230–231,
    236
  phenolic 149, 160, 230–231, 233,
    236
    in urine 189, 192–193, 199,
      201–202, 204
  phenylacetic 108, 148–149,
    187–188, 192–193, 234, 236,
    242
  uric 49, 175

Actinidia polygama 116, 118
Adsorbents 23–29, 31, 33
Affinity labelling 267
Agnosterol 69–70
Alces alces 77, 168
Alcohols, aromatic 36, 81, 99, 137,
  160, 236
Aldehydes 16, 27, 31–32, 36, 82–83,
  88, 97–99, 145, 155, 160, 162, 177,
  180, 231, 259
Alkanediols 32, 88, 177, 231
  moieties in sebum 63–64, 73
Alkanols 16, 27, 32, 36, 81–82, 87, 91,
  99, 105, 107, 127, 137, 177, 231,
  234
  moieties, families in sebum 64, 68,
    73, 79, 127
Alkyl glyceryl ethers 114–115, 149,
  218
Allantoin 175
Allelochemic 4
Allomone 4
Amines 32, 145, 149–150, 160, 228,
  232
  di- and poly- 149–150, 205–206,
    213, 228, 235
  phenylethyl- 150, 197–199
  trimethyl 145–146, 149, 212–213,
    259
  in urine 196–207
Amino acids 49, 145, 148–149, 235
  in urine 194–196
Ammonia 32, 50, 105, 149, 158, 175,
  240
Ampullae 211
Anal pocket/pouch 105–107, 115, 147,
  150–151
Anal sac 30, 115, 137–162

5α-Androst-16-en-3α-ol (androstenol) 114, 131–133, 186, 237–241
5α-Androst-16-en-3-one (androstenone) 93, 131–132, 186, 221, 237–241. 259–260
Angwantibo 121, 134
Anosmia, specific 213, 258–259
Anosmic 166, 229, 233
Ant, leaf-cutting 5
Antelope, klipspringer 91
Antidorcas marsupialis 88–89
Antilocapra americana 79, 86–89
Aotus trivirgatus 128
Apocrine glands 46–48
distribution in man 41, 131
effects of hormones 47, 51–54
Apocrine secretion, composition 48, 132–133
Apocrine sweating and emotional stress 48, 130
and thermoregulation 47
Apodemus flavicollis 107
Aquatic mammals 17, 44, 106–107, 160–162, 218–219, 222
Arctocebus calabarensis 121, 134
Aroma character impact compounds 35
Aroma value 37–39
Artiodactyla 74–93
Arvicola 107
Atta texana 5
Attraction to signal 92, 106, 112, 126, 144–146, 162–163, 165, 178, 222–224, 233
Axillary organ 26, 128–132
Axon 253

Baboon 61–62, 221
Badger 151
Balloon chamber 100–101
Bat 4, 213, 253
Bear 210
Beaver 160–162
Behaviour and chemical signals, aggressive 14, 93, 95–96, 105, 109, 112, 125, 127, 138, 150, 166, 201, 224–226, 233, 238
alarm 77, 85, 89, 121, 144, 164
appeasement 150
fear 164–165, 200
threat 77–78, 80, 83, 87
(*see also* Attraction, Confidence, Distress signalling, Mother–young interaction, Scent marking)

Behavioural assay 5–6, 77, 84–85, 87, 100–102, 127, 133, 229–230
Bile 164, 209
Bison 168
Blesbok 90
Bloch pathway 68, 217
Blood volatiles 26
'Boar Mate' 238
Body temperature and chemical communication 78, 83, 110–111
Bontebok 89–90
Bovid, tsetse fly attractants 91–92
Bovids, domesticated, *see* Cattle, Goat, Sheep
Breast milk volatiles 27, 29
Breast odour 41, 130
Breath volatiles, 23, 26, 235
Bruce effect 14, 172
Bulbourethral gland 211
Bushbabies 121–122, 167
Butyrate esters 99, 127

Caecum 139, 163–164
Callimico goeldii 126
Callithricidae 125–128
Callithrix jacchus 127
Camels 92–93, 168, 236
Camelus bactrianus 92–93, 168, 236
Camelus dromedarius 92–93
Canids 118, 138
Canis familiaris 30, 43, 50. 61–62, 64, 118, 137, 144, 147–149, 166, 210, 212, 215, 221, 233, 275
Canis latrans 30, 144, 192, 275
Capillary columns 26, 29, 281–283
Capreolus capreolus 53–54, 77, 162
Caprini 168
Capybara 106–107
Carbon dioxide 92
Caribou 79, 81
Carnivores 115–118, 137–139, 210
Carnosine 248
Carotenoid 118, 151
Carrion 144–146
Castor canadensis 160–161
Castor fiber 161–162
Castor sac 160–161
Castoramine 160
Castoreum 160–161
Cat, domestic 50, 61–62, 64, 115–116, 138–139, 144, 150, 183, 211, 221, 253
Catecholamines 199–204

Catmint (catnip) response   116–117
Cattle   17, 37, 61–62, 64, 91–92,
    133–134, 173, 192–193, 210,
    212–215, 221, 226, 229, 234
Cavia aperea   105–106
Cavia porcellus   13, 46, 50, 61–62, 64,
    105–106, 178–179, 194, 221, 252
Cebidae   128, 167, 236
Cebuella   127
Cephalophus maxwellii   80
Cercopithecoidea   128–129
Cerumen   41
Cervical mucus   173, 227
Cervus   162. 168
Cervus elaphus   76, 193–194
Chamois   168
Chemical analytical approach   10–13,
    278–279
Chemical communication   1
Chemical ecology   1
Chemical image strategy   6–7, 13
Chemical ionization   291–294
Chemical knowledge, total   11–13
Chemical signal, visual   44, 78
Chemoreception   243–277
Chenopodium vulvaria   146
Chimpanzee   48, 61–62, 129
Cholesterol, biosynthetic intermediates
    in skin lipids   68–71
Choline   212
Chromosorb   24–26, 178
Chrysocyon brachyurus   144
Civet, African   159
Civetone   159
Clitoral glands (rodent)   216, 224
Coagulating gland   210–211, 213,
    222–223
Cold exposure and urine chemistry
    203–204
Column efficiency   280–281
Confidence modification and chemical
    signals   101–102, 225
Conjugates   132, 161, 184, 186–187,
    193–194, 201, 240–241
Convergence in olfactory system   248
'Copulins'   229
Cowper's glands   211
Coyote   30, 144, 192, 275
Coypu   103
Creatinine   49, 175, 216
Cribriform plate   243–244
Crocuta crocuta   150
Cuis   59
Cynomys ludovicianus   104

Damaliscus dorcas   89–90
Deer, black-tailed   13, 15, 75–79, 162,
    193
    marsh   79
    mule   75–79, 162, 168
    musk   76, 85–86
    red   76, 162, 168, 193–194
    roe   53–54, 77, 162
    white-tailed   79
Deermouse, prairie   171
Defensive chemicals   2, 138, 153
Dendrite   253, 258
Dermis   44
Desquamation   42–43, 136, 140
Detector, gas chromatographic   282, 284
Diamine oxidase   212–213
Dicrostonyx groenlandicus   203–204
Diet, effects of   60, 108, 161, 174–176,
    178–179, 187–188, 191, 196
Dihydrothiazoles   182
Dimethyl disulphide   99, 178, 224,
    233–234
Distress signalling   78
Dodecenolide   77–78, 90, 193
Dog, African hunting   13, 40, 118, 216
    bush   144
    domestic   30, 43, 50, 61–62, 64, 118,
    137, 144, 147–149, 166, 210,
    212, 215, 221, 233, 275
Dolichotis patagonum   103–104
Dopamine   198–199
Drill   129
Duiker, Maxwell's   80

Eccrine glands   48–50
    gland distribution in man   41, 48, 130
    sweat, composition   48–50
    sweat, visualization   48
Ejaculate derived odour   228, 230
Electron impact source   290–291
Electro-olfactogram (EOG)   261–266,
    275
Elephants   13–14, 102–103
Elephas maximus   13–14, 102–103
Emotion correlates   48–49, 130, 184,
    196–197, 200–203
Enantiomeric composition   79, 86, 200,
    219
Enfleurage techniques   21–22
Epidermis   42
Epididymis   211, 221
Epinephrine (adrenaline)   47, 200–204
Ergothionine   215
Erinaceus europaeus   236

Esters, acetate   81, 99, 108, 217–218, 223
  alkyl glyceryl ether   114–115, 218
  butyrate   36, 99
  formate   91
  glyceride   48, 60–63, 65, 82, 95, 114, 133–134, 159, 218
  macrocyclic   159, 259–260
  volatile   87, 99, 145
  wax, *see* Wax esters
Exaltolide (ω-pentadecalactone)   259–260
Externalization of signal   163, 230

Faeces   26, 85, 94, 96–102, 120, 144, 152, 162–164, 168, 187, 222
Farnesene   88
Farnesol   68–69, 103
Felids   116, 138
Felinine   196
Felis catus   50, 61–62, 64, 115–116, 138–139, 144, 150, 183, 211, 221, 253
Fermentation hypothesis of chemical recognition   147–148
Fermentative scent source   139, 145
Fermented egg   22, 145–146, 162
Ferret   156–157
Field ionization/field desorption   294–295
Field studies   144–146
Flea, rabbit   183
Flehmen   14–15, 166, 168
Fox, red   23, 26–27, 58–59, 116, 138–150, 158, 164, 179–182
Freeze etching   244
Fungi, skin   136–137
Fur lipid   46

Galago crassicaudatus   121–122, 167
Galago senegalensis   167
Galea musteloides   59
Gas chromatography (GC)   17, 279–285
Generator potential   255–256
Genetic aspects   116, 165–166, 259
Geosmin   36
Geraniol   68–69
Geranylacetone   88, 179–180
Gerbil   51, 79, 107–111
Gland, apocrine   41, 46–48, 131
  Bartholin's   226–227
  clitoral (rodent)   216, 224

  coagulating   210–211, 213, 222–223
  Cowper's   211
  eccrine   41, 48–50, 130
  Harderian   94, 109–115
  Littre's   211
  periurethral   227
  salivary   59, 236–241
  sebaceous   41, 43–46, 136
  Skeen's   227
  skin scent glands, *see separate entry*
  urethral   169, 211
  vesicular   210–215, 219, 221
  vestibular   226–227
Glomerulus   174, 248
Glossina morsitans   91–92
β-Glucuronidase   212
Glucuronides   184, 186–187, 201, 240–241
Glycerides, in scent glands   82, 95, 114, 159
  in skin lipid   48, 60–63, 65, 82, 133–134, 218
Glycogen, vaginal   227–229
Glycoprotein   103, 152, 158, 237
Goat   62, 92, 168, 173, 192, 212
  black bedouin   47
Gorilla gorilla   48, 129
Group acceptance and chemical signals   95–96
Growth and chemical signals   169–170
Grysbok   91
Guinea pig   13, 46, 50, 61–62, 64, 105–106, 178–179, 194, 221, 252
Gymnemic acid   251, 267

Haematogenic smell   260
Hair   43–44, 50
Hamster   13–14, 26–27, 61, 109, 111–112, 229, 233–234, 252–253
Hapalemur griseus   124
Hare   95–96, 212
Headspace analysis   18–21
  volatiles 26, 97–99, 102, 105, 177–178, 181
Heart rate   10–11, 97–99, 101
Hedgehog   210–212, 236
Herpestes auropunctatus   144, 147–148
Heterocyclic compounds   5, 16, 27, 30, 32, 35, 38–39, 90, 99, 105, 117, 148–150, 154–155, 157, 164, 175, 177–178, 182, 189, 192–193, 196, 204–205, 212–213, 231–232, 235–236
  see also Porphyrins

High performance liquid
    chromatography (HPLC)   62, 286,
    300
High resolution mass spectrometry
    297–298
Hindleg–head contact (HHC)   83–84
Histamine   49, 206–207
Holocrine process   45–46
Homoeochemic   3
Hormone dependency, of accessory sex
    organs   211, 214–215, 217–222,
    224–226, 229
  of apocrine function   47, 51–54
  of hair growth   50
  of Harderian gland   111, 113
  of salivary glands   236–237
  of sebaceous function   51–52, 71, 113
  of skin glands   53, 80, 85, 92–96,
    102, 106–108, 111, 113, 120, 125
  of urine chemistry   169, 171–172,
    180–182, 194, 206–207
  of vaginal signals   228–233
Horse   47, 191–192, 212–215
Human, apocrine sweat composition
    48, 132–133
  attitudes modified   133
  foot odour   137
  meibomian glands   133–134
  mosquito attractants   92
  racial/cultural variations   130, 186
  scalp odour   136–137, 162
  sebum/skin surface lipid   48, 60–62,
    64–71
  semiochemicals, possible   129–133,
    173, 213–215, 229–232, 236
  skin glands   41, 43–50, 130
  skin hydroxysteroid dehydrogenase
    activity   57–58
  sweat, steroids in   54
  total volatiles   30–33
Hyaena brunnea   150
Hyaena hyaena   150–151
Hyaenas   150–151
Hyaluronic acid   44
Hydrocarbons isoprene   27, 32, 82,
    94–95, 99, 107, 133, 178, 215, 227,
    231
  see also Isoprene, Springene, Squalene
Hydrochoerus hydrochaeris   106–107
Hydrophobic interaction   270–271
Hydroxysteroid dehydrogenase (HSD)
    activity   57–59

Individuality, semiochemical   65, 77,

83, 85, 94–95, 104–105, 107–108,
    123, 125–128, 130, 146–148, 150,
    173, 179
Indoles   149–150, 157, 189, 193,
    204–205, 231, 235–236
Infrared (IR) spectroscopy   287
Insect chemoreception, contrasted
    mammalian   276–277
Insect semiochemistry   2, 5, 9, 23,
    30–31, 135
Insectivore   211
Interdisciplinarity   8, 11
Involatile compounds and
    semiochemistry   13–17, 165, 171,
    226, 234
Ion sources, in mass spectrometry
    289–294
Isopentenyl methyl sulphide 158,
    179–180
Isoprene (2-methylbuta-1,3-diene)
    32–33, 236
Ixodes matopi   91

Jacobson's organ (vomeronasal organ)
    14–15, 252–253

Kairomone   4
Kandutsch–Russell pathway   68, 217
Keratinization   42–44, 67
Ketones   16, 27, 32, 36, 80–81, 85–86,
    88, 90–92, 97, 99, 105, 145,
    159–160, 176–177, 179–181, 192,
    218–219, 236
Kidney   172, 174, 183, 206–207, 221
Koala   187

γ-Lactones   36, 77–79, 90, 93, 137,
    177, 187, 193
Lactones, macrocyclic   259–260
Lagomorphs   93–102
Lagothrix   236
Lanosterol   69–70, 82
Lathosterol   69–70
Latrine sites   97, 222
Lee-Boot effect   171–172
Lemming   203–204
Lemmus trimucronatus   203–204
Lemur catta   123–125, 173
Lemur fulvus   123–124
Lemur macaco   52, 124
Lemurs   43–44, 52, 123–125, 283
Lepus europaeus   95–96, 212
Lion   138–139, 144, 148–149

Lipid components, nomenclature 63, 71–72
Lipid extraction, natural 78–79
Lipoprotein 48, 103
Liquid chromatography 286
Littre's glands 211
Loligo pealei 135
Lophiomys imhausi 79, 102–103
Loris tardigradus 121, 166
Lorises 120–121, 166–167
Loxodonta africana 102–103
Lutra lutra 152–153
Lycaon pictus 13, 40, 118, 216

Macaca mulatta 129, 173, 229–230
Macaca speciosa 129
Macaque, stump-tailed 129
Macrocyclic compounds 85–86, 159, 218–220
Mandrillus leucophaeus 129
Mara 103–104
Marmoset 127
Marmota monax 160
Marsupials 48, 210
Mass analysers (magnetic deflection, quadrupole, time of flight) 295–297
Mass fragmentography 298
Mass spectrometry (MS) 288–298
Meibomian glands 41, 123, 133–134
Melanin 52, 123
Melanocyte stimulating hormone (α-MSH) 52, 113, 219, 224
Meles meles 151
Menstrual odour 163, 230–232, 235
Menstrual synchrony 173
Mephitis mephitis 153–155
Meriones lybicus 109
Meriones tristrami 108–109
Meriones unguiculatus 51, 79, 107–111
Merocrine process 46–47, 49
Mesocricetus auratus 13–14, 26–27, 61, 109, 111–112, 229, 233–234, 252–253
Microbial hydrolysis, in anal sac 148–149
of sebum glycerides 60–61, 137
Microcebus murinus 123
Microdistillation methods 22–23
Microflora, of anal sacs/anal pockets 139–142, 147–148, 160, 162
of the axillae 131
of castor sacs 161
of the oral cavity 235–236

of the skin 136–137
of the vagina 135, 228–229, 234
Microorganisms 45, 135
as odour producers 96, 105, 130–131, 135–136, 139
Microtus 172
Mink 30, 144, 149, 151–153, 156–159, 229, 232–233
Mitral cells 247–248
Mole 211
Mongoose 144, 147–148
Monkey, Goeldi's 126
owl 128
rhesus 129, 173, 229–230
squirrel 128, 130, 173–174
woolly 236
Monotremes 210
Moose 77, 168
Moschus moschiferus 76, 85–86
Mosquito 92
Mother–young interaction and chemical signals 41, 83, 85, 130, 163–164
Mouse, field 107
laboratory 14, 26, 62–64, 66–67, 69–70, 112, 114, 165, 168–172, 182, 206, 217–218, 221, 224–225
wild house 170
Mucopolysaccharide 49, 93, 153
Mus musculus, *see* Mouse
Muscone 85–86
Muscopyridine 86
Mustela erminea 156–157
Mustela putorius furo 156–157
Mustela vison 30, 144, 149, 151–153, 156–159, 229, 232–233
Mustelids 151–159, 210
Myocastor coypus 103

Nepeta cataria 116
Nepetalactone 116–117
Nerve impulse, propagation 256–257
recording 262–263
Neurohumoral control, apocrine function 47
eccrine function 49
Harderian gland 110
Neuron 253
Nipple, odour 41
release of scent via 104
Nitriles 27, 32, 122, 236
Nomenclature, lipid components, 63, 71–73
Norepinephrine (noradrenaline) 200–204

Nuclear magnetic resonance (NMR)
   spectroscopy 287–288
Nycticebus coucang 121, 167

Ochotonidae 93
Odocoileus dichotomus 79
Odocoileus hemionus columbianus 13,
   15, 75–79, 162, 193
Odocoileus hemionus hemionus
   75–79, 162, 168
Odocoileus virginianus borealis 79
Odour, complexity of 35–39
   descriptors 37
   image, chemistry of 34–39
   primary 258
   thresholds 16, 35–36, 39, 132, 214,
      259–260
   unit values 37–39
Odourogram 232
Oestrus, detection 226
   synchrony 171, 173
Olfaction 243–248, 258–277
   chemical modification studies of
      267–268
   range of sense of 275
Olfactometer 261
Olfactory bulb 109, 243–244, 246–247
   cilia 243–245
   data, variability of 260
   epithelium 243–246, 248, 266
   marker protein 248
   system, a model for 268–277
Ondatra zibethica 218–220
Orang-utan 129
Oreotragus oreotragus 91
Oryctolagus cuniculus 10–11, 25–26,
   48, 50, 53, 61–62, 64, 93–102,
   114–115, 137, 168, 183, 212, 221,
   284
Osmetrichia 43–44, 78–79, 103–104,
   106, 108
Otter 152–153
Ox, musk 216
Oximes 194

Pan satyricus 48, 61–62, 129
Panthera leo 138–139, 144, 148–149
Panthera tigris 138–139, 144, 150
Peptide 14, 169, 172, 248
Periglomerular cells 248
Perissodactyla 74
Perodicticus potto 120–121
Peromyscus maniculatus bairdii 171

Pest control 2, 22, 91–93, 103,
   145–146, 162–163, 222
Petaurus breviceps 75
pH, of anal sac 143, 158
   of body surface 50
   of eccrine sweat 50
   effect on perceived odour 105, 228,
      235
   of reproductive tract 228, 233
   of urine 174–175
Phalanger, gliding 75
Phascolarctos cincreus 187
Phenols 81, 90–91, 99, 105, 121, 160,
   177, 182, 186, 191, 193–194,
   201–202, 204, 231–232, 236,
   240–241
Pheromone 2–4
   aversive 225–226
   maternal 163–164
   mounting 234
   primer 3, 14, 92, 111, 125, 168–174
   puberty accelerating 14, 168–171,
      173, 182
   releaser 3
Photoaffinity labelling 267
Pig 48, 59, 61, 173, 186, 192, 210–212,
   214–216, 237–242
Pikas 93
Piloerection 43–44
2-Piperidone 30, 148–149, 231–232
Pituitary hormones 51–52, 113, 164,
   172, 216, 219, 224
Pongo pygmaeus 129
Porapak 18, 27, 30, 33, 178
Porphyrins 109–113, 196, 208–209
Potto 120–121
Prairie dog 104
Preen gland lipids 64, 161
Preorbital sac 75–76, 80–81, 91, 109
Preputial glands (rodent) 211,
   216–225
Preputial sac (diverticulum) 193, 216,
   240–242
Primary odour 258
Primate, non-human 43, 47–48,
   119–130, 173, 205, 213
Prolactin 52, 164, 216
Pronghorn 79, 86–89
Propithecus verreauxi 124
Prostaglandins 49, 214–215
Prostate 210–213, 215, 219, 221
Protein 14, 16, 49, 95, 103, 144, 153,
   194, 234, 248
Psychophysical methods 258–261

Putrescine   149–150, 158, 205–206, 213, 235
Pyrazines   16, 35, 38–39, 99, 178, 192
1-Pyrroline   213–214

Quinoline, 2-methyl-   150, 154–155, 180

Rabbit   10–11, 25–26, 48, 50, 53, 61–62, 64, 93–102, 114–115, 137, 168, 183, 212, 221, 284
Radioimmunoassay   132, 216, 299–300
Rangifer tarandus caribou   79, 81
Rangifer tarandus tarandus   76–77, 79–85, 168
Raphicerus melanotis   91
rat, crested   79, 102–103
    laboratory   26, 46, 50–52, 56, 62–64, 66–67, 69–70, 109, 112–114, 135, 163–164, 169–170, 173, 203–204, 210–213, 215–217, 219, 221–224, 228–229
    musk   218–220
Receptor, olfactory   244, 268–275
    taste   251
Reindeer   76–77, 79–85, 168
Repellant, deer   162–163
Reproduction and chemical signalling   13, 14, 125, 168–174, 210, 213–215, 226, 229–230, 233–234, 237–239, 253
Resolution of mass spectrum   295–296
Response guided strategy   5–6, 13
Resting potential 254–255
Rhinoceros   47
Rodents   103–114, 159–165, 211, 213
Rub-urination   78–79
Rupicapra rupicapra   168

Saguinus fuscicollis   126–128
Saguinus geoffroyi   126
Saguinus oedipus   126–127
Saimiri sciureus   128, 130, 173–174
Saliva   17, 29, 75, 110–111, 120, 128, 235–241
Salivary gland   59, 236–241
Salting out   23
Scent mark, aging of   84, 97, 122, 150, 173
    model   15
Scent marking, behaviour   74, 119, 172, 233
    conspecific   75, 94–95, 108, 115–116, 120–121, 123–124, 126, 167–168, 212

environment   75–76, 87, 93–95, 97, 102, 105–106, 108, 120–123, 125–126, 138, 144, 150–152, 161, 166, 174, 222
self   78–79, 83–84, 89, 93, 106, 110–111, 125–126, 161, 167–168, 174, 193
Scent precursor   16, 135, 213
Sciuridae   104
Sebaceous glands   43–46, 136
    distribution in man   41, 45
    effects of hormones   51–52, 71, 113
    hydroxysteroid dehydrogenase activity in   57–59
Sebum   45–46
    biosynthesis 72–73
    compound classes present in   60–64
    effects of diet   60, 108
Secretory process, apocrine   46
    holocrine   45–46
    merocrine   46–47, 49
Self-anointing   236
Semen, chemistry of   211–216
Seminal vesicles   210–215
Semiochemical   2
Semiochemical message value, context and   94
    learned response   230, 234, 277
Semiochemistry   1
    and thermoregulation   78, 83, 110–111
Serotonin   204–205
Sexual arousal   227, 232, 237
Sheep   43, 46, 61–64, 69–71, 92, 168, 173, 212, 214–215
Shrew, house   51, 54
    tree   119–120
Sifaka   124
Skin, uniqueness of   40
Skin glands, see Apocrine glands, Eccrine glands, Sebaceous glands, Skin scent glands
Skin lipid, acid moieties in   64–67
    alcohol moieties in 64, 68
    cholesterol biosynthesis intermediates in   68–71
    relative epidermal and sebaceous contributions to   67–68, 71
Skin scent glands, hydroxysteroid dehydrogenase activity in   57–59
    abdominal (including epigastric and ventral)   51, 79, 85, 107–108, 119–120, 122–123, 128
    anogenital (anal, circumanal,

circumgenital, labial, perianal, perineal, scrotal, suprapubic, *but not* anal sacs/pockets) 26, 53, 75, 94, 96, 102–107, 118, 120–121, 123–124, 126, 128, 227

arm/leg (antebrachial) 59, 124–125; (axillary) 26, 128–132; (brachial) 121, 124–125; (carpal) 240; (interdigital) 76; 81–84, 89–90; (metatarsal, tarsal) 76–77, 79, 82

back (dorsal) 87–89, 104

chest (sternal) 75, 119, 122, 126, 128–129, 236

flank (hip, ischiadic, side) 51, 88, 103, 107, 109

groin (inguinal) 53, 94–96

head 92; (cheek, maxillary, postauricular, subauricular) 54, 80, 87, 93; (chin, submandibular) 53, 59, 94–95, 106; (frontal) 53, 75–76, 129; (labial, oral, perioral) 104, 121, 123; (nasal, perinasal) 79, 106, 121; (occipital) 92–93; (preorbital) 75–76, 80–81, 91, 109

tail (caudal, subcaudal, supracaudal) 23, 58–59, 76, 82–83, 85, 103, 106–107, 116, 118, 128, 151, 194

throat (gular) 54, 119–120, 124, 128

Skunk 153–155

Specific anosmia 213, 258–259

Specific ion monitoring 298

Speothos venaticus 144

Spermine 205–206, 213

Spider, bolas 4

Spilopsyllus cuniculi 183

Springbok 88–89

Springene 88

Squalene 48, 60–62, 68–69, 71, 89, 127, 217, 231

Squames 42, 136

Squid 135

Squirrel, ground 104

Stercobilin 208–209

Sternotherus odoratus 149

Steroids
  in axillary sweat 130–132
  biosynthetic relations between 239, 301
  metabolism in various tissues 54–57, 114, 221, 239
  in scent glands 85–86, 93

semiochemical function 131–133, 183–186, 237–241
  structural formulae 55, 132, 185, 239, 301
  in sweat 54
  in urine 183–186, 192, 241
  see also Adrostenol, Androstenone, Testosterone

Sterols and sterol esters 48, 60–64, 68–71, 74, 79, 82, 86, 93, 95, 103, 107, 133–134, 148–149, 159–160, 216–218, 236
  structural formulae 69

Stoat 156–157

Strategy, chemical image 6–7, 13
  response guided 5–6, 13
  sensory 276–277

Stratum corneum 42, 45, 136

Stress, emotional 48–49, 130, 196–197, 200–203
  social 170, 220

Structure–activity relationships 273–275

Sweat, pubic 213
  in schizophrenia 49–50
  steroids in 54, 130–132

Sweating, apocrine 46–48, 130–133
  eccrine 48–50, 130

Subcaudal pouch 151

Sugar 49, 215

Sulphur compounds, organo- 16, 32, 99, 145, 150–159, 164, 178–183, 196, 224, 233–235

Suncus murinus 51, 54

Sus scrofa, see Pig

Synapse 254, 258

Tamarins 126–128

Tarsiers 122–123, 134

Tarsius syrichta 122–123

Taste 249–251

Tenax 18, 24–26

Testis 211, 221, 239–241

Testosterone, excretion 184, 186, 236
  5α-dihydro- 52, 55–57, 130, 221, 240

Thermoregulation 47–49
  and semiochemistry 78, 83, 110–111

Thietanes 155–158

Thin layer chromatography (TLC) 286

Thiols 154–155

Tick, parasitic and mammalian semiochemicals 91

Tiger 138–139, 144, 150

360

Total ion current   289
Trails, chemical   5, 84, 166–167
Transevaporator   28–29
Transmitter substance   254, 257
Trapping methods   23–34
Trigeminal sense   248–249
Tryptamine   197
Tsetse fly   91–92
Tupaia belangeri   119–120
Turbinate bones   243–244
Turtle, stinkpot   149

Ungulates   74–93, 168
Urea   49, 103, 174–175, 231
Urethral glands   169, 211
Uric acid   49, 175
Urine   13–14, 24–27, 75, 78, 81, 94,
    105, 109, 120, 126, 128, 144, 150,
    152, 158, 165–209, 224–226,
    240–242, 298
  accessory sex organs contribution to
    211–212, 216, 221–225, 241
  bladder   165, 168–169, 172,
    225–227
  body marking with   78–79, 83, 89,
    94, 106, 121, 128, 167–168, 193,
    212
  fluorescence   194, 209
  pigments   207–209
  volatiles   176–183
  washing   121, 167, 174

Vagina   215, 226–234
  permeability of   215, 227, 232

Vaginal secretion   13–14, 26–27, 125,
    128, 144, 226–234
Vas deferens   211, 221
Vernix caseosa   60–61, 64, 66–67
Vesicular gland   210–215
Viverra civetta   159
Viverrids   159
Volatile profiles, technique dependency
    of   17–18, 25, 39
  whole organisms   30–34
Volatiles, interaction with non-volatile
    substances   15–16, 20–21
Vole, field   172
  water   107
Vomeronasal organ   14–15, 252–253
Vulpes vulpes   23, 26–27, 58–59, 116,
    138–150, 158, 164, 179–182

Wax diesters, type 1   62–64, 79, 114
  type 2a   62–64, 133
  type 2b   63–64
Wax esters   48, 60–62, 71, 86, 95, 113
Wax monoesters   62–64, 79, 107,
    133–134, 157, 161–162, 218
Whitten effect   14, 171–172
Wolf   144
Wolf, maned   144
Woodchuck   160

X-ray energy probe micronalysis   152,
    158

Zebra   215